NOBEL LECTURES

PHYSICS

2011 – 2015

Nobel Lectures

Including Presentation Speeches
and Laureates' Biographies

Physics

Chemistry

Physiology or Medicine

Economic Sciences

Nobel Lectures

Including Presentation Speeches and Laureates' Biographies

PHYSICS

2011 – 2015

Editor

Lars Bergström

Stockholm University, Sweden

NEW JERSEY • LONDON • SINGAPORE • BEIJING • SHANGHAI • HONG KONG • TAIPEI • CHENNAI • TOKYO

Published by

World Scientific Publishing Co. Pte. Ltd.
5 Toh Tuck Link, Singapore 596224
USA office: 27 Warren Street, Suite 401-402, Hackensack, NJ 07601
UK office: 57 Shelton Street, Covent Garden, London WC2H 9HE

NOBEL LECTURES IN PHYSICS (2011–2015)

ISBN 978-981-124-552-7 (hardcover)
ISBN 978-981-124-680-7 (paperback)
ISBN 978-981-124-553-4 (ebook for institutions)
ISBN 978-981-124-554-1 (ebook for individuals)

For any available supplementary material, please visit
https://www.worldscientific.com/worldscibooks/10.1142/12508#t=suppl

Printed in Singapore

PREFACE

Hardly anything can inform us about the development of science better than to follow the sequence of Nobel Prizes and Laureates, to get to know their work and sources of inspiration through first-hand accounts. In this volume, we will do this for the physics Laureates for the years from 2011 to 2015.

Physics was the first field of science mentioned in the will of Alfred Nobel from 1895, to be awarded with a yearly Prize "to the one who in the field of physics has made the most important discovery or invention." This was in the general spirit of sharing dividends of the prize fund equally in five parts to "those who during the past year have conferred the greatest benefit to mankind", where the other fields beside Physics are Chemistry, Physiology or Medicine, Literature, and Peace. The Laureates of the Physics and Chemistry Prizes are chosen by vote in the full assembly of the Royal Swedish Academy of Sciences, with suggestions prepared by its respective Nobel Committee and the Academy's classes of Physics and Chemistry. A Foundation, the Nobel Foundation, was created in 1900 to manage Alfred Nobel's fortune in a manner that ensures a secure financial standing for the Nobel Prize over the long term and that the prize-awarding institutions are guaranteed independence in their work of selecting recipients. The first set of Nobel Prizes was given in 1901, with Wilhelm Conrad Röntgen as the first Physics Laureate, for the discovery of X-rays, and with an impressive list of Laureates ever since.

The will of Alfred Nobel is maintained to the fullest degree possible, even if the requirement of "during the past year" has been reinterpreted in the bylaws of the Prize to mean that the awarded work should have gained recent actuality, even if performed earlier.

One of the most important aspects of the Nobel Prize is its international character, as the will states that the Prize should be awarded to the worthiest, irrespective of nationality. Although Alfred Nobel was a very productive inventor with more than 350 patents, the most well-known being for dynamite in 1867, the best invention of all, in retrospect, seems to be the large, international Prize which bears his name and now is the most prestigious Prize in the world — the Nobel Prize.

There are several things that are not specified in the will, and therefore for the Nobel Prizes in Physics and Chemistry this has been addressed in the bylaws of the Royal Swedish Academy of Sciences. For the Physics Prize, the sentence about "most important discovery or invention" is of course the crucial one, and unchanged. However, the number of Laureates each year has been set to at most three. The Academy has so far chosen not to give a Physics Prize to an institution or collaboration, in contrast to the Peace Prize of the Norwegian Parliament where this has been done several times.

We now give a short overview of the theme of this volume, the Physics Laureates between 2011 and 2016.

The Nobel Prize in Physics for 2011 was awarded for one of the most surprising discoveries in the field of cosmology, where careful studies of the emitted light from a certain type of exploding stars, supernovae, indicated that the expansion of the Universe is in fact accelerating! This enigmatic situation has changed the starting point of the cosmological standard model to one which besides matter (dark and visible) and radiation also needs an additional component — dark energy. Thus, this discovery, in 1998, has meant a paradigm shift, to something which is often referred to as the ΛCDM model. Here CDM stands for Cold Dark Matter, meaning slowly moving electrically neutral particles of non-zero mass. The letter Λ (Greek Lambda) is the symbol Albert Einstein introduced for his "cosmological constant" in 1917. This corresponds to an energy density of the vacuum, but also a related negative pressure, i.e., a repulsion. Einstein wanted a static Universe and realized that all mass in the Universe would cause a shrinking thanks to the mutual attraction of massive bodies such as stars. To counter this attractive force, he noticed that by introducing the repulsion given by a cosmological constant and fine-tuning its value, he could balance the attraction with the repulsion and indU niverse. Unfortunately, such a fine-tuned model is not stable, and when G. Lemaître and E. Hubble by the end of the 1920's discovered that the Universe is in fact expanding, Einstein abandoned the model. From then on, most scientists in the field for simplicity assumed that Λ was zero, even with no convincing reason.

The two research teams, one led by Perlmutter, and the other by Schmidt and Riess, raced each other to find the most distant supernovae. By establishing the distance to these far-away supernovae and the speed at which they are moving away from us, the scientists hoped to indicate the future fate of the Universe. They expected to find signs that the expansion of the Universe was slowing down. The discovery awarded in 2011 surprisingly showed, on the other hand, that the expansion of the Universe is accelerating. It is therefore needed to add a component of dark energy, the simplest of which is indeed an energy density Λ of the vacuum, that is a cosmological constant, and with it a negative pressure.

The awarded discovery "for the discovery of the accelerating expansion of the Universe through observations of distant supernovae" was made by two teams, the Supernova Cosmology Project led by Saul Perlmutter, and the High-z Supernova Search Team with Brian Schmidt and Adam Riess the awarded scientists.

In the memoirs and the Nobel lectures the reader can follow the sometimes-thorny road to this important and unexpected discovery. When the dust had settled it was reassuring that both teams had compatible results, with a value of the energy density of the vacuum giving a sizeable, even dominant, fraction of the total energy density of the Universe. (The current value, using several sets of observational data, also from measurements of the cosmic microwave background, is close to 70%.)

In 2012, the Physics Prize was awarded for brilliant, enabling technologies, that made possible measurements and manipulation of individual quantum systems. The Laureates were from both sides of the Atlantic, with the French scientist Serge Haroche trapping individual light quanta, microwave photons, in a cavity with perfectly reflecting walls, and American scientist David J. Wineland working instead with charged ions using an ion trap (the technique of which was awarded with a Physics Nobel Prize to Hans G. Dehmelt and Wolfgang Paul in 1989). There are several similarities between the two methods, both using two-level quantum states. In either case quantum non-demolition measurements can be performed where two-level systems are coupled to a quantized harmonic oscillator. The two-level system consists of either a trapped ion in the setup of Wineland (with two levels coupled by laser light) or a highly excited Rydberg atom in the setup of Haroche (with two atomic levels coupled by a microwave field). The quantized harmonic oscillator can then describe either the motion of the ion in the trap or the microwave field in the cavity.

One of the secrets behind the breakthrough of Wineland was the use of laser beams to minimize the motion of the ion, so that it enters the ground state (in quantum mechanics, the allowed energy states of an atom or ion are quantized, and the state of lowest energy is called the ground state). Once there, additional laser pulses may put it in a so-called superposition state, for example superposition of the ground state and the first excited state. In such a superposition, one may say that the ion is simultaneously in both states — something which is conceptually different from anything we can experience in our everyday, macroscopic world. This was noted by Erwin Schrödinger (Physics Prize 1933), who famously in 1935 said that if this would work also in our macroscopic would, we could construct a so-called entangled state where for example a cat is simultaneously dead and alive — a clear paradox.

So why does Schrödinger's cat not exist, even if an ion in "cat states" consisting of single trapped ions entangled with coherent states of motion, or in the setup of Haroche, a superposition of cat-like microwave field states entangling a Rydberg atom with the cavity field, can be in such a state? It turns out that the measurements of the two groups illustrate the solution, which is an effect called decoherence. By interacting with the environment, the delicate state of superposition where the two states are perfectly entangled with each other, becomes gradually more and more like classical states, i.e., either "dead" or "alive" but not a quantum superposition of the two. The two groups devised creative experiments to show in detail how the act of measuring actually causes the quantum state to finally collapse and lose its superposition character.

Of course, Schrödinger's cat is a stimulating philosophical problem, but the main applications of the results of Haroche and Wineland are the ability to use the new quantum systems for, e.g., atomic clocks with exquisite precision, or creating building blocks for future quantum computers, which would radically transform computer science.

The 2013 Nobel Prize in Physics was special in several ways. It was a good illustration of how the early adopters of the Prize modified the "impossible" requirement in the will of Alfred Nobel that only a discovery or invention during the "preceding year" should be eligible for a Prize. In the bylaws of the Nobel Prize this was reinterpreted to mean that those activities which are the most current in the field of award are eligible, whereas older work can be considered if their importance has been clarified only during recent times. Another section of the bylaws states that the Prize can only be given to living persons, and not posthumously.

In particle physics, the origin of mass for some mediators of forces, was long a complete mystery. It seemed that the successful formulation of the action of these mediators pointed to certain mathematical theories, gauge theories, for their interactions. However, gauge theories seemed to forbid explicit mass terms for the mediators. In this situation, in 1964, two independent scientific articles appeared which found an ingenious solution. The articles were by F. Englert and R. Brout (the latter deceased in 2011), "Broken Symmetry and the Mass of the Gauge Vector Mesons", and P.W. Higgs, in the two papers "Broken symmetries, massless particles and gauge fields" and "Broken Symmetries and the Mass of the Gauge Bosons". These authors used a property of some non-relativistic condensed matter theories, developed in particular by Y. Nambu (Physics Prize, 2008) and P.W. Anderson (Physics Prize, 1977). This property is that an effective mass of a mediator can be obtained by so-called spontaneous breaking of a symmetry. In the fully relativistic models that Brout, Englert, and Higgs constructed, this is obtained by giving one component of hypothetical spin-0, scalar fields a non-zero expectation value for the field in the state of lowest energy, the ground state. One way to view the working of this mechanism, now frequently called the BEH mechanism, is that the otherwise massless mediator "eats" one of the scalars by which it gets heavy. In the minimal model with two scalar fields, Higgs took the analysis one step further and showed that the remaining degree of freedom in the scalar sector corresponds physically to a massive, spin-0, particle. This is the famous Higgs particle, for which particle physics experiments were searching for several decades to come. In fact, one of the main motivations for building the Large Hadron Collider (LHC) at CERN was to enlarge the mass range for the Higgs search, as this was the last remaining particle missing to make the Standard Model of particle physics complete.

The LHC was ready in 2008, and two large teams each of thousands of scientists, ATLAS and CMS, started a very careful search that resulted in a joint, statistically significant discovery of the Higgs particle in 2012. This was the piece missing to show that the BEH mechanism is the one realized in nature. In line with the bylaws of the Nobel Prize, the Royal Swedish Academy of Sciences could announce the two living members of the BEH trio, François Englert and Peter Higgs, as the recipients of the 2013 Nobel Prize in Physics almost 50 years after their fundamental discovery. The Royal Swedish Academy of Sciences has, in contrast to the Peace Committee of the Norwegian Parliament, not awarded Prizes for institutions or collaborations.

The citation for the 2013 Physics Nobel Prize is, "for the theoretical discovery of a mechanism that contributes to our understanding of the origin of mass of subatomic particles, and which recently was confirmed through the discovery of the predicted fundamental particle, by the ATLAS and CMS experiments at CERN's Large Hadron Collider".

The Nobel Prize in Physics is given for a discovery or invention, and the three Prizes in the years 2011–2013 had a clear component of discovery. In 2014, however, the Prize was given for an invention: the blue light emitting diode (LED), by the Japanese trio Isamu Akasaki, Hiroshi Amano, and Shuji Nakamura.

The history of the invention was that energy efficient light emitting diodes for both red, and green light had been developed since the 1960s. However, to create white light one needs the triad of red, green, and blue, something known for example from color photography, and color TV. A solid-state device such as a light emitting diode needs very little power compared to the previously used incandescent light bulbs. Therefore, it was realized that finding a blue LED could bring quite a revolution of light sources in our modern society, with lamps built on red, green, and blue LEDs being at least a factor of 10 more energy efficient than regular light bulbs (which get warm when light is on, meaning they radiate a lot in the invisible, infrared part of the spectrum). Moreover, LEDs last some 10 times longer, which means that, for example, streetlights do not have to be changed so often. As we know, the revolution has indeed come, and the switch to LEDs for both indoor and outdoor lighting has been very fast internationally.

With only red and green LEDs existing, and blue light LEDs remaining a dream for three decades, the big challenge was clear for the Japanese scientists, to find a way to construct a blue LED, to enable energy efficient white light sources. This turned out to be a daring and very complicated task, however, and many research laboratories around the world had tried, but failed. For instance, which material should one use to obtain the blue light? Akasaki was convinced by previous experience that the choice of material he had made, gallium nitride, was the correct one, and worked effortlessly with Amano, who was his doctoral student at Nagoya University, to prove this. Nakamura at Nichia Chemical corporation also chose gallium nitride instead of the alternate, zinc selenide, which was used and considered more promising by competing groups.

In 1992, the long-sought bright blue light beams appeared from their semiconductors, with Akasaki and Amani's method of producing blue light using an electron beam to sensitize the light-emitting region being slightly different from that of Nakamura, who in his independent work used a simpler and cheaper method, by heating the material. From then on, both research groups succeeded in further improving their blue LEDs, making them more efficient. They had thus completed one of the most important inventions of the last century — the blue LED, which has enabled bright, and energy-saving white light sources used all over the world. The citation for the Physics Nobel Prize of 2014 reads "for the invention of efficient blue light-emitting diodes which has enabled bright and energy-saving white light sources".

In 2015, the Physics Nobel Prize was awarded for a very important discovery in basic science, the fact that neutrinos have a very small, but non-zero, mass. The citation for Takaaki Kajita and Arthur B. McDonald was "for the discovery of neutrino oscillations, which shows that neutrinos have mass."

Neutrinos are some of the most interesting elementary particles, but also the hardest to catch. The existence of the neutrino was first hypothesized by Wolfgang Pauli in 1930 but was not experimentally discovered until 1956 by F. Reines and C.L. Cowan Jr. (Reines shared the Nobel Prize in Physics 1995 with M.L. Perl). Neutrinos do not interact through electromagnetism as they are electrically neutral, but they play an important part in weak interactions. They participate for instance in the decay of free neutrons, but more importantly in the nuclear reactions in the interior of stars like our Sun — this is where the energy of the Sun comes from (H.A. Bethe, Physics Prize 1957). These energy-giving, exothermic reactions cause effectively fusion of two protons and two neutrons into helium nuclei, and in the interconversion of protons to neutrons through the weak interactions in the first stage of the reaction chain, neutrinos are also produced (Willy Fowler, Physics Prize 1987). Thus, a huge flux of neutrinos was predicted to emanate from the Sun, and this was verified in experiments, started already in 1968, by Raymond Davis Jr (Physics Prize, 2002). However, there was a problem with the measured neutrino rate, which was named the "solar neutrino problem". This was a mismatch of the theoretically predicted flux, pioneered by John Bahcall and collaborators, and the measured one by Davis. There was a factor of roughly 2–3 smaller flux measured than predicted.

The mystery of the missing neutrinos could have different explanations, for instance relating to the temperature at the center of the Sun. However, when data were verified in several other experiments, the only solution which fits was the existence of neutrino oscillations, a hypothesis first put forward in a similar context by Bruno Pontecorvo in 1957 (the first quantitative theory of neutrino oscillations was developed by Maki, Nakagawa, and Sakata in 1962). In the Standard Model of particle physics, the constituent particles are grouped in three so-called families, where each family contains its own neutrino species, besides the electron neutrino also the muon neutrino and the tau neutrino. If the electron neutrinos produced in the Sun would mix with the muon and tau neutrinos, then the measurements could be explained if the electron neutrino oscillates between all three species. Most of the detections used so far were mainly sensitive to the interaction of the electron neutrinos with the electrons in the target, so this could explain the factor of three lower rate. However, it was realized that neutrino oscillations can only take place if there is a difference in mass of the neutrinos. Thus, if one demonstrates that oscillations occur, one has also shown that neutrinos have mass.

In 1998, Takaaki Kajita reported results from the Super-Kamiokande Collaboration in Japan, who could use directional information to find that cosmic ray muon neutrinos were depleted depending on their length of travel through the earth — a clear signature of oscillations. It was not until 2002, however, when Arthur MacDonald leading a collaboration at the Sudbury mine in Canada, the Sudbury Neutrino Observatory, could independently show that the other types of neutrinos are still present in the neutrino flux from the Sun — it is just that they do not interact as strongly with electrons. This was shown by using a detector filled with heavy water, i.e., where the hydrogen nucleus is replaced by deuterium. By comparing the rate for so-called charged current reactions, where the deuterium nucleus breaks up into two protons and one electron (sensitive only to electron neutrinos) with the neutral-current reaction giving a breakup of the deuterium in one proton and one neutron (where all three neutrino species contribute), the oscillation hypothesis was finally verified.

As we have seen, during the five years covered in this volume, outstanding work has been rewarded in many subfields of physics, from the smallest particles like neutrinos to the whole Universe with its unexpected dark energy, the manipulation and control of quantum systems, the origin of mass through the Brout–Englert–Higgs mechanism in particle physics. Also, the invention of an amazing new light-source, the solid-state blue LED has been awarded. And, of course, physics is now as much as ever before a developing field with new discoveries and new inventions appearing all the time, and some of the worthiest will be awarded Nobel Prizes in Physics "for conferring the greatest benefit to humankind".

Lars Bergström

CONTENTS

Physics 2011

Saul Perlmutter, Brian P. Schmidt and Adam G. Riess

"for the discovery of the accelerating expansion of the Universe through observations of distant supernovae"

The Nobel Prize in Physics

Speech by Professor Olga Botner of the Royal Swedish Academy of Sciences.

Your Majesties, Your Royal Highnesses, Ladies and Gentlemen,

Allow me to start in English by citing a short poem by the Danish scientist, poet and designer Piet Hein, called *Nothing is indispensable – grook to warn the Universe against megalomania*:

The Universe may
be as big as they say.
But ... it wouldn't be missed
if it didn't exist!

Well ... If the universe didn't exist, we would definitely not be sitting here today!

Humans are a part of the universe – just like the planets, the stars and the galaxies. The building blocks of every cell in our bodies – carbon, oxygen and other atoms – were formed inside ancient stars that exploded in the Milky Way perhaps 10 billion years ago, long before the solar system came into existence, just like the elements that the planets were formed from. Galaxies, stars and planets are affected by the same forces of nature as we are – especially gravity, which determines the orbits of planets and affects the life cycles of stars – and makes sure we have our feet on the ground. In order to understand ourselves, we must try to understand the universe!

This year's Nobel Prize in Physics is about star explosions and about gravity – indeed, about the whole universe.

By observing distant galaxies with the help of telescopes, nearly 100 years ago scientists discovered that our universe is getting bigger and bigger. The distances between the galaxies are constantly increasing, like the spaces between raisins in a raisin cake swelling in the oven. If we extrapolate backwards in time, we realise that this expansion must have begun about 14 billion years ago – in a primordial explosion that British astrophysicist Fred Hoyle called the Big Bang. During the past five billion years, roughly since the formation of our solar system, the universe has doubled in size, and it is continuing to expand. But – all matter in the universe is attracted to all other matter because of gravitation. The distances between the galaxies should not be able to continue increasing as rapidly for all eternity – the expansion should eventually slow down. It can be calculated that if the universe contains more than six atoms per cubic metre, this expansion must stop. The universe should start shrinking and eventually end with a Big Crunch, the opposite of the Big Bang. Is that the fate of the universe?

In order to answer this question, we must study whether the expansion rate of the universe is changing over time. Luckily we have access to a "time machine": The light reaching our telescopes from distant stars has travelled through space for millions and even billions of years and has gradually become redder, since space itself has expanded. If we compare the redshift that is measured for many different objects using the models for the expanding universe permitted by Einstein's General Theory of Relativity, we can discover the model that describes the real universe. The redshifts can be interpreted as distances, but this interpretation depends on various assumptions in the models about the balance between different forms of energy, for instance the quantity of radiation and matter. If we can compare these model-based distances with other, independent distance measurements of the same objects, we can determine the energy balance of the universe. But then we need objects that are visible across enormous distances – sources of light billions of light years away.

The scientists who are being awarded the Nobel Prize in Physics today have studied distant star explosions – supernovae. These events release enormous quantities of energy: during a few weeks, a single supernova can outshine all the hundreds of billions of stars in a galaxy. Stars can be ripped apart in many different ways, but it turns out that in special cases, explosions occur that always emit the same amount of light. These explosions can be recognised by carefully studying observed starlight. Since the quantity of light released is always the same, the distance can be determined from the intensity of observed starlight. The further away the supernova – the fainter it seems!

But the universe is big, and finding the right kind of supernovae is a challenge to scientists. The Laureates and their research teams used digital technology and invented efficient methods for repeatedly searching patches of sky and comparing thousands of images.

Dozens of supernovae were discovered, and with the help of the world's largest telescopes, their type and brightness could be determined, as well as the distance indicated by their redshift.

To everyone's utter amazement, the supernovae seemed to be much too dim – weaker than expected if the expansion of the universe had slowed down. Instead, the universe has been expanding faster and faster.

What, then, is causing this expansion rate to increase? The answer might be: a special form of energy, called dark energy, which is pushing the universe "outward". A similar form of energy – the cosmological constant – was discussed by Einstein as early as 1917 but was later removed from his theory. Dark energy seems to make up about 73 per cent of all energy in the universe, which has been confirmed during the past decade by the distribution of galaxies across large distances and by studies of cosmic background radiation.

The discovery of accelerating expansion through studies of distant supernovae has thus changed our image of the universe in an unexpected, dramatic way. We have realised that we live in a universe which largely consists of components that are unknown to us.

Understanding dark energy is a challenge for scientists all over the world – in keeping with Piet Hein's motto:

Problems worthy of attack
prove their worth by hitting back.

Professor Perlmutter, Professor Schmidt, Professor Riess,

You have been awarded the Nobel Prize in Physics for the discovery of the accelerating expansion of the Universe through observations of distant supernovae. On behalf of the Royal Swedish Academy of Sciences it is my honour and my pleasure to convey to you the warmest congratulations. I now ask you to step forward to receive your Nobel Prize from the hands of His Majesty the King.

Saul Perlmutter. © The Nobel Foundation. Photo: U. Montan

Saul Perlmutter

My four grandparents all immigrated as young adults to the United States from Eastern European Jewish towns and villages early in the twentieth century. This was a generation of poor but optimistic intellectuals, who expected that the newly rationalist world would use education and creativity to leave behind boundaries, borders, and religion to build a better world. (My mother's father, for one, although he earned a living welding in the shipyards during the Second World War and later ran a sandwich stand, was by avocation a self-educated scholar and teacher of Yiddish literature and history.) It is perhaps not surprising then that their only children, my parents, both became professors, my mother, Felice Davidson Perlmutter, in Social Work and Social Administration, and my father, Daniel D. Perlmutter, in Chemical Engineering.

My mother's work life was full of research collaborations that she organized and enjoyed professionally and personally. The social work field appeared full of warm, friendly people who took great pleasure in working with each other. My father's work involved careful, accurate calculations with slide rule and graphs (both attractive to me as a child), and experiments with lab equipment – and graduate students whom he taught patiently and mentored. On weekends our home was full of my parents' friends, discussing politics and movies, books and arts late into the night. Concerns about social and political conditions of the world were a constant theme.

In this atmosphere, I grew up wanting to know all the "languages" – music, literature, math, science, symbols, architecture, psychology – particularly those that seemed most universal. I thought *everybody* needed to know what was in the missing "Owner's Manual" of the Universe, since we lived here and "used" it. The Quaker schools my two sisters, Shira and Tova, and I attended encouraged creativity, critical thought, and social concerns; there were excellent teachers in the basic math and science courses but (at the time) not much beyond the fundamentals to encourage further creativity in these areas. Instead, I was excited to learn to write clearly and expressively, and to become familiar with the non-science approaches to the world.

As with many scientists, music was an important part of my life. I don't know if there is a documented relationship between musical and scientific thinking, but there certainly seem to be shared elements, perhaps starting off with the delayed gratification inherent in practicing an instrument. In my case, my violin teacher Frances Duthie (the only teacher who was a constant throughout my school years) certainly taught a strong ethic of perfectionism, but in the service of a warm, humanistic tradition of interpersonal communication, and the chamber music ideal of collaborative, shared listening and contributing. The pleasure of group music-making continued though string-quartet playing and group singing throughout my life – and my favorite science group experiences share some of the same feeling. (Many years later, as a professor at Berkeley, I was glad to have the opportunity to design a "Physics & Music" course for undergraduates.)

By the time I reached college, I thought I would follow my fascination with the big Philosophical Questions. As I saw it, the two big mysteries are: How does the world work? and How does the mind work? (The latter was also relevant to the former, since we only perceive the world through our minds.) So starting Harvard, in 1977, I considered double majoring in Philosophy and Physics – until I realized that if I did this there would be no time for courses in Humanities and Social Sciences. I decided to major in Physics first, since it seemed more likely that I would later come back to Philosophy than the other way around. (We'll see.)

College then became a chance to take courses in many fields. I discovered a strong interest in the then relatively-new area of Cognitive Sciences. I found out that Biology was not the boring field full of memorization that I had taken it to be, but in fact new exciting concepts were being discovered every day. I learned that Physics is a very social activity, with groups of friends and roommates meeting every week to solve the math and physics problems sets. I also found that there is barely any time to digest the implications of the physics that you learn as an undergraduate – and when I graduated in 1981 I was curious to go on to graduate school and have a chance finally to focus on the physics.

I chose to go to the University of California, Berkeley, for graduate school because I wanted a wide variety of strong experimental research groups to choose from, and this was something that Berkeley appeared to offer well beyond the other top physics departments. When I started grad school, my goal was to find a research project with real data (not just theory) that would address a deep philosophical question. The most likely route for this was a particle physics accelerator experiment, and I expected to end up doing this; but first I thought I should look to see if I could learn on a more small-group project. In my second year I was lucky enough to find an unusual, dynamic, eclectic research group led by Professor Richard Muller.

Rich followed – perhaps outdid – the tradition of his mentor Luis Alvarez to pursue interesting research topics wherever they arose, with a can-do experimentalist's attitude – and rigorous skepticism of all claims. When I joined the group, the group members were reporting at the weekly meeting on their work: (1) an idea for a new fundamental physics experiment, measuring the gravitational deflection of starlight by Jupiter; (2) a robotic-telescope supernova search to measure the Hubble constant; (3) a Raman scattering measurement of atmospheric carbon to study the carbon cycle; (4) a table-top cyclotron for radioisotope dating; (5) cosmic microwave background measurements searching for anisotropy; and (6) sundry topics, including implications of the impact theory that Luis had recently developed with his son to explain the extinction of the dinosaurs. This was the tradition of physicists at their best: playing with ideas, building toys. (Caution: physicists at play!)

I soon focused on the robotic-telescope supernova project, since it seemed to offer the possibility of a fundamental measurement, the Hubble constant. I ended up developing the software (and a little hardware) that made it possible to automatically identify the supernovae in the images, and rule out sources of confusion such as asteroids and cosmic rays. I had always enjoyed playing with computers, and the latest computers were just getting fast enough and with enough memory to make possible the near-real-time image analysis that triggered the follow-up of the supernovae. By the time I graduated with my PhD in 1986 the automated supernova search was successfully up and running, and I was asked to stay on as a postdoc to get the results from this project. (Along the way, I had of course been drawn into the myriad topics of Rich's group, and my thesis actually used the same robotic telescope and image analysis techniques to search for a possible companion-star of our Sun, which had been proposed to explain *periodic* mass extinctions on the earth every 26 million years – the extinction of the dinosaurs was just one of these.)

In 1980 when Rich had begun the Hubble constant project, it looked likely that the so-called Type II supernovae would be used as a distance indicator for the measurement, with a calibration based on the Baade-Wesselink expanding photosphere method (later studied by my co-Laureate Brian Schmidt). But by the time the project was up and running in 1986, there was evidence (presented particularly strongly by Gustav Tammann and his student Bruno Leibundgut) that the new sub-classification of Type "Ia" supernovae could be used as an alternative (perhaps better) distance indicator. This news prompted Dr. Carl Pennypacker (a more senior researcher in Rich's group) and me to think about possible implications for new projects. The Type Ia supernovae were significantly brighter than the average Type II, so they could in principle be studied at much further distances. Ever since the first supernovae were studied in the 1930's there had been the hope that they could someday be used to measure the deceleration of the universe's expansion – now the uniformity of the new Type Ia subclassification opened up the possibility that this idea could be revisited. And now we also had several new tools at our disposal to exploit the Type Ia supernovae: for the robotic supernova search we had worked with the very first generation of "charge-couple-device" (CCD) detectors to be used for ultra-sensitive astronomical imaging, and we had developed the image analysis tools that could sift through the large amounts of digital data that the CCD's produced.

Late in 1987 Carl and I thus proposed a new project: we would design and build a new wide-field camera, the widest ever with a CCD on a 4-meter class telescope, and develop the software to search through tens of thousands of galaxies in one night. Unlike our previous robotic nearby supernova search, which studied one galaxy in each image, this approach would allow us to look at thousands of much more distant galaxies at a time. We estimated that in several years we could in this way discover sufficient numbers of supernovae at redshifts as high as ~0.3 that we could make an excellent measurement of the deceleration parameter. The project got off the ground in 1988 with the support of our skeptical mentor, Rich, and was one of the founding projects of Berkeley's new Center for Particle Astrophysics.

The project began slowly with only two-and-a-half nights of good weather out of more than a dozen nights scheduled over almost two years at the Anglo-Australian 4-m Telescope in Siding Springs, Australia. Still, by 1992, when I was asked to take over from Rich as leader of the supernova research group, we had found a Type Ia supernova at $z = 0.45$ – doubling the world's high-redshift sample. This was at the time the highest redshift supernova known. (The other high-redshift type Ia SNe was the one found in a several year search by a Danish team led by Norgaard-Nielson.) In the next two years, we parlayed this success into access to other large telescopes (with somewhat more reliable weather).

Two key problems stood in our way: relating brightnesses of high- and low-redshift supernovae (measured in different filters); and guaranteeing distant supernova discoveries in advance – and in time to measure their peak brightness. Without such a guarantee, one could not obtain time on the large telescopes needed to study them. By 1994, we had solved these problems and we were able to guarantee entire "batches" of multiple high-redshift supernovae, all still brightening, and all found on a pre-selected date, perfect for scheduling the measurements of brightness and spectrum. Such "guarantees" led us to propose a novel use of the Hubble Space Telescope: precision measurements of distant supernovae, particularly important for the ultra-far $z \sim 1$ supernovae that collaborator Ariel Goobar (then a postdoc in Berkeley, now a Professor at Stockholm University) and I had shown could be used to distinguish among cosmological theories.

Meanwhile, between 1990 and 1993 several approaches had been developed to further calibrate the Type Ia standard candle: David Branch showed that selecting by color could give a standardized set, while Mark Phillips showed that a relationship could be established between the peak luminosity and the timescale of the brightening and fading lightcurve. A beautiful dataset of low-redshift supernovae had been found in the Calan/Tololo supernova search (led by Mario Hamuy, Jose Maza, Mark Phillips, and Nick Suntzeff) that allowed these improved calibrations to be made. So, by late 1994, after establishing that we had an effective approach with our batch discovery and multi-band follow-up of high-redshift supernovae, our now-international team of scientists was working together round-the-clock, collecting new batches of high-redshift supernova data using the best telescopes in the world for the purpose. And so was a new, competing team of experienced supernova researchers, organized by co-Laureate Brian Schmidt.

Finally, in 1997, we were analyzing our haul of 42 Type Ia supernovae at redshifts about $z \sim 0.5$ and finding an odd result: the universe's expansion was apparently dominated by a cosmological constant, or more generically a "dark energy" pervading all space, so it was actually *speeding up* – this didn't fit with known models of physics! We announced this startling evidence for a cosmological constant at the American Astronomical Society January 1998 meeting. Because both our team and Brian's team – including co-Laureate Adam Riess – independently announced matching results at conferences in the beginning of the year, by the end of the year most of the scientific community had accepted the extraordinary findings.

When we started the project we thought that whatever answer we found would be exciting: if the universe were decelerating enough we would know that the universe is finite and is coming to an end in a Big Crunch; if not then we could establish that the universe is likely infinite in space and time, and the inflation theory would have a successful prediction. We could not have hoped for the actual outcome, a surprise that presents a new puzzle to fundamental physics. This is the sort of conclusion to a project that in turn initiates many new projects. We now have the fun of trying to figure out what it is that causes the universe to accelerate. Since 1999 I have been working with colleagues on such new projects, including the development of a new space telescope that can make a much more precise measurement of the expansion history of the universe.

Perhaps when my wife, Laura Nelson, and I send our now eight-year-old daughter, Noa, to college, science will have the next installment of answers – or, better yet, new surprising questions about our world.

MEASURING THE ACCELERATION OF THE COSMIC EXPANSION USING SUPERNOVAE

Nobel Lecture, December 8, 2011

by

SAUL PERLMUTTER

University of California at Berkeley and Lawrence Berkeley National Laboratory Institute, Baltimore, MD, USA.

INTRODUCTION

The "discovery of the acceleration of the expansion of the Universe" does not flow trippingly off the tongue – which is fitting, since the work that led to it was comparably long and tortuous. In this Lecture, I would like to give you a feel for some of the science issues we were facing over the ten years leading up to the discovery. I will primarily use the graphics from the original overhead projector transparencies we were using in those years. Although they are not as beautiful as our modern-day graphics, I hope they will help give some of the texture of what was going on during that period.

AN ANCIENT QUESTION

The question that motivated all this work is something you can imagine that the very first humans might have asked when they walked out of their caves at night and found themselves looking up at the starry sky. Do we live in a Universe that goes on forever in space and will it last forever in time? I think it almost defines what it means to be the very first humans – that they could ask such questions.

For most of human history, this sort of question was a truly philosophical question. It wasn't until the 20th century that we began to have a scientific version of this question. This is partly because Einstein's theory of general relativity gave us some new conceptual tools that made it possible to think about this topic in a more rigorous way. But it's also because Edwin Hubble (1929) measured an expansion of the Universe, which meant that we started to see in more concrete terms what we could mean by the fate of the Universe.

Hubble's observations indicated that we don't live in a Universe that is standing still, but rather one in which all of the distances between galaxies are growing with time. You can then immediately start asking yourself whether it will continue to grow with the same speed over time, or might it slow down because gravity would attract all stuff to all other stuff in the Universe. In fact, you might wonder if it could be slowing enough so that someday it could come to a halt, and then collapse into a Big Crunch that could be an end of the Universe.

This is a question about the future of the Universe that you can address by looking into the past of the Universe, by looking to see what was happening billions of years ago and how much the Universe was slowing down back then. If it was slowing enough, you could predict that it is slowing enough to collapse in the future. This was understood, even back in the 1930s, in the decade following Hubble's discovery, by astronomers like Walter Baade and Fritz Zwicky who were studying supernovae. They saw that in principle, you could use a very bright exploding star, a supernova, to perhaps answer this question.

SUPERNOVAE AS PROBES OF THE HISTORY OF THE UNIVERSE

You could take the brightness of a supernova as an indicator of how far away it is: The fainter it is, the further away it is from us – and hence its light has taken more time to reach us. So with the fainter supernovae, you are looking further and further back in time. You can also use the colors of the spectral features of a supernova: a supernova would look blue if it were seen nearby, but when you see it very, very far away it looks red. How red it gets tells you how much the Universe has stretched since the supernova exploded, because while the light is traveling to us, its wavelength stretches by the exact same proportion as the Universe stretches. So, this is a very direct way of plotting how much the Universe has stretched as a function of time. In principle, if you observe enough supernovae and plot them on such a graph, you can see just how much the Universe has decelerated in the past and make a prediction about how much it will slow down in the future.

Baade (1938) wrote about the possible supernova measurement back in the 1930s. The problem was that the supernovae that they knew about at that time were not quite good enough "standard candles" – they weren't all quite the same brightness. They varied by more than a factor of two or three in brightness, so while it looked like a good idea, it was not really practical to do at that time. That's essentially where the problem stood for about the next 50 years (but see Kowal 1968 for a benchmark along the way).

HOW OUR WORK BEGAN

Two things happened in the mid-1980s that got me and others interested in the problem: First, there was the realization that the supernovae could be subdivided into subclasses. It was in the mid-1980s that the "Type Ia" subclass was identified (Panagia 1985, Uomoto and Kirshner 1985, Wheeler and Levreault 1985). It began to be clear that this subclass really was a better standard candle than the others. For example, a histogram of supernova brightnesses in a paper by Nino Panagia (1985) showed that supernovae found in spiral galaxies varied greatly, while those found in elliptical galaxies had only a small dispersion (Fig. 1). The implication was that the ellipticals might host a subclass that was a better standard candle, and also that the spiral galaxies are known to have dust that would add further dispersion. The possibility that we now had a good standard candle suggested that we now might be ready to go back to the original idea of Baade and Zwicky.

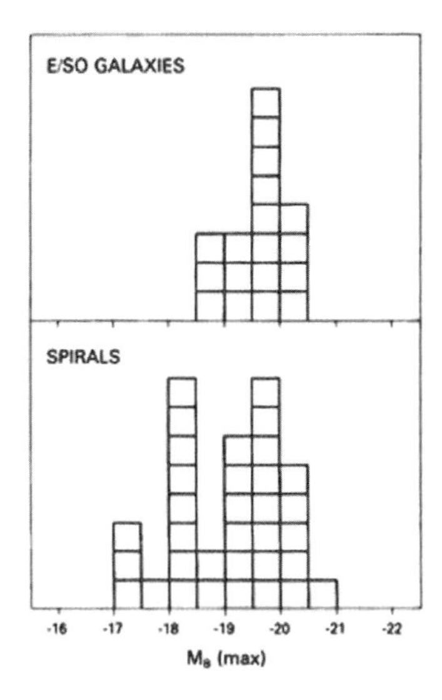

Figure 1. Histogram of supernova absolute magnitudes from Panagia (1985).

The other development of the mid-1980s was the introduction into astronomy of the new sensors, the CCD detectors, which are like the detectors in the back of the digital cameras that most people have today. These were just becoming available in the beginning of the 1980s, and I worked on one of the first astronomy projects to use CCDs. It also included the new computers that were just then becoming fast enough to analyze the large amounts of data that came out of these detectors. This was a project led by my inspiring thesis advisor, Richard Muller. It was doing a search for nearby supernovae using a small robotic telescope and automatic detection of the supernovae by the computers (Fig. 2). I worked on the software that made it possible to subtract the image of a galaxy from the image of a supernova plus the galaxy, and thus find these supernovae (Fig. 3). With the automatic

Figure 2. Saul Perlmutter and Richard Muller with a new telescope the group was automating to replace the Leuschner Observatory telescope (though it was later used for a different purpose at another site).

searching technique we had found 20 supernovae by the time the project stopped (Muller et al. 1992; Fig. 4).

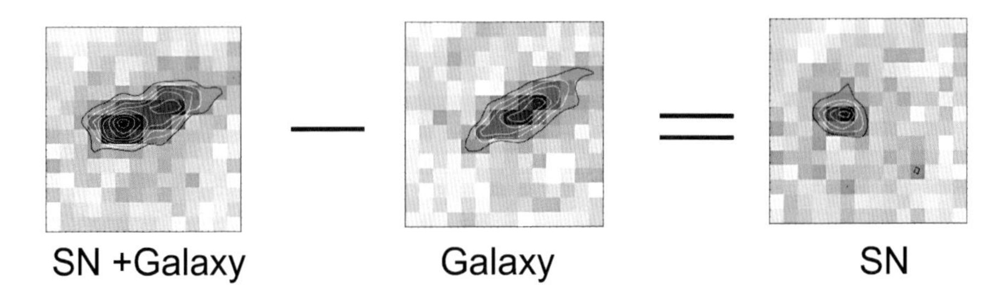

Figure 3. Example of digital image subtraction. From the CCD image of a supernova and its host galaxy, we subtract an image of the galaxy before the supernova appeared (or after it disappeared), leaving an image of just the supernova.

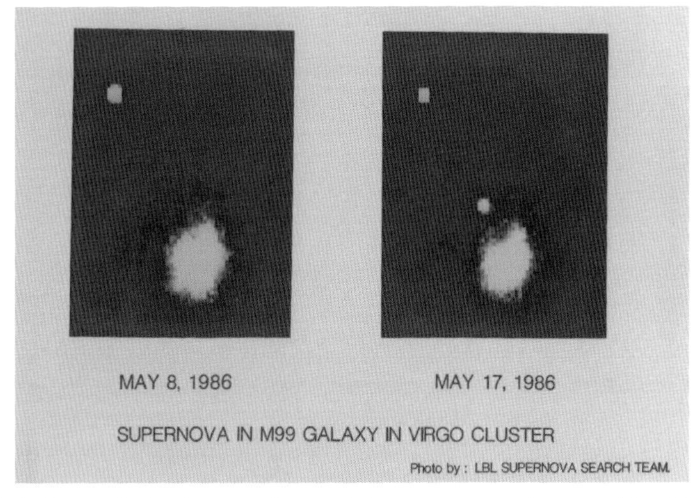

Figure 4. Before-and-after images of one of the supernovae discovered in 1986 by our Berkeley Automated Supernova Search, led by Professor Richard Muller.

These two developments led Carl Pennypacker and myself, both of us researchers in Rich Muller's group at that time in 1987 (I had stayed on as a postdoc after just completing my PhD), to decide to try out the original idea of Baade and Zwicky to measure the deceleration of the universe's expansion using supernovae. We began working on the project with Rich's support, and it looked very promising. However, we were aware that it was not going to be easy. For a number of reasons we knew there was a lot of work to be done to make this possible. So, when we proposed the project we did not get an immediately enthusiastic response from the referees and reviewers. But it's Carl's nature to be absolutely undaunted and to be optimistic that we can do anything, and so we carried on. There were the practical problems of trying to find the more distant supernovae. There were some specific technical problems concerning how you would analyze those very faint, distant supernovae and compare them to nearby examples. There were issues about the

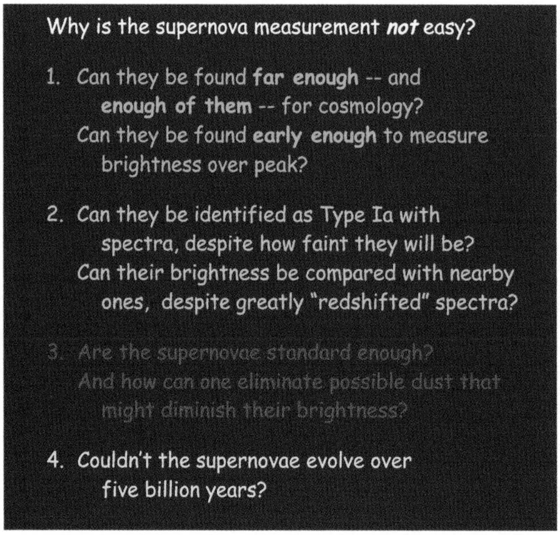

Figure 5. Some of the issues that we recognized as hurdles to be crossed in order to use supernovae to measure the deceleration of the Universe's expansion.

standardness of the supernovae themselves and their consistency over time that needed to be addressed. These were the specific details that we saw as the hurdles that had to be jumped over in order to do the project (Fig. 5).

HOW CAN WE FIND SUPERNOVAE ON DEMAND?

I will try to describe how we addressed the problems, because what is interesting about this particular measurement is that it is so simple to describe that it is possible to explain the difficulties in some detail, and most people can then understand what it would take to do this particular project.

First of all, there are the questions of:

- Can you find these supernovae at all?
- Can you find them far enough away?
- Can you find enough of them?
- Can they be found early enough so that they can brighten, fade away and you can measure the peak brightness?

The peak brightness is what we knew was standard, so we couldn't use them if we found them weeks after they had already peaked. These were very difficult problems, because if you had to choose a research tool, you would never choose supernovae – they are a real pain in the neck to do research on (Fig. 6). They are rare – the Type Ia supernovae explode only a couple of times per millennium in a given galaxy. They are random – they don't tell you when they're going to explode. They are short lived – they brighten and fade away in time scales of weeks, so they don't stick around so that you can study them. So, they are just very inconvenient to do research with.

The difficulties were illustrated when we later saw the results from another supernova project, which we learned had started shortly before ours. Hans Nørgaard-Nielsen and a Danish team searched several years for very distant Type Ia supernovae but found just one, several weeks after it had already

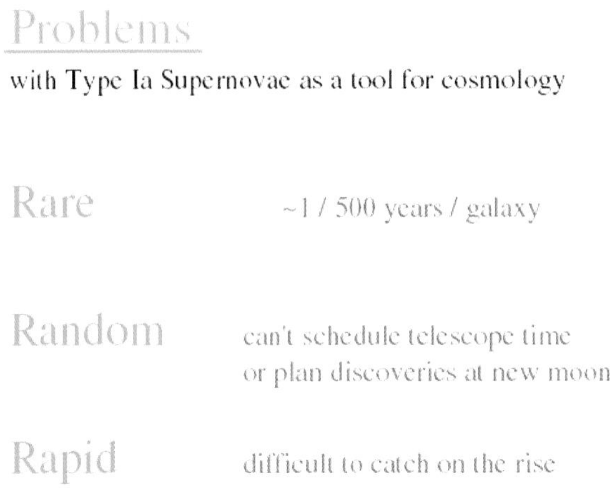

Problems

with Type Ia Supernovae as a tool for cosmology

Rare ~1 / 500 years / galaxy

Random can't schedule telescope time or plan discoveries at new moon

Rapid difficult to catch on the rise

Figure 6. The characteristics of Type Ia supernovae that make them difficult to find and study over their peak brightness.

passed its peak brightness (Nørgaard-Nielsen *et al.* 1989). So, while it was encouraging that such distant supernovae existed, the critics of our project said that this did not look like a very viable program. This was the first concern.

The approach that we took to this problem was to develop the capability to look at more than one galaxy at a time (since looking at one galaxy would have meant waiting the 500 years for a Type Ia supernova to explode), and even more than a small cluster of galaxies at a time (as the Danish program had done). We decided to build a wide-field camera that would allow us, with each exposure, to look at 10 to 20 times as many galaxies as you would find even in a cluster of galaxies. We had to develop an unusual optical system (Fig. 7) that would bring light from a very big field of view onto a small CCD detector. This novel instrument went onto the Anglo-Australian 4-m telescope which meant that we were able to work with a large enough telescope with a large enough field of view to be able to search for supernovae at great distances in thousands of galaxies at a time.

Figure 7. The novel F/1 wide-field CCD camera we developed for the Anglo-Australian 4-m telescope (AAT) to collect a wide enough field to search for $z > 0.3$ Type Ia supernovae in hundreds of galaxies in each image.

With our wide-field camera on the Anglo-Australian Telescope you obtain images like the one shown in Fig. 8a, in which all the small specks of light are the distant galaxies in which we were searching for supernovae. You then take an image another time in the year (Fig. 8b) and subtract it from the first one. You are left with an image that shows just the spot that got bright – and that's your supernova. Of course we did this with software, so we had to go back to our image analysis software and develop that much further. Fig. 8c shows what the computer subtraction looked like, and Fig. 9 shows some of the members of the group at that time.

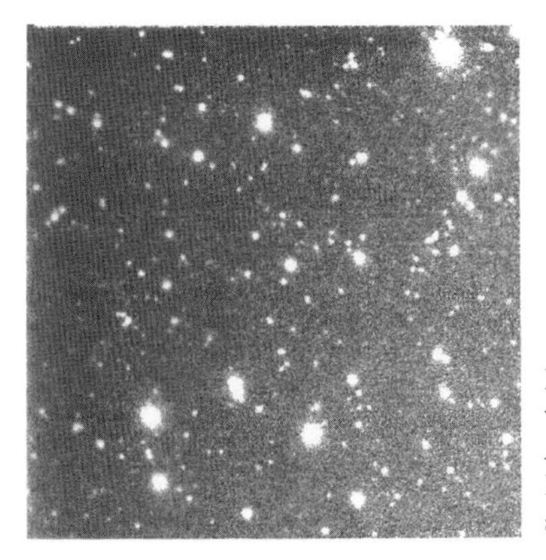

Figure 8a. An image obtained November 1989 with our wide-field camera on the Anglo-Australian 4-m telescope. The small specks of light are the distant galaxies in which we were searching for supernovae.

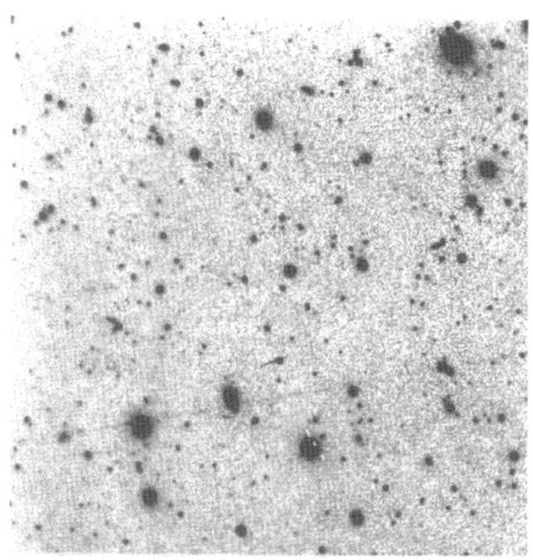

Figure 8b. The same field as in Fig. 8a, but observed January 1990. It is reversed in grayscale to indicate that it will be subtracted from the first.

Figure 8c. Computer subtraction of the Fig. 8b image from the Fig. 8a image. The spot remaining is what a supernova would look like.

Figure 9. Members of the group in 1992 discuss images with many distant galaxies obtained with a wide-field camera for the distant supernova search. Left to right: Carl Pennypacker, Saul Perlmutter, Heidi Marvin Newberg, Gerson Goldhaber, Rich Muller.

This method of finding supernovae seemed to work. However, the remaining problems were still haunting us, because even if you could find the supernova you couldn't prove that you had a supernova, since you would need to schedule the largest telescopes in the world to obtain the identifying spectrum. You would also need to schedule the largest telescopes in the world to follow it as it brightens and fades away, to measure its brightness at peak. Of course, no telescope time assignment committee would give you the time to schedule a telescope six months in advance for a supernova that may or may not appear on the proposed observing date, say March 3rd. As Mario Hamuy *et al.* (1993a) put it, in their discussion of the Calan/Tololo Supernova Search at much lower redshifts, "Unfortunately, the appearance of a SN is not predictable. As a consequence of this we cannot schedule the follow-up observations a priori, and we generally have to rely on someone else's telescope time. This makes the execution of this project somewhat difficult."

So, we had to figure out a way to make the whole operation more systematic, and what we came up with was a new search strategy that I developed to make this possible. Fig. 10 shows the strategy on a timeline of new moon to full moon to new moon. I realized that if you collect all of your first images just after new moon, wait 2 1/2 weeks or so, and collect all of the second set of images just before the next new moon (and then subtract all the second set from the first set), you now have enough galaxies collected with a wide field imager to guarantee not just one supernova discovery, but more than half a dozen supernova discoveries. Once you reach this statistically significant sample size of supernovae, then you can be sure that you always have some new supernovae to observe by the second new moon. The other advantage is that with this short timescale between the two sets of images, you can guarantee that the supernovae won't have enough time to reach maximum and then start fading away; since they rise in a couple of weeks and fade in a few months, this timescale ensures that you always catch them while they're still brightening. You can then guarantee that right before that second new moon (which is the time you need to do the spectroscopy) you will have new supernovae, they will be on the rise, and there will be more than one. Now you can schedule the follow-up spectroscopy and photometry on the following nights.

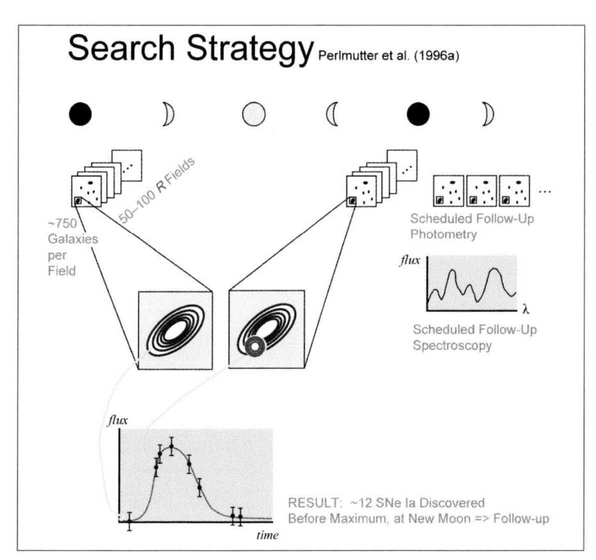

Figure 10. The "batch" observational strategy that made it possible to guarantee multiple new supernova discoveries at high redshift. They would all be on a pre-specified date (in particular just before a new moon) and all while still brightening (Perlmutter et al. 1995a). This in turn made it possible to propose many months in advance for scheduled telescope time at the largest telescopes to study the supernova over the peak of their light curves with both photometry and spectroscopy.

Once we demonstrated this, we were able to start applying for telescope time at the best telescopes in the world – the telescopes in Chile, where the weather is good enough that you can usually follow this whole time series without getting hit by clouds and rain. The first time that we tried this new "batch" observational strategy we called up the International Astronomical Union telegram service, which notifies astronomers around the world when a new supernova is discovered so it can be studied. They're very careful about what they allow in the telegram and I wanted to be sure they would be able – and willing – to accept our results. I warned them that we would be sending them a half a dozen new supernova discoveries two weeks from now. They laughed, because nobody had ever predicted a supernova discovery before, and certainly nobody ever found more than one at a time before, so it sounded a little unusual. In fact, we then sent out such telegrams semester after semester for the following years, as this new observing technique worked, producing batches of supernovae. So, this did work and we had surmounted the problem of making the supernova a systematic tool.

MORE DIFFICULTIES TO OVERCOME

The next problem was that even if you found the supernovae, there was the question of whether you could identify the subtype, i.e. the Type Ia's. It wasn't clear that they would be bright enough to obtain a reasonable spectrum. In fact the first spectra that we obtained looked mostly like a lot of noise. We realized, however, that we could take advantage of the fact that the spectral features of supernovae are very broad on these wavelength scales, unlike the narrow spectral features of the underlying host galaxy and the noise. (This is because the supernova explosion spreads the spectral features out over a wide range of wavelengths, due to the Doppler shifts.) This meant that we could cut out all the sharp lines, and then smooth the whole spectrum, in order to bring out the broad features that then did look like a supernova (Fig. 11).

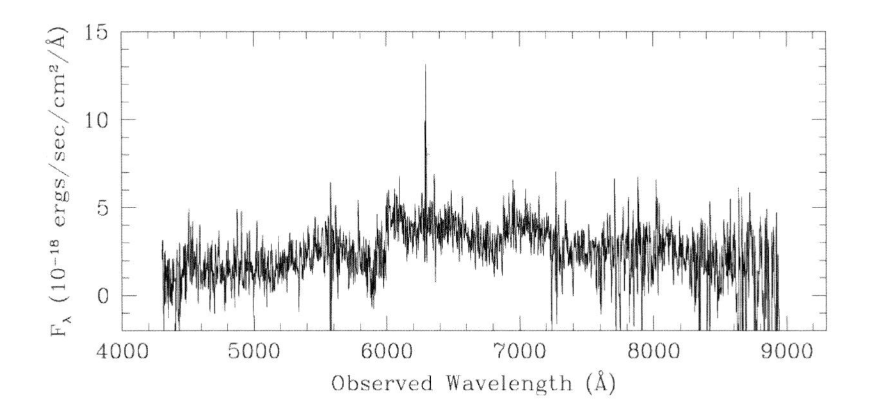

Figure 11a. Spectrum of a high redshift Type Ia supernova (Lidman et al. 2005).

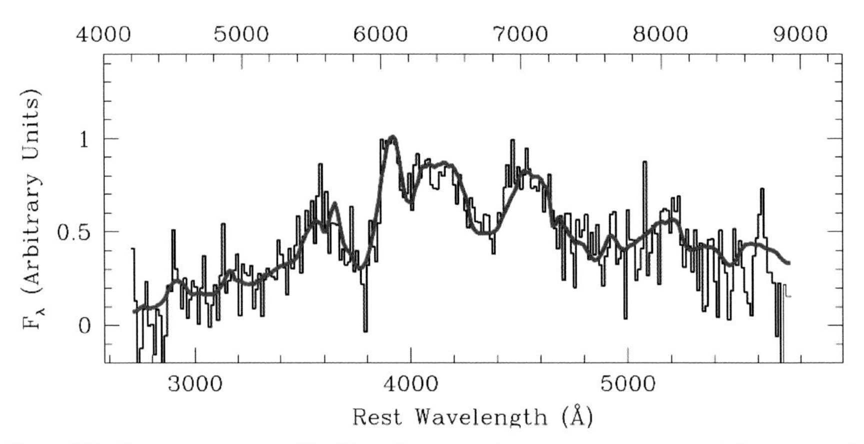

Figure 11b. Same spectrum as Fig. 11a, after removing very narrow spectral features, and smoothing to bring out the broad supernova features. The red curve shows the excellent match with a spectrum of a low-redshift Type Ia supernova, as it would appear redshifted to z = 0.55.

Once we were able to show this, it became believable that we could identify the supernovae at these great distances. It also helped that the larger telescopes came online just at this time, e.g. the Keck telescope that was being developed just upstairs from me at Lawrence Berkeley Laboratory while I was a graduate student. At that time, I had no idea that I would be using it, but just a few years later – as it was being commissioned – it was exactly the telescope that we needed to do this kind of work.

Another problem arises because the spectrum of a low-redshift supernova gets shifted to the redder wavelengths when you look at supernovae at greater distances – that's what we mean by "redshift". Seen nearby, most of the light of the supernova is emitted in the blue wavelengths, so typically at low redshifts we study the supernova using the B-band filter ("B" for "blue"). But, at high redshift the same blue filter would be looking at the very faint UV tail of the spectrum (Fig.12). The question is, how are you going to compare a supernova seen in this faint tail of the spectrum with a supernova seen at the peak of the spectrum. This kind of comparison is called a K correction, and whether it could be done well enough to compare low- and high-redshift supernovae was far from clear. A paper by Bruno Leibundgut (1990) tried to show what the correction would have to be to account for the difference between the different parts of the two spectra. He did a very careful job, but the uncertainties in the correction looked like they would present a significant problem.

In 1992 we found the first high-redshift supernova that we were able to follow throughout its entire light curve. It was at a redshift of $z = 0.45$, (z stands for redshift). When we started trying to analyze it, we saw that the K correction everybody thought we'd have to use was the wrong way to go about it. Instead of trying to find what part of the spectrum we were looking at based on what the blue filter saw, what we really had to do was look at the high redshift supernova with a red filter, the R-band filter. The same part of the spectrum that comes through the R-band filter at high redshift is what comes through the B-band filter at low redshift (Fig. 13). Alex Kim, Ariel Goobar, and I wrote a paper describing this "cross-filter K correction" approach, which made us confident that we could control uncertainties that would otherwise be introduced by the single-filter K correction (Kim, Goobar and Perlmutter 1996; see also Nugent, Kim and Perlmutter 2002).

The next set of issues had to do with how standard these supernovae were. Could you rely on them as a marker of distances? The papers I mentioned earlier, e.g. Panagia (1985), were certainly suggestive of the possibility that the Type Ia supernovae could make a good standard. Bruno Leibundgut (1988) and Gustav Tammann (Tammann and Leibundgut 1990) were really responsible for making the community aware of how consistent the Type Ia supernovae were with each other, during the late 1980s and into the 1990s, so it's important to mention their names in this context. It's also important to note David Branch's group (Branch and Tammann 1992, Miller and Branch 1990, Branch and Miller 1993) in Oklahoma, who were compiling sets of published supernovae, and finding that they could come up with ways

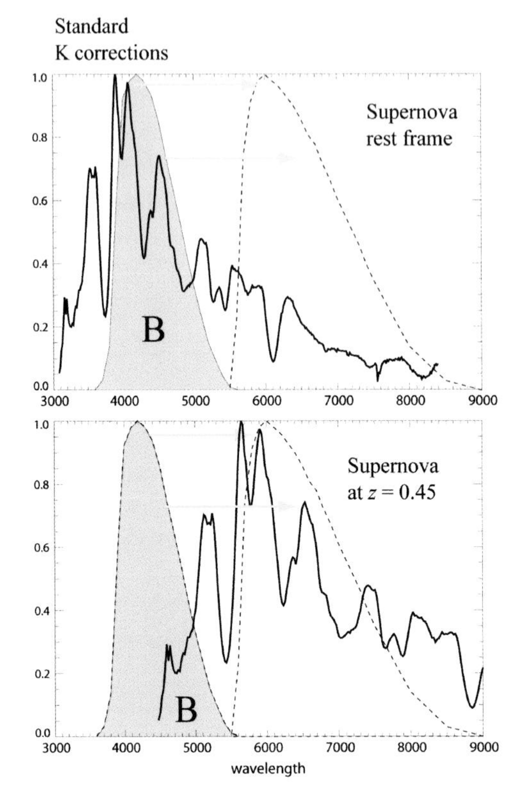

Figure 12. The standard "K correction" was intended to capture the difference between the amount of light in a given filter (here the "B" filter) seen at zero redshift and at high redshift, due to the different parts of the spectrum that would be observed in that filter. This was calculated for high-redshift supernovae by Leibundgut (1990) and by Hamuy et al. (1993b).

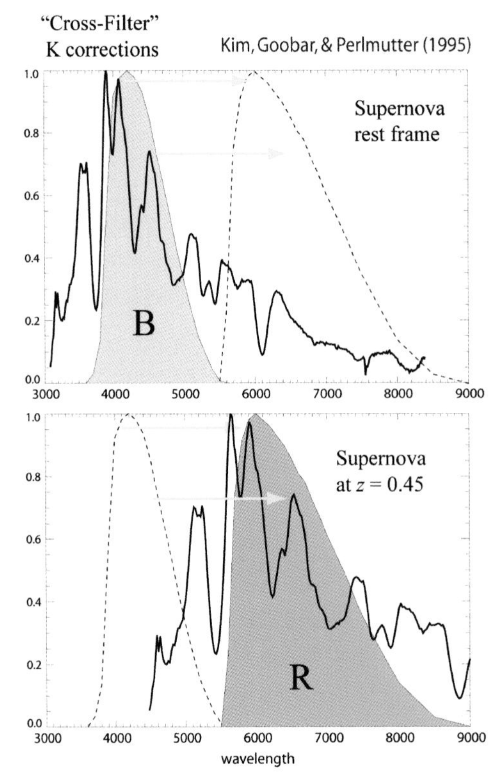

Figure 13. The new "Cross-filter K correction" developed in Kim, Goobar, and Perlmutter (1996) worked by observing the high redshift supernova with a red filter, in this case the R-band filter, so that the same part of the supernova spectrum is coming through this filter at high redshift as comes through the B-band filter at low redshift. This approach makes possible a much smaller uncertainty for the K correction.

to make them even more standard than the earlier papers indicated. One good example of this was the paper by Vaughn *et al.* (1995), which I worked on with the Oklahoma group, that showed the improvement in the dispersion of Type Ia supernovae that was possible by simply rejecting the redder supernovae (Fig. 14). This would remove supernovae that were dimmed by dust (making them appear redder), and also peculiar supernovae that were intrinsically redder. This meant that you could take the dispersion of brightness that you observed from the 40 to 50% range that was seen in the earlier dataset down to something like a 30% dispersion by removing the redder supernovae. The Vaughn *et al.* paper pointed out that this dispersion included measurement errors, and since these were not very well measured data sets the measurement errors could, in fact, already account for most of this 30% dispersion. So, it was possible that the Type Ia supernovae were perfectly good standard candles but, given the quality of data available, we just couldn't tell.

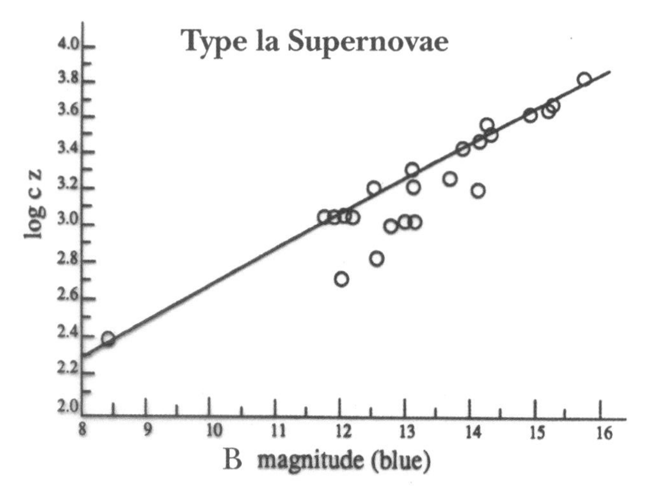

Figure 14. One approach to improving the standardization of the Type Ia supernovae suggested by David Branch's Oklahoma group (Branch et al. 1993, Vaughan et al. 1995) was to remove all the redder supernovae (plotted here as red-filled circles). This apparently removed both the dust-reddened supernovae and many peculiar Type Ia supernovae, thus greatly improving the dispersion about the Hubble line.

CALAN/TOLOLO: A NEW AND IMPROVED LOW-REDSHIFT DATASET

In the early 1990s, the Calan/Tololo supernova search joined in the game. This was a key moment for this field. I should particularly mention the names of Mario Hamuy, José Maza, Mark Phillips, and Nick Suntzeff who were leaders in this work, because they were responsible for setting the next stage. The dataset they developed contained relatively nearby supernovae that were just far enough out into the Hubble flow so that you could measure the relative distances very well (Hamuy *et al.*, 1993a, 1996). Using the exact same color cut from the Vaughn *et al.* paper, you could bring the brightness dispersion down to about 18%, just by using this better measured dataset. Now, this was good enough for the cosmological measurement we wanted to make.

There was another approach to this that Mark Phillips proposed in the early 1990s: You could use the timescale of the supernova event, e.g. how rapidly it decayed from its peak brightness, to estimate what its brightness had been at the peak of its light curve (Fig. 15) (Phillips 1993). Mark saw this in the previously published supernova data, but even more clearly in the new Calan/Tololo data. Adam Riess then came up with a sophisticated statistical analysis of this, adding and subtracting template light curve shapes (Riess, Press and Kirshner 1995). Our group used a third method of capturing this supernova timescale that I had developed: Simply stretching or compressing the time axis of the light curve by a single "stretch factor" (Perlmutter *et al.* 1995a, 1997, 1999; see also Goldhaber *et al.* 1995, 2001).

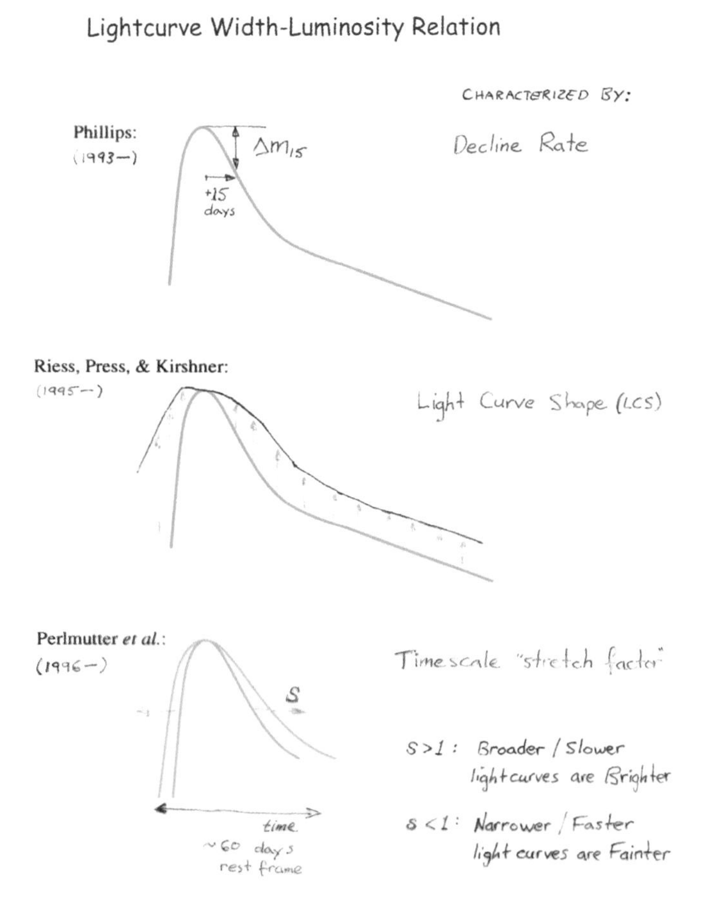

Figure 15. Three alternative approaches to using the time scale of the supernova event (or the shape of the light curve) as an indicator of how bright the supernova reached at peak. These methods followed Phillips' (1993) recognition that the faster the supernova's decline the fainter its peak magnitude.

With this approach, you could take the range of beautifully measured Calan/Tololo data that includes brighter/slower light curves as well as fainter/faster light curves, and for each one measure the stretch factor that would be needed to make them all match each other – and this stretch factor would predict the brightness of each one. In fact, if you "correct" each light curve by appropriately brightening the faster ones and dimming the slower ones while stretching or compressing their timescales, you can standardize all the light curves so that they all lie right on top of each other (Fig. 16). These techniques brought the field another step forward, further improving the brightness dispersion from 18% down to the 12-15% range.

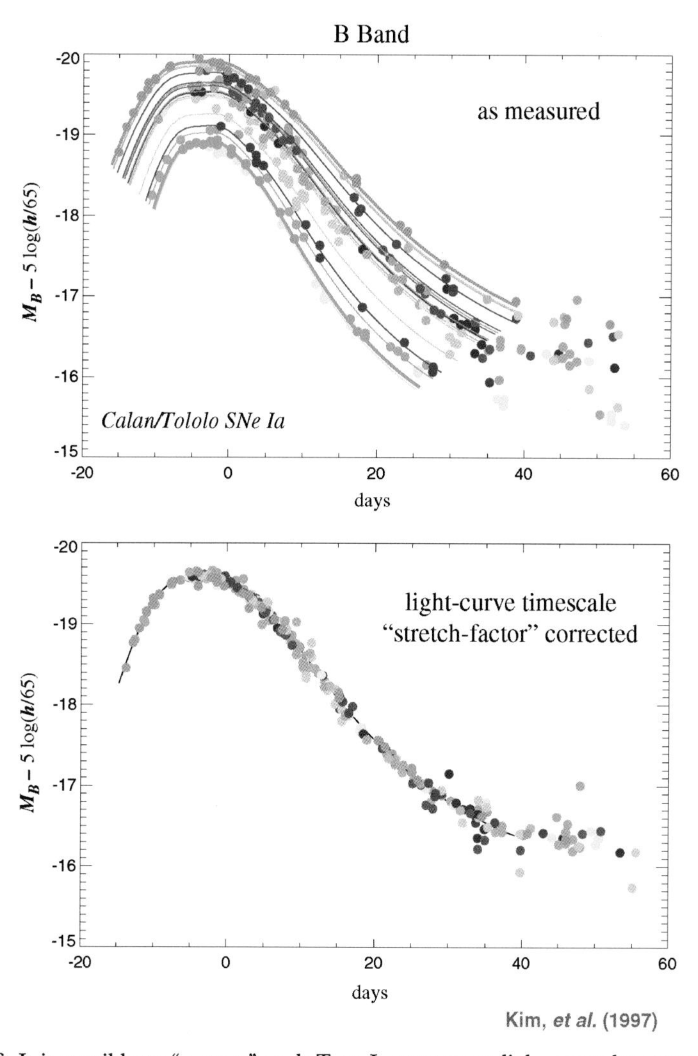

Figure 16. It is possible to "correct" each Type Ia supernova light curve by appropriately brightening the faster ones (low "stretch factor") and dimming the slower ones (large "stretch factor") while stretching or compressing their timescales. The lower panel shows the result of this, using a linear relation between the stretch of the light curve timescale and its peak luminosity.

With respect to the problem of dust dimming the supernova, I mentioned earlier that the color cut proposed by the Oklahoma group in the Vaughan *et al.* paper not only removed intrinsically redder supernovae, but also supernovae that were dimmed (and reddened) by dust. Another way to handle dust is to measure how red each supernova is, and then correct the brightness for that amount of dust. It was also found that such a color correction can account for some color-brightness relation intrinsic to the supernovae themselves, since it appears to follow a very similar trend to dust's (Tripp 1998, Tripp and Branch 1999). As I'll discuss later, once we had proven the batch-discovery-and-follow-up technique with that first batch of discoveries we were able to start getting the telescope time to make better supernova measurements, including the good color measurements needed for this purpose.

An important element of this dust story is that apparently most supernovae are not suffering much dimming by dust. At the time, you could already look at the range of relative brightnesses and colors of nearby supernovae and see that most supernovae were in a very narrow range – it would be spread more if dust were prevalent. And, of course, what you really care about if you are comparing nearby and distant supernovae is whether the range of supernova colors is the same for both groups, indicating essentially the same dust-dimming. So we developed two approaches to this color-study of dust: You could compare the nearby and distant distributions of supernova color, or you could correct each individual supernova's brightness using the color as an indication of how much correction is needed. (We used both of these approaches in the results that I'll be discussing, and we were thus able to robustly account for the dust with our measurements.)

There is also a third handle on dust that we began presenting at meetings, as an option for the future if we could find even more distant supernovae. At these much higher redshifts it becomes possible to differentiate the dimming due to dust from the cosmological effects. At great distances the supernova events are so far back in time you would not expect dust dimming to increase with distance the same way it does at closer distances. The farther back you go, the more the dimming should be due to the cosmology and not to dust (Fig. 17). In 1998 (Aldering *et al.* 1998), using the Hubble Space Telescope, we discovered, spectroscopically confirmed, and measured the light curve of a supernova with a redshift of 1.2. It is the first supernova ever found at a redshift above $z = 1$. We nicknamed it Albinoni (Fig. 17), and it was the kind of supernova needed to perform these tests. This approach was then taken to its conclusion by Adam Riess and colleagues, in some beautiful Hubble Space Telescope work (Riess *et al.* 2004).

Finally, there is the concern that the supernovae are not guaranteed to act the same over billions of years. Since we are looking so far back in time with these studies, the worry would be that the supernovae back then might not be the same as supernovae today. It would then be meaningless to compare the brightnesses if they were not identical. Our group realized that there was a nice route to addressing this problem: Just as the low-redshift supernova sample had been separated into those hosted by elliptical galaxies or spiral galaxies, the same study could be performed for the high-redshift supernovae. Then the cosmology could be determined separately for the low- and high-redshift supernovae found in elliptical galaxies and for those in spirals (Fig. 18). These different host galaxy environments have very different histories, so if the cosmology results from these different environments agree we have a strong indication that the results are not strongly distorted by the environmental histories changing the behavior of the supernovae.

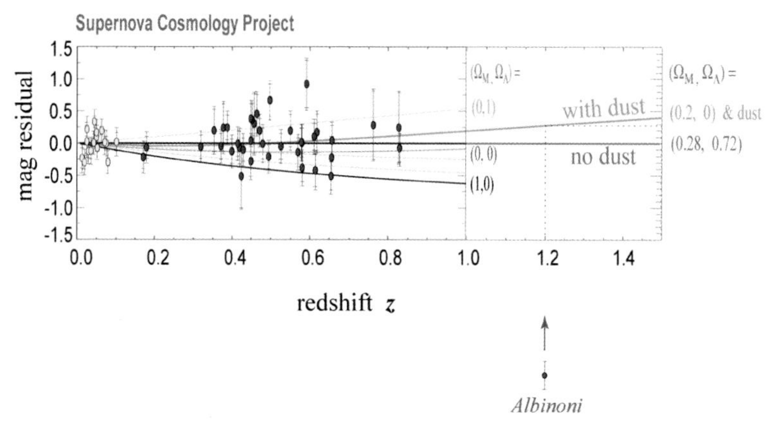

Figure 17. With the discovery (Aldering et al. 1998) of SN 1998eq (nicknamed "Albinoni"), we showed that it was possible to discover and spectroscopically confirm a Type Ia supernova well beyond redshift 1. We suggested that measurements at these very high redshifts would make it possible to separate dust from cosmology in the dimming seen on the Hubble diagram.

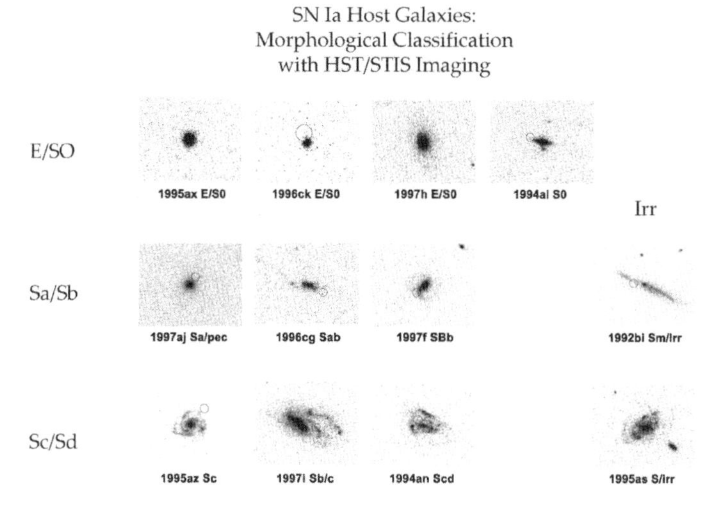

Figure 18. By performing the cosmology measurement with separate subgroups of supernovae found in different host-galaxy environments, it is possible to test that the measurements are not strongly distorted by the different evolutionary histories at low- and high-redshift. We proposed such tests in Perlmutter et al. (1995b, 1997, 1999) and implemented them in several stages, leading to the HST-morphology-based study shown here from our Sullivan et al. (2003) paper.

A NEW CONCERN

By mid-1994, we had answers in hand to the series of concerns listed in Fig. 5 about using distant supernovae for the cosmological measurement – but then we added one further new concern (Fig. 19). We began thinking about this when we were analyzing the cosmological implications of that very first high-redshift supernova we found in 1992. Of course, that supernova by itself did not give you a very definitive measurement. But, it happened to come out with a very low value for the slowing of the expansion of the Universe, and hence the mass density of the Universe responsible for this slowing. In fact, it was so low that we started to think about a mathematical term called the "cosmological constant" that Einstein had put into his equations describing the Universe's expansion. Einstein early on rejected this term, once it was learned that the Universe was expanding. If it were there, however, it would have the effect of fighting against gravity's slowing of the expansion. We realized that our first supernova's very low values for the slowing could be due to some of this cosmological constant fighting against gravity – and then how could you tell what you were seeing? Is it less mass density, that is, less slowing due to gravity, or is it perhaps more of the cosmological constant? The concern was that we wouldn't be able to tell these apart.

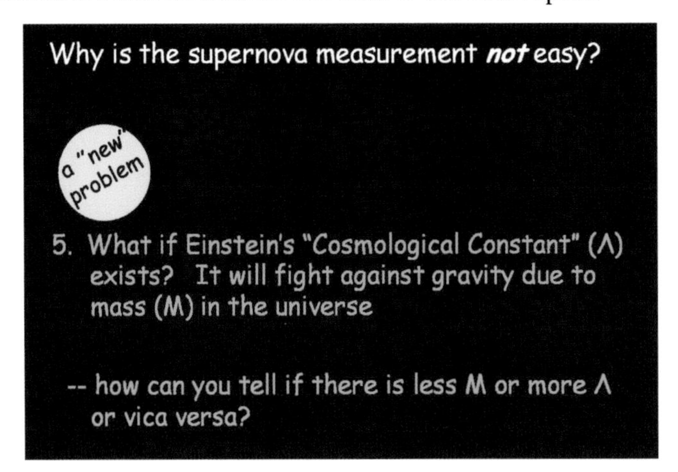

Figure 19. By 1994 there was a new concern about the supernova measurement – in addition to the four problems described in Fig. 5.

The problem distinguishing between less mass or a stronger cosmological constant is illustrated in Fig. 20a (from a paper that Ariel Goobar and I wrote together). Say you have a supernova with redshift $z = 0.5$ (which we didn't yet, so this was still hypothetical) and plot the combinations of mass density and the density of the cosmological constant consistent with its brightness. You get a steadily rising strip of possible values. With just this plot, you wouldn't know if the true values describing our Universe were at the lower-left part of this strip, with low mass density and low cosmological constant density, or at the upper-right with high values for both densities. What Ariel and I realized was that if you go out to a larger range of redshifts, that is, if you study much farther supernovae at redshifts as high as $z = 1$ and beyond, then this plotted strip of allowed cosmological values starts to rotate (Fig. 20b). Then the intersection of these strips for different redshift supernovae allows us to separate out these two effects – the mass density and the cosmological constant density on the expansion history of the Universe. Fig. 21 shows the plot we used in the paper to show that for a given mass density you could distinguish between a zero cosmological constant – no cosmological constant at all – and a significant cosmological constant. Interestingly, the values we chose for this example of a significant cosmological constant turned out to be very close to the final answer we found. Of course, we were expecting to find the zero cosmological constant result shown. (We pointed out in the paper that it would be easier to make the measurement if there were a cosmological constant, because the error bars are smaller in that region of this plot.)

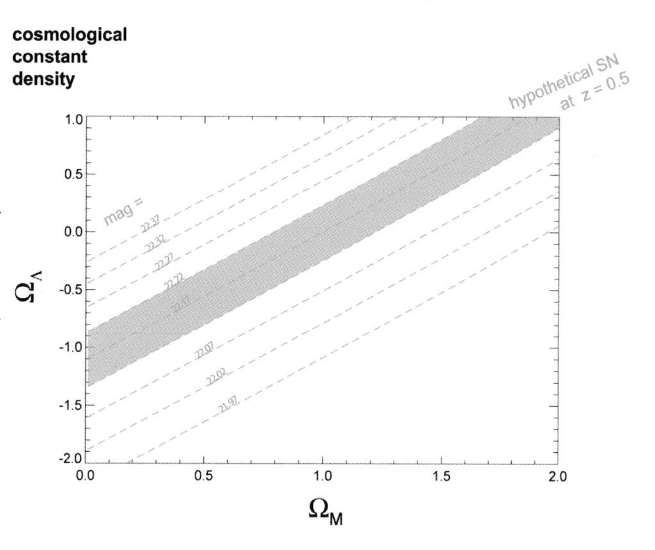

Figure 20a. Goobar and Perlmutter (1995) used this figure to show that a measurement of a supernova at z = 0.5 (hypothetical at the time) would constrain the possible values of the mass density and the cosmological constant density, but they could be traded against each other along the strip of values shown here as a green band.

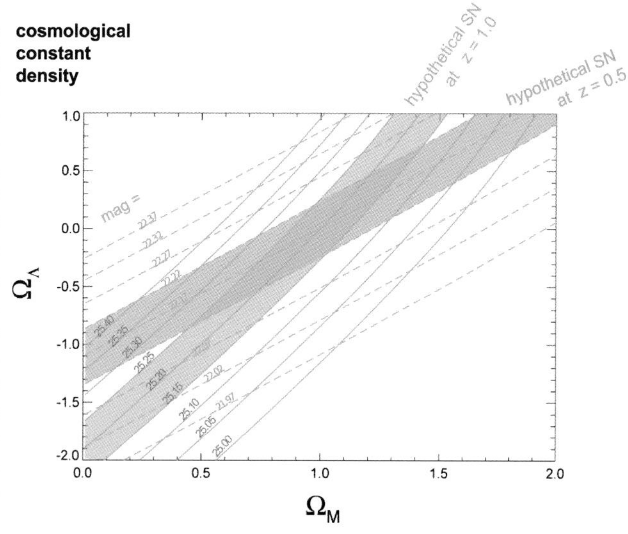

Figure 20b. The strip of possible values rotates in the plane of mass density versus cosmological constant density when we observe a supernova at higher redshifts. The results for a then hypothetical z = 1 supernova are shown here with the blue band. The intersection of such measurements at different redshifts allows the two densities to be distinguished. (Also from Goobar and Perlmutter 1995.)

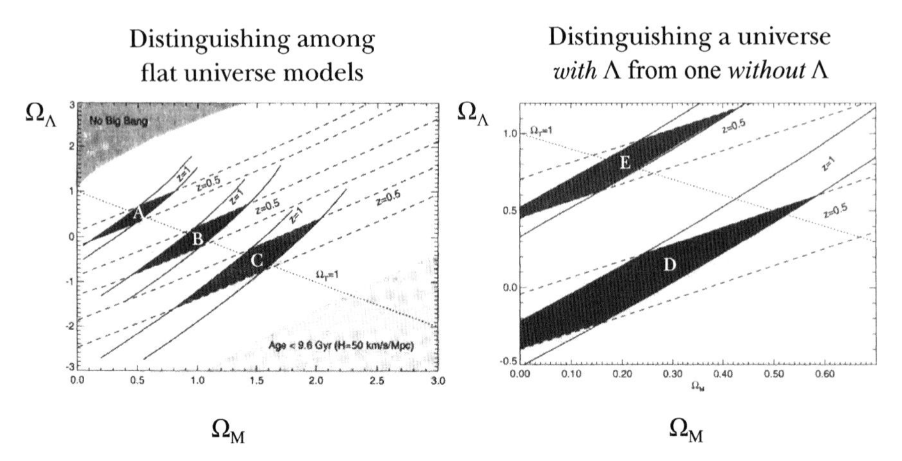

Figure 21. In Goobar and Perlmutter (1995), we gave a range of sample cosmologies to show how supernovae at, e.g., z = 0.5 and z = 1 could distinguish mass density from cosmological constant density. The examples in the right panel turned out to be particularly prescient, since they showed that in a universe with significant cosmological constant the smaller error bands make it easier to confirm the cosmological constant's existence.

AN UNEXPECTED OUTCOME

Now we were in a position at the end of 1994 to use the new on-the-rise/on-demand "batch" discovery strategy over and over again during each of the following semesters at the telescopes. It became a production job: We wanted to turn out enough of these supernovae and make these delicate high-precision measurements in multiple colors so that we could have the strong statistical sample needed to measure how much the Universe is slowing down. We were by now applying to telescopes all around the world to do this. We had to use telescopes to find the supernovae (at the Cerro Tololo 4-meter telescope in Chile), to follow the supernovae with spectroscopy (at the Keck 10-meter telescope in Hawaii), to follow the supernova with photometry (at the Isaac Newton telescope in the Canary Islands and the WIYN telescope in Tucson).

So it was a pretty dramatic scene during the weeks that we would conduct one of these supernova campaigns. One team would be flying down to Chile and then returning to Chile three weeks later for the "discovery images," while another team back in Berkeley would be pulling the data in near-real-time over the then-fledgling internet to analyze the data. Meanwhile teams would head out to Hawaii, the Canary Islands, and Tucson, with email going back and forth updating everything we knew about each supernova in our batch of discoveries. By this time, the whole Supernova Cosmology Project (SCP) team was more than the three or four people that we had at the beginning. Fig. 22 shows much of the team at that time. I want to emphasize what a capable, creative, and dedicated group of people this was – and it made for a collaboration that was a great example of teamwork.

Figure 22. A collaboration meeting of the Supernova Cosmology Project in the late 1990s. *Top row:* Greg Aldering, Saul Perlmutter, Isobel Hook; *Second row:* Sebastian Fabbro, Alex Kim; Third row: Rob Knop, Pilar Ruiz-Lapuente; Bottom row: Peter Nugent, Ariel Goobar, Gerson Goldhaber. (*Not shown:* Reynald Pain, who was taking the picture.)

Semester by semester, we started to build up this increasingly larger sample of supernovae, covering an increasing redshift range. In Fig. 23 the color coding shows the supernova redshifts for each batch of supernova discoveries from a given semester's search-and-follow-up. First a half-a-dozen, and then a dozen, and then another dozen... and by 1997 we had enough supernovae in hand to get results that were statistically significant.

As we added the new batches of high-redshift supernovae to the Hubble diagram, publishing the results at each step, the history of the Universe's expansion slowly began to be apparent. The very first data (the red points around a redshift of $z = 0.4$ in Fig. 24a) appeared to favor a slowing Universe with no cosmological constant, but with only 7 supernovae the uncertainties were large. These were the very first high-redshift supernovae we had studied. After we had used them to show that the batch-discovery-and-follow-up method worked, we were able to request and obtain sufficient follow-up time on major telescopes to make more comprehensive measurements of the following batches of supernovae. So the next supernovae were all much better measured.

In a *Nature* article that appeared 1 January 1998, we then reported that even one very-well-measured supernova – it had Hubble Space Telescope observations – at twice the redshift (the red point at $z = 0.83$ in Fig. 24a) already began to tell a different story, pointing to a Universe with a cosmological constant! But the evidence really became strong almost immediately afterwards with 42 supernovae (the red points in Fig. 24b). Now there was a clear bulk of the supernova data indicating a Universe that is dominated by a cosmological constant, not ordinary matter.

Plotting the same data as an Expansion History of the Universe (Fig. 25a), we see that it doesn't match any of the entire range of possible slowing histories shown (the curved lines on the figure). Apparently we don't live in a Universe that is currently slowing in its expansion, but rather a Universe with one of the more interesting histories shown in the panel of Fig. 25b. Its expansion rate used to be slowing, but has been speeding up for the last half of its history, and presumably could speed up forever.

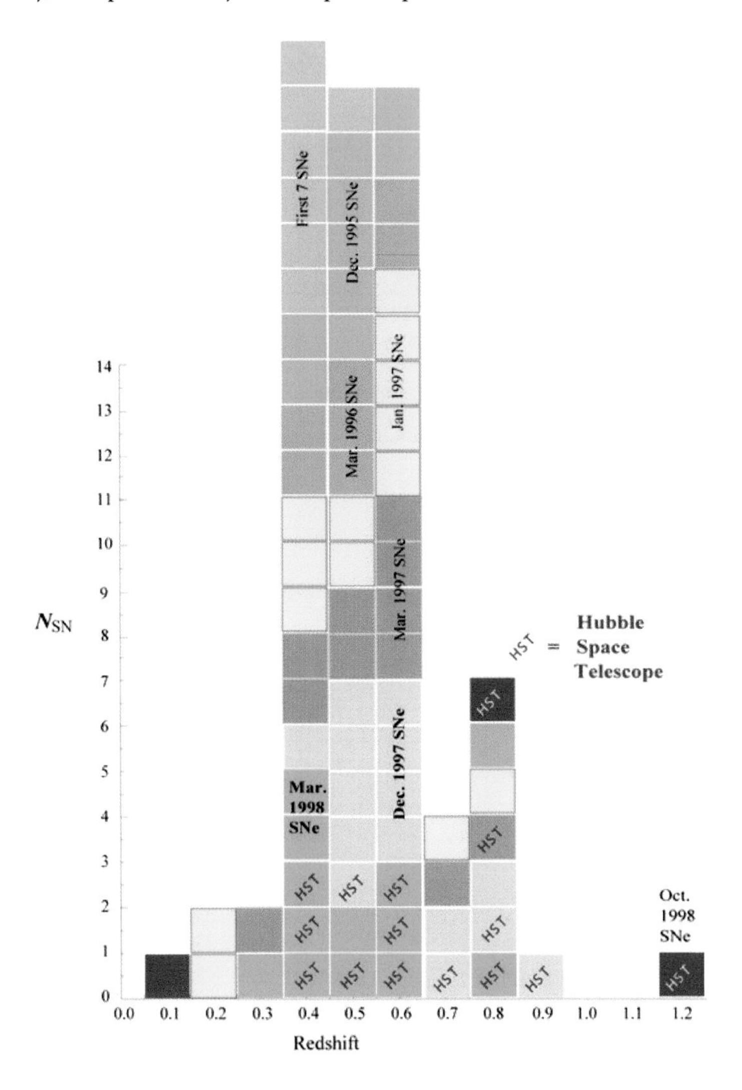

Figure 23. Each semester (shown in different colors) between 1994 and 1998 we employed the "batch search-and-follow-up" method to add another set of Type Ia supernovae. The distance (redshift) range increased, and we began to use the Hubble Space Telescope to follow the SNe to obtain the most precise measurements.

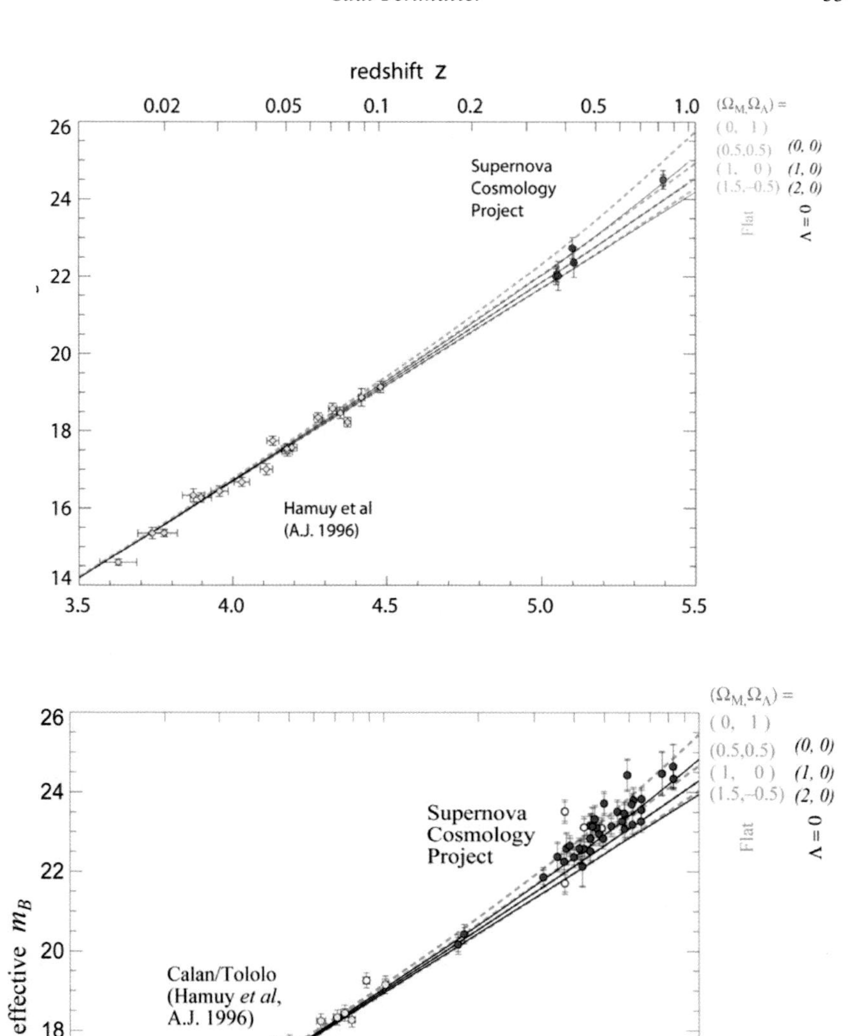

Figure 24. Fig. 24a top – Fig 24b bottom. As the SCP added the new batches of high-redshift supernovae to the Hubble diagram the history of the universe's expansion slowly began to be apparent (Perlmutter et al. 1998). The very first data (the red points around a redshift of z = 0.4 in Fig. 24a) appeared to favor a slowing Universe with no cosmological constant, but with only 7 supernovae the uncertainties were large. Even one very-well-measured supernova – it had Hubble Space Telescope observations – at twice the redshift (the red point at z = 0.83 in Fig. 24a) already began to tell a different story. But the evidence really became strong with 42 supernovae (the red points in Fig. 24b). Now there was a clear bulk of the supernova data indicating a Universe that is dominated by a cosmological constant, not ordinary matter. Its expansion is apparently *speeding up*.

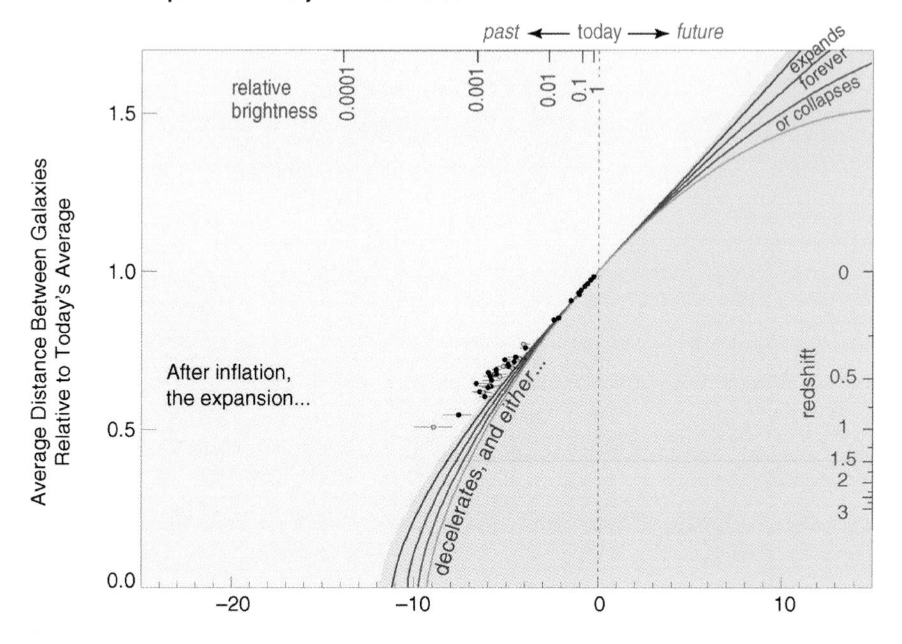

Expansion History of the Universe

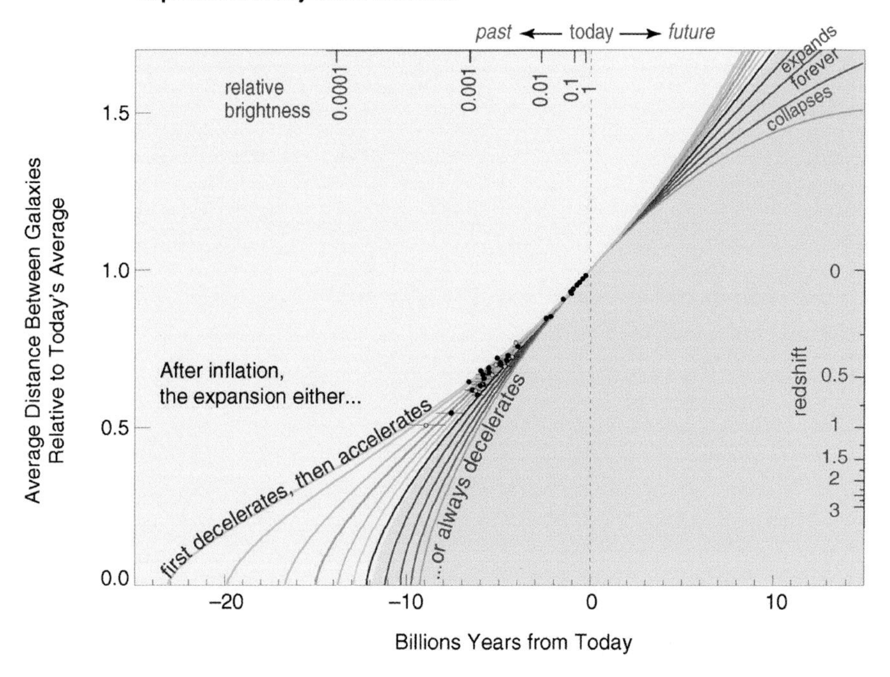

Figure 25. Fig. 25a top – Fig 25b bottom. Our supernova data clearly didn't fit with any of the decelerating options shown in Fig. 25a. To fit the data, we now had to add curves that are currently accelerating, as shown in the blue region of Fig. 25b. The best fit curve was decelerating for about the first seven billion years, and then accelerating for the most recent approximately seven billion years. This was the surprising result the supernovae were showing us. (Based on figure from Perlmutter 2003.)

Apparently we have a Universe that is dominated by some new ingredient, some previously unknown "Dark Energy" that makes the Universe expand faster and faster. It is so rare that you come across something that is not part of our current physics model! This is one of the best outcomes that you could ever get in a project like this. I feel very lucky to be able to work on this at all, because here was a project where any result you could discover would be exciting: We might have found that the Universe was infinite, or that it was finite and going to come to an end. Either of these results would have been great. Instead we found an answer that was even better than "great", in that it was a surprise. This is not something that you can even wish for in science.

This outcome is a perfect example of how science can so often be two-headed. On the one hand, it was only possible to discover what came as a surprise to all of us because our field, physics, had already made such great progress in understanding the Universe. Less than a century ago we had no idea that there was more to the Universe than our own Milky Way. The immense size of the Universe, the fact that it is expanding, the fact that it is populated with such things as exploding stars – all this and more had to be discovered before we could do the work that led us to contemplate an unknown form of energy that accounts for more than two-thirds of everything there is.

It's amazing how much we figured out, but on the other hand it's amazing how big a mystery has opened up as a result, and how much we still have left to discover. One of the real pleasures of doing science – which will continue to be true, I believe, on any given day for the next few centuries – is that we have so much knowledge to build upon, yet there is still be so much for us to discover.

These two aspects of science remind us that science is a method, not a finished product. We don't know where it will lead or what new, seemingly magical powers it will give us in the future. We never know whether what we find will turn out to be useful, but we do know that in the past, whenever we made a major step forward in our understanding of how the world works, we've ultimately been able to solve more problems, including very practical problems. I think that's the only way we can proceed as basic scientists: We try to see what we can understand, and we hope it opens more possibilities for what we can do in the world.

SCENES FROM A DECADE: AN IMPRESSIONISTIC SKETCH

So far, you may have the impression of this science being a very cut-and-dried activity: We identify each problem, and solve it, and then see the results. But the actual experience of this work is completely different: It's a non-stop whirlwind of activity and people. Unfortunately there are almost no photographs of these 10 year project to show this.

But a few years ago I tried to convey this with the following very fast, impressionistic, verbal sketch of scenes from the decade leading up to the discoveries. (I'll supplement this with the few photos that we do have, Figs. 26–71.)

It begins with brainstorming in Berkeley with Carl Pennypacker (in 1987) as we first batted around hardware and software plans for a new high-redshift SN project in Rich Muller's group, which Rich soon embraced. And then the consequence: The mountaintop observatory cafeteria at Coonabarabran as Carl, graduate student Heidi Newberg, former-graduate student Shane Burns, and I got to know our pioneering Australia-based colleagues, Warrick Couch and Brian Boyle. They were installing and then using our weird crystal-ball of a wide-field corrector and camera at the AAT 4-meter telescope – which led to our first high-redshift (but unconfirmed!) SN.

Figure 26. Carl Pennypacker (left) and Saul Perlmutter.

Figure 27a and b. Rich Muller then and now.

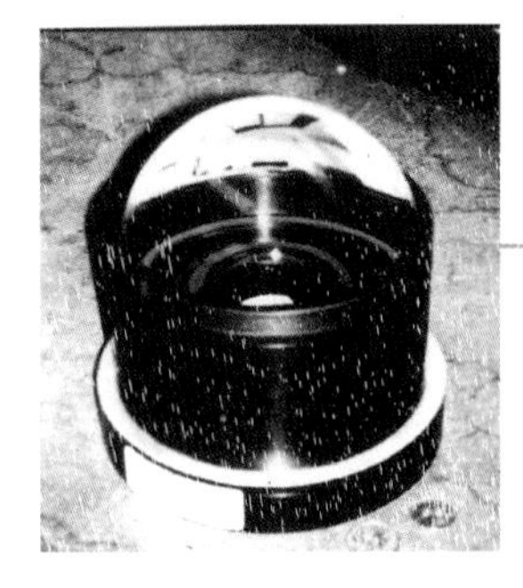

Figure 28a and b. The F/1 wide-field CCD camera being mounted at the Anglo-Australian telescope.

Figure 29. (left to right) Heidi Marvin Newberg, Warrick Couch, Carl Pennypacker, and Shane Burns then.

Figure 30. Shane Burns now.

Figure 31. Brian Boyle.

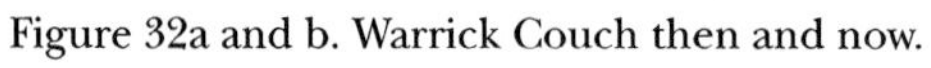

Figure 32a and b. Warrick Couch then and now.

Figure 33. Heidi Marvin Newberg.

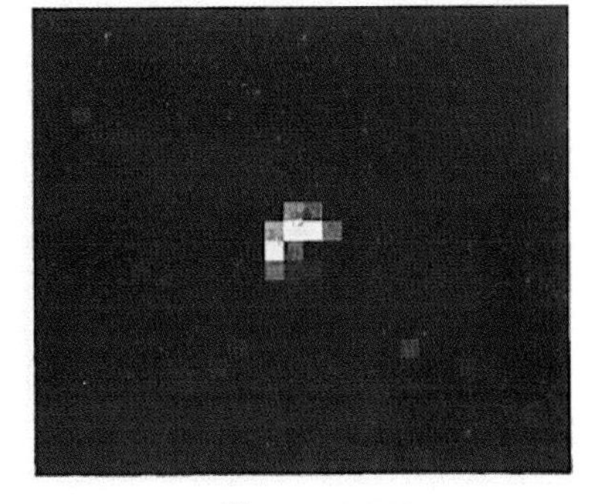

Figure 34a and b. The group's first (unconfirmed) high-redshift supernova, observed at the Anglo-Australian telescope.

Back in Berkeley, I have an image of Gerson Goldhaber overlaying transparencies with negative and positive images of fields full of galaxies – image analysis for the days when the computers were down! I should pause here to say that we are very sad that Gerson isn't here today – he died just over a year ago. He was a warm heart – and sharp eyes! – of our team. He was the host of the collaboration parties, and he should have been here to celebrate with us. We miss him.

Figure 35. Gerson Goldhaber.

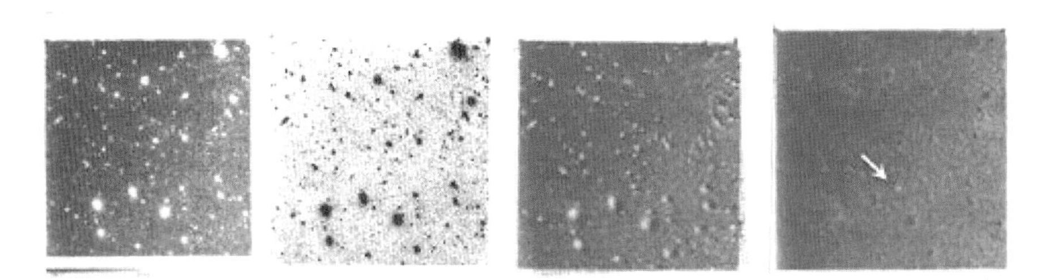

Figure 36a–d. Transparencies showing successive images of distant galaxies with computer subtraction to isolate possible supernova.

In Australia, we had rain, rain, rain, and more clouds – and then the sunny relief of beautiful La Palma where our new Cambridge colleagues, Richard McMahon (working with Mike Irwin) studied the most distant quasars. Long nights debugging a new instrument for La Palma, and tense phone calls to the Isaac Newton telescope while the data was sent to Berkeley for analysis. And then, our first "official" high-redshift supernova – and a crucial La Palma spectrum from the Herschel telescope as Richard Ellis has by now joined the team (after being one of those Danish group pioneers).

Figure 37. Anglo-Australian Telescope in Coonabarabran, Australia (Fig. 37a) and Isaac Newton Telescope on La Palma in the Canary Islands (Fig. 37b).

Figure 38a and b. Richard McMahon then and now.

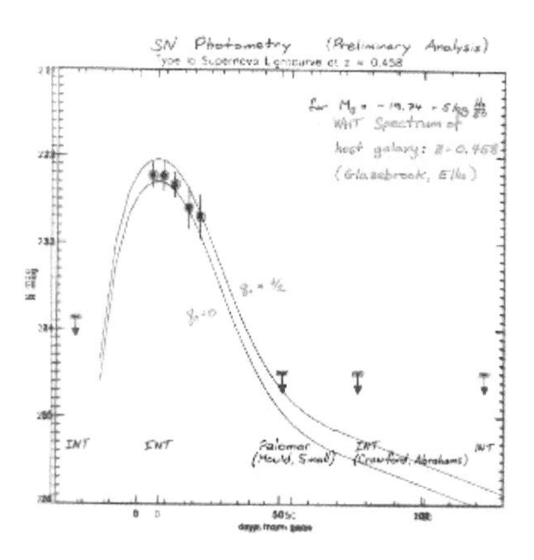

Figure 39a. One of the preliminary graphs showing the brightening and fading light of the group's first official high-redshift supernova, SN 1992bi. The plot is on a transparency that is overlaid with another transparency showing the expected light curve, based on nearby Type Ia supernovae.

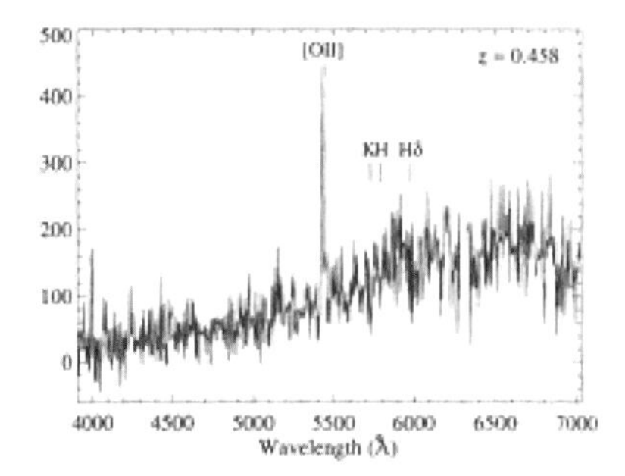

Figure 39b. Spectrum, obtained at the William Herschel telescope on La Palma, of the galaxy that hosted SN 1992bi.

Figure 40. Richard Ellis then and now.

By this time the Europeans had arrived full force in Berkeley – Ariel Goobar from Stockholm kicked it off by developing new analyses with our then grad student Alex Kim, and brainstorming with me about the cosmological measurements. A glimpse of Reynald Pain from Paris assessing the damage (and successes!) at the end of a complex telescope run – the first of the so-called "batch discoveries." This epoch ends in my mind with a celebratory party at Gerson Goldhaber's in the Berkeley hills, where we have a bottle of champagne for each of the half dozen SNe discovered in a batch.

Figure 41a. Ariel Goobar *(left)* and Carl Pennypacker in 1998.

Figure 41b. Ariel Goobar now.

Figure 41c. Alex Kim.

Figure 41d. Reynald Pain.

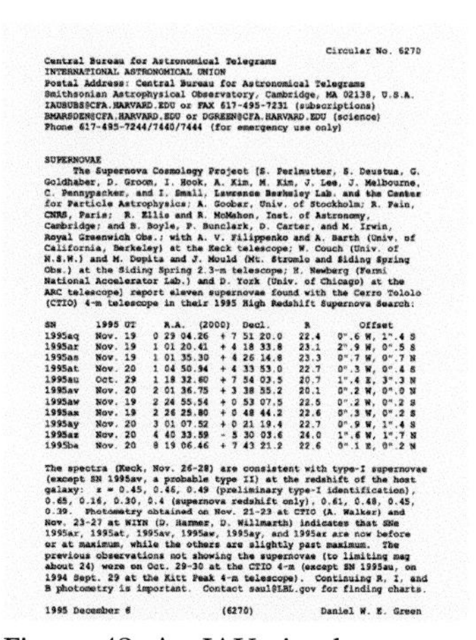

Circular No. 6270

Central Bureau for Astronomical Telegrams
INTERNATIONAL ASTRONOMICAL UNION
Postal Address: Central Bureau for Astronomical Telegrams
Smithsonian Astrophysical Observatory, Cambridge, MA 02138, U.S.A.
IAUSUBS@CFA.HARVARD.EDU or FAX 617-495-7231 (subscriptions)
BMARSDEN@CFA.HARVARD.EDU or DGREEN@CFA.HARVARD.EDU (science)
Phone 617-495-7244/7440/7444 (for emergency use only)

SUPERNOVAE

The Supernova Cosmology Project [S. Perlmutter, S. Deustua, G. Goldhaber, D. Groom, I. Hook, A. Kim, M. Kim, J. Lee, J. Melbourne, C. Pennypacker, and I. Small, Lawrence Berkeley Lab. and the Center for Particle Astrophysics; A. Goobar, Univ. of Stockholm; R. Pain, CNRS, Paris; R. Ellis and R. McMahon, Inst. of Astronomy, Cambridge; and B. Boyle, P. Bunclark, D. Carter, and M. Irwin, Royal Greenwich Obs.; with A. V. Filippenko and A. Barth (Univ. of California, Berkeley) at the Keck telescope; W. Couch (Univ. of N.S.W.) and M. Dopita and J. Mould (Mt. Stromlo and Siding Spring Obs.) at the Siding Spring 2.3-m telescope; H. Newberg (Fermi National Accelerator Lab.) and D. York (Univ. of Chicago) at the ARC telescope] report eleven supernovae found with the Cerro Tololo (CTIO) 4-m telescope in their 1995 High Redshift Supernova Search:

SN	1995 UT	R.A. (2000)	Decl.	R	Offset
1995aq	Nov. 19	0 29 04.26	+ 7 51 20.0	22.4	0".6 W, 1".4 S
1995ar	Nov. 19	1 01 20.41	+ 4 18 33.8	23.1	2".9 W, 0".5 S
1995as	Nov. 19	1 01 35.30	+ 4 26 14.8	23.3	0".7 W, 0".7 N
1995at	Nov. 20	1 04 50.94	+ 4 33 53.0	22.7	0".3 W, 0".4 S
1995au	Oct. 29	1 18 32.60	+ 7 54 03.5	20.7	1".4 E, 3".3 N
1995av	Nov. 20	2 01 36.75	+ 3 38 55.2	20.1	0".2 W, 0".0 N
1995aw	Nov. 19	2 24 55.54	+ 0 53 07.5	22.5	0".2 W, 0".2 S
1995ax	Nov. 19	2 26 25.80	+ 0 48 44.2	22.6	0".3 W, 0".2 S
1995ay	Nov. 20	3 01 07.52	+ 0 21 19.4	22.7	0".9 W, 1".4 S
1995az	Nov. 20	4 40 33.59	- 5 30 03.6	24.0	1".6 W, 1".7 N
1995ba	Nov. 20	8 19 06.46	+ 7 43 21.2	22.6	0".1 E, 0".2 N

The spectra (Keck, Nov. 26-28) are consistent with type-I supernovae (except SN 1995av, a probable type II) at the redshift of the host galaxy: z = 0.45, 0.46, 0.49 (preliminary type-I identification), 0.65, 0.16, 0.30, 0.4 (supernova redshift only), 0.61, 0.48, 0.45, 0.39. Photometry obtained on Nov. 21-23 at CTIO (A. Walker) and Nov. 23-27 at WIYN (D. Harmer, D. Willmarth) indicates that SNe 1995ar, 1995at, 1995av, 1995aw, 1995ay, and 1995az are now before or at maximum, while the others are slightly past maximum. The previous observations not showing the supernovae (to limiting mag about 24) were on Oct. 29-30 at the CTIO 4-m (except SN 1995au, on 1994 Sept. 29 at the Kitt Peak 4-m telescope). Continuing R, I, and B photometry is important. Contact saul@LBL.gov for finding charts.

1995 December 6 (6270) Daniel W. E. Green

Figure 42. An IAU circular reporting the results of a "batch" SN discovery observing run.

The sociology changes a little as we move to mass production, with new outposts at telescopes around the world, typically manned by a lone team member tenuously connected by a stream of email, phone, and fax. Pilar Ruiz-Lapuente and Nic Walton are at La Palma, Chris Lidman is the voice at the VLT in Chile, Brad Schaefer at Kitt Peak. Larger expeditionary forces head to Cerro Tololo in Chile where all the SN discoveries now are generated (I picture Don Groom and Susana Deustua on one such trip), immediately followed by another team of us rushing with the new list of likely SNe to the oxygen poor mountaintop of the then new Keck Telescope. (I have a memory video-clip of Alex Filippenko – then on our team, his student Tom Matheson, and our new Cambridge PhD Isobel Hook and I crowding round the computer screen as SN after SN proved itself. And at the control center in Berkeley the graduate students working round the clock: I picture Matthew Kim and – from France – Sebastian Fabbro in the cramped room full of students, postdocs, and computers, where our former undergrad intern, Ivan Small, now presides over the growing search software.

Figure 43a and b. Pilar Ruiz-Lapuente then and now.

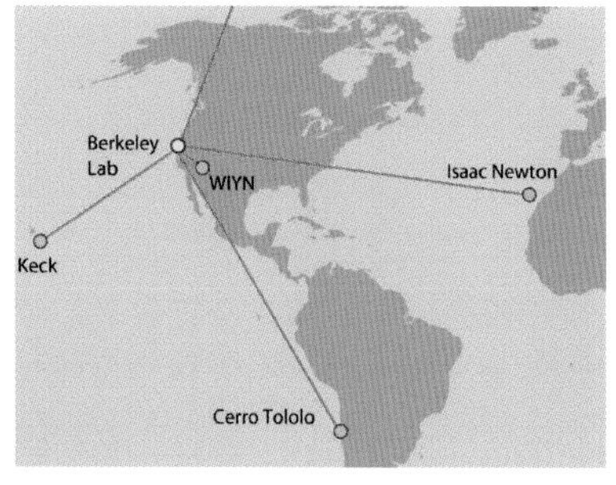

Figure 44. Telescopes around the world.

Figure 45a and b. Chris Lidman then and now.

Figure 46.
Brad Schaefer.

Figure 47.
Don Groom.

Figure 48.
Susana Deustua.

Figure 49.
Alex Filippenko.

Figure 50.
Tom Matheson.

Figure 51a and b. Isobel Hook then and now.

Figure 52.
Matthew Kim.

Figure 53.
Sebastien Fabbro.

Figure 54. Ivan Small.

I then imagine the calm of space as the Hubble Space Telescope quietly does its part of the job, but of course down on Earth the same flurry of humans – here Andy Fruchter and Nino Panagia – make it possible to use this robot effectively.

Figure 55. The Hubble Space Telescope.

Figure 56. Andy Fruchter.

Figure 57. Nino Panagia.

The last act begins with a view of the end-of-night clean up after the next collaboration party, but this time there are entire cases of bottles of champagne left un-drunk – we're lightweights – all labeled with the names of the now-scores of new SNe to be analyzed. A fresh relief team of scientists is now on the field at Berkeley: Rob Knop, who thinks, types – and programs – faster than I talk, Peter Nugent, juggling SN theory and practice, and Greg Aldering pulling together all the strains of the analysis ...and the search.

Figure 58. Matthew Kim *(left)* and Rob Knop, relaxing afterwards.

Figure 59. Rob Knop.

Figure 60a and b. Peter Nugent then and now.

Figure 61a and b. Greg Aldering then and now.

A final push of analyses has all of our Berkeley-based pre-graduate-school interns working nights and weekends (parallel computing at its finest): First Julia Lee, and then Patricia Castro, Nelson Nunes, and Robert Quimby – all of whom continued careers in the field.

Figure 62a and b. Julia Lee then and now.

Figure 63. Nelson Nunes and Patricia Castro then (Fig. 63a), Nelson Nunes now (Fig. 63b) and Patricia Castro now (Fig. 63c).

Figure 64a and b. Robert Quimby then and now.

And we all fall gasping in a metaphorical heap with the surprising discovery about the Universe that you just heard about. ... And that's just our science team!

I think it's pretty clear from all this that the popular image of the lone scientist in a lab looks nothing like our experience: Science is – at least for us – an extremely social activity. This particular work was the product of an amazing community of scientists. In fact, between our two teams we include a large fraction, but not all, of the community of scientists studying supernova. I show here several of the key players that I mentioned in the talk, just *representing* the rest of this supernova community. It was an honor and a pleasure to work with all of you on this.

Figure 65. The Supernova Cosmology Project team in 2007.

Figure 66. The High-Z Supernova Search team in 2007.

Figure 67. José Maza.

Figure 68. Mario Hamuy.

Figure 69.
Craig Wheeler.

Figure 70.
Gustav Tammann.

Figure 71. David Branch.

A WORD OF GRATITUDE

None of this whirlwind of human choreography happens without the constant support of our families and friends, our teachers and mentors – and our staff at the universities, labs, and observatories – and also, in many cases, the really courageous administrators and funders who took risks on things when it wasn't obvious they were going to work. They are represented today by the family and friends who are here and we all thank you for helping – and putting up with – all this.

But the work is not done! We look forward to joining in with the next tag-teams of scientists as we delve into the mystery that we are currently calling Dark Energy.

And, finally, we are grateful for the Nobel Prize committees and foundation, who have found a way to encourage this human activity of science.

Thank you.

REFERENCES

1. G. Aldering *et al.*, *IAUC* **7046**, 1A (1998)
2. W. Baade, *Astrophys. J.* **88**, 285 (1938)
3. D. Branch, G. Tammann, *Annu. Rev. Astron. Astrophys.* **30**, 359 (1992)
4. D. Branch, D. Miller, *Astrophys. J. Lett.* **405**, L5 (1993)
5. D. Branch, A. Fisher, P. Nugent, *Astron. J.* **106**, 2383 (1993)
6. G. Goldhaber *et al.*, in: "Four Papers by the Supernova Cosmology Project", ed. S. Perlmutter, *Lawrence Berkeley Laboratory Report*, **LBL-38400** (1995); also in Thermonuclear Supernovae, ed. P. Ruiz-Lapuente, R. Canal, J. Isern, *NATO ASI Series C* (Kluwer, Dordrecht), Vol. **486** (1997)
7. G. Goldhaber *et al.*, *Astrophys. J.* **558**, 359 (2001)
8. A. Goobar and S. Perlmutter, *Astrophys. J.* **450**, 14 (1995)
9. M. Hamuy *et al.*, *Astron. J.* **106**, 2392 (1993a)
10. M. Hamuy *et al.*, *Pub. Astron. Soc. Pacific* **105**, 787 (1993b)
11. M. Hamuy *et al.*, *Astron. J.* **112**, 2398 (1996)
12. E. Hubble, *Proc. Nat. Academy Sci.* **15**, 168 (1929)
13. A. Kim, A. Goobar, S. Perlmutter, *Pub. Astron. Soc. Pacific* **108**, 190 (1996)
14. C. T. Kowal, *Astron. J.* **73**, 1021 (1968)
15. C. Lidman *et al.*, *Astron. Astrophys.* **430**, 843 (2005)
16. B. Leibundgut, PhD Thesis, University of Basel (1988)
17. B. Leibundgut, *Astron. Astrophys.* **229**, 1 (1990)
18. D. Miller, D. Branch, *Astron. J.* **100**, 530 (1990)
19. R. Muller *et al.*, *Astrophys. J. Lett.* **384**, L9 (1992)
20. H. U. Nørgaard-Nielson *et al.*, *Nature* **339**, 523, (1989)
21. P. Nugent, A. Kim, S. Perlmutter, *Pub. Astron. Soc. Pacific* **114**, 803 (2002)
22. N. Panagia, *Lecture Notes in Physics* (Springer Berlin/Heidelberg) **224**, 14 (1985)
23. S. Perlmutter *et al.*, in: "Four Papers by the Supernova Cosmology Project", ed. S. Perlmutter, *Lawrence Berkeley Laboratory Report*, **LBL-38400** (1995a); also in Thermonuclear Supernovae, ed. P. Ruiz-Lapuente, R. Canal, J. Isern, *NATO ASI Series C* (Kluwer, Dordrecht), Vol. **486**, (1997)
24. S. Perlmutter *et al.*, *Astrophys. J. Lett.* **440**, L41 (1995b)
25. S. Perlmutter *et al.*, *Astrophys. J.* **483**, 565 (1997)
26. S. Perlmutter *et al.*, *Nature* **391**, 51 (1998)
27. S. Perlmutter *et al.*, *Astrophys. J.* **517**, 565 (1999)
28. S. Perlmutter, *Physics Today* **56**, 53 (2003)
29. M. Phillips, *Astrophys. J. Lett.* **413**, L105 (1993)
30. A. Riess, W. Press, R. Kirshner, *Astrophys. J. Lett.* **438**, L17 (1995)
31. A. Riess *et al.*, *Astrophys. J.* **607**, 665 (2004)
32. M. Sullivan *et al.*, *Mon. Not. R. Astron. Soc.* **340**, 1057 (2003)
33. G. A. Tammann, B. Leibundgut, *Astron. Astrophys.* **236**, 9 (1990)
34. R. Tripp, *Astron. Astrophys.* **331**, 815 (1998)
35. R. Tripp, D. Branch, *Astrophys. J.* **525**, 209 (1999)
36. A. Uomoto, R. P. Kirshner, *Astron. Astrophys.* **149**, L7 (1985)
37. T. Vaughan, D. Branch, D. Miller, S. Perlmutter, *Astrophys. J.* **439**, 558 (1995)
38. J. C. Wheeler, R. Levreault, *Astrophys. J. Lett.* **294**, L17 (1985)

Brian P. Schmidt. © The Nobel Foundation. Photo: U. Montan

Brian P. Schmidt did not submit an autobiography. See https://www.nobelprize.org/prizes/physics/2011/schmidt/facts/

THE PATH TO MEASURING AN ACCELERATING UNIVERSE

Nobel Lecture, December 8, 2011

by

BRIAN P. SCHMIDT

Australian National University, Weston Creek, Australia.

INTRODUCTION

This is not just a narrative of my own scientific journey, but also my view of the journey made by cosmology over the course of the 20th century that has led to the discovery of the Accelerating Universe. It is completely from the perspective of the activities and history that affected me, and I have not tried to make it an unbiased account of activities that occurred around the world.

20TH CENTURY COSMOLOGICAL MODELS

In 1907 Einstein had what he called the 'wonderful thought', that inertial acceleration and gravitational acceleration were equivalent. It took Einstein more than 8 years to bring this thought to its fruition, his theory of General Relativity [1] in November 1915. Within a year, de Sitter had already investigated the cosmological implications of this new theory [2], which predicted spectral redshift of objects in the Universe dependent on distance. In 1917, Einstein published his Universe model [3] – one that added an extra term called the cosmological constant. With the cosmological constant he attempted to balance the gravitational attraction with the negative pressure associated with an energy density inherent to the vacuum. This addition, completely consistent with his theory, allowed him to create a static model consistent with the Universe as it was understood at that time. Finally, in 1922, Friedmann, published his family of models for an isotropic and homogenous Universe [4].

Observational cosmology really got started in 1917 when Vesto Slipher observed about 25 nearby galaxies, spreading their light out using a prism, and recording the results onto film [5]. (I am indebted to the family of Slipher for helping fund my undergraduate education through a scholarship set up at the University of Arizona in his honour.) The results confounded him and the other astronomers of the day. Almost every object he observed had its light stretched to redder colours, indicating that essentially everything in the Universe was moving away from us. Slipher's findings created a conundrum for astronomers of the day: Why would our position as observer seemingly be repulsive to the rest of the Universe?

The contact between theory and observations at this time appears to have been mysteriously poor, even for the days before the internet. In 1927, Georges Lemaître, a Belgian monk who, as part of his MIT PhD thesis, independently derived the Friedmann cosmological solutions to general relativity, predicted the expansion of the Universe as described now by Hubble's Law. He also noted that the age of the Universe was approximately the inverse of the Hubble constant, and suggested that Hubble's and Slipher's data supported this conclusion [6]. His work, published in a Belgian journal, was not initially widely read, but it did not escape the attention of Einstein who saw the work at a conference in 1927, and commented to Lemaître, "Your calculations are correct, but your grasp of physics is abominable." [7]

In 1928, Robertson, at Caltech (just down the road from Edwin Hubble's office at the Carnegie Observatories), predicted the Hubble law, and claimed to see it when he compared Slipher's redshift versus Hubble's galaxy brightness measurements, but this observation was not substantiated [8]. Finally, in 1929, Hubble presented a paper in support of an Expanding Universe, with a clear plot of galaxy distance versus redshift – it is for this paper that Hubble is given credit for discovering the Expanding Universe [9]. Assuming that the brightest stars he could see in a galaxy were all the same intrinsic brightness, Hubble found that the faster an object was moving away in Slipher's measurements, the fainter its brightest stars were. That is, the more distant the galaxy, the faster its speed of recession. It is from this relationship that Hubble inferred that the Universe was expanding.

With the expansion of the Universe as an anchor, theory converged on a standard model of the Universe, which was still in place in 1998, at the time of our discovery of the accelerating Universe. This standard model was based on the theory of general relativity, and two assumptions. Assumption one was that the Universe is homogenous and isotropic on large scales, and assumption two that it is composed of normal matter, i.e. matter whose density falls directly in proportion to the volume of space, which it occupies. Within this framework, it was possible to devise observational tests of the overall theory, as well as provide values for the fundamental constants within this model – the current expansion rate (Hubble's constant), and the average density of matter in the Universe. For this model, it was also possible to directly relate the density of the Universe to the rate of cosmic deceleration and the geometry of space. It stated that the more material the faster the deceleration, that above a critical density the Universe has a finite (closed) geometry and below this critical density a hyperbolic (open) geometry.

In more mathematical terms: If the Universe is isotropic and homogenous on large scales, the geometric relationship of space and time is described by the Robertson-Walker metric,

$$ds^2 = dt^2 - a^2(t)\left[\frac{dr^2}{1-kr^2} + r^2 d\theta^2\right]. \tag{1}$$

In this expression, which is independent of the theory of gravitation, the line element distance s between two objects depends on the coordinates r and θ, and the time separation, t. The Universe is assumed to have a simple topology such that if it has negative, zero, or positive curvature, k takes the value {–1,0,1}, respectively. These Universes are said, in order, to be open, flat, or closed. The Robertson-Walker Metric also requires the dynamic evolution of the Universe to be given through the evolution of the scale factor $a(t)$, which gives the radius of curvature of the Universe — or more simply put, tracks the relative size of a piece of space over time. This dynamic equation of the Universe is derived from General Relativity, and was first given by Friedmann in the equation which we now name after him:

$$H^2 \equiv \left(\frac{\dot{a}}{a}\right)^2 = \frac{8\pi G\rho}{3} - \frac{k}{a^2}. \tag{2}$$

The expansion rate of the Universe, H, called the Hubble parameter (or the Hubble constant, H_0, at the present epoch), evolves according to the content of the Universe. Through the 20th century, the content of the Universe was assumed to be dominated by a single component of matter with density, ρ_i, compared to a critical density, ρ_{crit}. The ratio of the average density of matter compared to the critical density is called the density parameter, Ω_M, and is defined as

$$\Omega_i = \frac{\rho_i}{\rho_{crit}} \equiv \frac{\rho_i}{\left(\frac{3H_0^2}{8\pi G}\right)}. \tag{3}$$

The critical density is the value where the gravitational effect of material in the Universe causes space to become geometrically flat (k=0 in Eq. 1). Below this density the Universe has an open, hyperbolic geometry (k=–1) and above, a closed, spherical geometry (k=+1).

As experimentalists, what we need are observables with which to test and constrain the theory. Several such tests were developed and described in detail in 1961 by Allan Sandage [10] and are often described as the classical tests of cosmology. These tests include measuring the brightness of an object as a function of its redshift. The redshift of an object, z, indicates the amount an object's light has been stretched by the expansion of the Universe and is related to the scale factor such that

$$1 + z = \frac{\lambda_{obs} - \lambda_{emit}}{\lambda_{emit}} = \frac{a(z=0)}{a(z)}. \tag{4}$$

The redshift is measured from the observed wavelength of light, $\lambda_{obs,}$ and the wavelength at which it was emitted, $\lambda_{emit.}$ The luminosity distance, D_L, is defined from the inverse square law of an object of luminosity, L, and observed flux, f,

$$D_L \equiv \sqrt{\frac{L}{4\pi f}} \quad (5)$$

This was traditionally solved as a Taylor expansion to Eq. (1) and (2),

$$D_L \approx \frac{c}{H_0}\left[z + z^2 \frac{(1-q_0)}{2}\right], \quad (6)$$

where c is the speed of light. H_0, the current cosmic expansion rate, has units of velocity over distance, and the deceleration parameter, q_0, is defined as

$$q_0 \equiv -\frac{\ddot{a}_0}{\dot{a}_0^2} a_0 = \frac{\Omega_M}{2} \quad (7)$$

The equivalence of Ω_M and q_0 is provided through solutions of the Friedmann equation assuming a Universe consisting solely of normal matter. The Taylor expansion is accurate to a few percent over the region of interest of the day (z<0.5), but was perfected by Mattig (1958), who found a closed solution,

$$D_L = \frac{c}{H_0 q_0^2}\left[q_0 z + (q_0 - 1)\left(\sqrt{1 + 2q_0 z} - 1\right)\right] \quad (8)$$

These equations provide one of the classic tests of cosmology – the luminosity distance versus redshift relationship. For an object of known luminosity, a single measurement of its redshift and brightness will yield an estimate of H_0. The measurement has to be made at a moderate redshift (not so low that gravitationally induced motions, typically z~0.002, are important, and not so high that the second order term in Eq. (6) is important). By measuring a standard candle's (an object of fixed luminosity) brightness as a function of redshift, one can fit the curvature in the line, and solve for q_0.

In principle, from Eq. (6), measuring H_0 does not appear to be difficult. An accurately measured distance and redshift to a single object at a redshift between 0.02<z<0.1 is all that it takes, with their ratio providing the answer. But making accurate absolute measurements of distance in astronomy is challenging – the only geometric distances that were typically available were parallax measurements (measurement of the wobbles in the positions of nearby stars due to the Earth's motion around the Sun) of a handful of nearby stars. From these few objects, through a bootstrapping process of comparing the brightness of similar objects in progressive steps, known as the extragalactic distance ladder, researchers came to conclusions which varied by more than a factor of two, a discordance which persisted until the beginning of the new millennium.

Measuring q_0 required making accurate measurements of the relative distances (absolute distances are not required since the Hubble constant can be normalised out of Eq. (6) and (8)). Attempts made in the 1950s [11], based on the brightest objects in the sky, giant galaxies in the centre of clusters, provided a range of answers. Ultimately, Tinsley [12] showed these galaxies should change dramatically in brightness as we look back in time, making them problematic cosmological probes. Progress in measuring q_0 required a precise standard candle bright enough to be seen to $z>0.3$, where curvature in the luminosity distance and redshift relationship could be accurately measured.

SUPERNOVAE AND MY EARLY CAREER

My astronomical career started in 1985 when I arrived as a bright-eyed freshman at the University of Arizona studying physics and astronomy. In my first astronomy class I felt daunted by all of the astronomy majors, many of whom seemed to me to have encyclopaedic knowledge of everything from white dwarf stars to quasars. I understood physics, but I knew nothing of all of these things, so I looked around for something to do at Steward Observatory to increase my knowledge, and started working for John McGraw on his CCD Transit Instrument (CTI).

This instrument was 15 years ahead of its time, and made the first large digital maps of the sky. Employing Charged Coupled Devices (CCDs), CTI did not track the sky, instead it let the night sky pass overhead, and followed the motion caused by the Earth's rotation electronically, using a technique known as drift scanning. The Sloan Digital Sky Survey applied this technique with its highly successful survey starting in 2000. In 1985, CCDs were still very young, and the data rates that this telescope achieved in the mid 1980s were staggering. This data rate pushed the software and computational hardware capabilities of the day to the detriment of the telescope's overall scientific impact. As with all undergraduates, my progress was slow, but by the end of my 3rd year I had a real job within the group, to try to come up with ways to discover exploding stars known as supernovae in this data set. With a newly minted classification, Type Ia supernovae were reputed to be good standard candles, and the CTI instrument had the opportunity to obtain the first digital light curves of a set of objects at redshifts greater than $z>0.01$ where they could be tested as standard candles. The task was hard because the data set was enormous and, for computational reasons, we only had the ability to

Figure 1. CTI Telescope at Kitt Peak, Arizona.

search catalogues of objects. Supernovae, though, usually occur in galaxies, and when making catalogues it is difficult to discern new objects in the complex structure of a galaxy. By the time I finished my undergraduate degree I had managed to discover a possible object. Unfortunately, it was in data that was more than a year old, and was therefore never confirmed.

Supernovae: Supernovae (SN), the highly luminous and physically transformational explosions of stars, show great variety, which has lead to a complex taxonomy. They have historically been divided into two types based on their spectra. Type I supernovae show no hydrogen spectroscopic lines, whereas Type II supernovae have hydrogen. Over time, these two classes have been further divided into sub-classes. The Type I class is made up of the silicon rich Type Ia, the helium rich Type Ib, and the objects which have neither silicon nor helium in abundance, Type Ic. The Type II class is divided into II-P, which have a ~100 day "plateau" in their light curves, II-L which have a "linear" decline in their light curves, and II-n which have narrow lines in their spectrum [13].

Massive Star Supernovae: Massive stars typically undergo core collapse as the last amount of silicon is burned to iron in their cores. As pressure support is removed by the loss of heat previously supplied by nuclear reactions, their interiors collapse to neutron stars, and a shock wave is set up by neutrino deposited energy outside of the neutron star region. A massive star that has a substantial, intact hydrogen envelope produces a SN II-P. Other variants are caused by different stages of mass loss. SN Ib represents a massive star which has lost its hydrogen envelope, and SN Ic are objects which have, in addition, lost their helium envelope.

Thermonuclear Detonations: These explosions are the result of the rapid burning of a white dwarf star. The entire star is burned, mainly to ^{56}Ni, but also to intermediate mass elements such as Sulfur and Silicon. The actual mechanism has long been assumed to occur when a white dwarf star accretes mass from a companion, and approaches 1.38 $M_{\odot}$, i.e. 1.38 times the mass of our Sun. In 1931 Chandrasekhar showed that at this point a white-dwarf's self-gravity will exceed the pressure support supplied by its electron degenerate gas [14]. As the star approaches this critical juncture, the high pressure and density in the star's core initiates carbon burning near its centre, which eventually leads to the entire star being consumed by a rapidly expanding thermonuclear burning front. We now suspect that it is maybe possible to ignite such an explosion in a variety of ways. These include sub-Chandrasekhar explosions initiated by a surface helium detonation which compresses the star's centre to its nuclear flash point, and super-Chandrasekhar explosions involving the merger of two white-dwarfs via gravitational radiation.

GRADUATE SCHOOL AT HARVARD

Late in 1988 I applied to a number of universities with the hope of receiving a scholarship to work on my PhD – I was not particularly optimistic as I had heard horror stories from others about how competitive the process was. To my surprise, on my 22nd birthday (24 Feb 1989), I received a call from Bob Kirshner at Harvard University, telling me of my acceptance to Harvard's PhD Astronomy programme. It was the best birthday gift of my life. This call was followed up with several more offers in the coming hours and days, and I had a hard choice of deciding where to study. Either I could stay in the western United States where I was comfortable, or move east, which was tantamount to a foreign country to me. After visits to several campuses, Harvard had risen to the top of my list, a decision which I finalised when Bob Kirshner visited Tucson to give the first Aaronson Memorial Lecture by asking if I could work with him on my PhD.

When I arrived at Harvard to work with Bob Kirshner, I decided to focus on studying supernovae rather than discovering them. The idea of measuring the Hubble constant appealed to me, and so we took the tack of building on my supervisor's thesis, to calibrate the luminosity of Type II supernovae and use them to measure the extragalactic distance scale [15]. SN 1987A, the nearest observed supernova to the Earth in almost 400 years had created a frenzy of activity in the subject, and Bob's finishing PhD student, Ron Eastman, had developed a sophisticated computer code to model how radiation emerged from this supernova. My thesis involved applying Ron's theory to several supernovae at sufficient distances so that we could reliably estimate the Hubble constant. Type II-P supernovae are well suited for this purpose because they have simple hydrogen based atmospheres, whose emergent flux is close to a blackbody. In addition, their expansion is unaffected by gravity, enabling us to infer their radius by making measurements over time using absorption lines in their spectra to indicate the velocity of the material from which the supernova's flux is emerging. Put together, the emergent flux calculations and expansion rate allow the distance to a supernova to be determined on a purely physical basis. We named the method the Expanding Photosphere Method (EPM).

In addition to observations, this method has as an essential ingredient: Atmospheric models to calculate the correction to the blackbody assumption. Ideally, these calculations would be hand-crafted for each SN, but the calculations took weeks to run, and instead, we used an approximation where we found that the blackbody correction depended almost entirely on the SN's temperature, and not on other factors. For my thesis I used this technique to measure the distances to 14 SN II at redshifts between $0.005>z>0.05$, and found a 95 % range for the value of the Hubble constant to be $61<H_0<85$ km/s/Mpc [16]. This result was completely independent of the cosmic distance ladder – the bootstrapping of distances from our solar system to the nearest galaxies, but was in almost perfect accord to galaxies which distances were determined using Cepheid variable stars as part of Hubble Key Project. The accepted value today is $67<H_0<75$ km/s/Mpc. Work on using Type II SN to measure distances continues, and while some of the approximations made during my thesis have been challenged, the fundamental technique remains in place.

After my thesis, the next step was obviously to use these objects to measure the deceleration parameter, q_0, but SN II and the expanding photosphere method have three significant drawbacks in measuring the global properties of the Universe. The first is that SN II are difficult to observe beyond $z>0.3$ with current instrumentation – they are too faint. The second is that they require significant observations to obtain each distance – multi-epoch high quality spectra with simultaneous photometric observations, making them observationally prohibitively expensive for measuring q_0. The final difficulty is that the EPM distance precision, while not poor at about 15%, means many objects need to be observed to make a sufficiently precise measurement of q_0 to be interesting. The principal advantage of EPM, that objects were calibrated in an absolute sense, while essential for H_0 measurements, was irrelevant in q_0 measurements. Fortunately, during my PhD, I was exposed to the rapidly emerging work directed at measuring distances to Type Ia supernovae. More importantly, I had got to know and work with the worlds experts on these objects, and these relationships were ultimately the basis of forming the High-Z SN Search Team.

THE FOUNDATIONS OF THE HIGH-Z TEAM

When I arrived at Harvard in 1989, I arrived with Bob's newest postdoc, Swiss national Bruno Leibundgut. Bruno, rather than studying SN 1987A and its

Figure 2. Bob Kirshner examining my Thesis results at Harvard in 1993.

sibling Type II supernovae like most of the world was doing at the time, had concentrated on understanding just how standard of candles Type Ia supernovae were. Type Ia supernovae, and their antecedents Type I supernovae, had developed a reputation from less than ideal data for being essentially identical, making them potentially very good cosmological probes.

For his thesis, Bruno spent many a night on telescopes in Chile, taking photographic images to discover objects in a project lead by his supervisor Gustav Tammann, and collaborator Allan Sandage. While this project successfully discovered supernovae, the search was unable to deliver a dataset useful for testing the veracity of SN Ia as standard candles. So Bruno used the entirety of data collected over the previous 5 years, and by other groups previously, to develop a standard template of the average SN Ia light curve which could be used as a reference to test the homogeneity of the SN Ia family. The results were extremely encouraging – all of the SN Ia seemed to fit a single template [17]. Now at Harvard, Bruno was able to use Harvard facilities, the new 1.2m telescope equipped with a CCD to monitor the light curves of nearby SN Ia as they were discovered, and the huge Multiple Mirror Telescope to obtain their spectra. Our first observing trip together, soon after we both arrived to Harvard, resulted in what I believe are the only ill-feelings ever between Bruno and myself. We had trouble understanding each other's enthusiasm for thinking we knew the right way to observe. The fact that Bruno was the postdoc and I the student didn't occur to me at the time as being a key factor in the discussion. Within a few months, though, we grew to know and respect each other – and to this day, if Bruno challenges anything I say or do, I listen first, and ask questions later.

Bruno's first scientific big break at Harvard came with SN 1990N, an object that was discovered in the summer of 1990, just as Bob and I were off to Europe for a summer school on Supernovae at Les Houches in the French Alps. This object was discovered extremely soon after explosion, and its spectrum showed some funny features that persisted and were different to other SN Ia. But SN 1990N's light curve was well matched by Bruno's template [18].

In Les Houches I realised just how lucky I was to be an astronomer. Situated in a gorgeous village at the base of Mount Blanc, the summer school immersed me for 5 weeks in a group of students from around the world, tutored by the greats of the field. I consider it to be the greatest 5 weeks of my life. There I met a young Chilean, Mario Hamuy, who was working at Cerro Tololo Inter-American Observatory (CTIO) as a research assistant for CTIO staff astronomer Nick Suntzeff. I was familiar with Mario by reputation, for the photometric data he and Nick had amassed on SN 1987A in the Large Magellanic Cloud, which I was using to measure this supernova's distance as part of my thesis.

Mario told us of a new project, the Calan/Tololo survey, which would use the Curtis Schmidt telescope at CTIO to discover objects at redshifts more distant than the objects we were all studying. By discovering SN at $0.02<z<0.1$, the Calan/Tololo survey aimed to test rigorously SN Ia as standard candles, using the redshift as an accurate proxy for relative distance. The members of this group, Mario Hamuy, Nick Suntzeff, and Mark Phillips, at CTIO, and Jose Maza at University of Chile, were starting their programme that year. In addition to Nick and Mario's work on SN 1987A, Jose Maza had lead a highly successful SN search from Calan in the 1980s, while Mark had made an impact in the field by observing SN 1986G in the nearby Centaurus A galaxy. SN 1986G was one of the first objects to be observed with a CCD, and showed a light curve that was ultimately accepted as being unusual compared to the Leibundgut template.

Partially as a result of the Les Houches school, and mainly due to subsequent work that Bob Kirshner was doing with Mark Phillips and Nick Suntzeff on SN 1987A, a 5 week trip for me to visit Cerro Tololo was planned for the end of 1991. There I would use data from the Calan/Tololo survey on Type II SN for my thesis – and learn the techniques that were being used at CTIO to accurately measure the light curves of SN Ia using CCDs, and apply them to my SN II. To ensure the cultural shock wasn't to great, my trip was sequenced with the arrival to CTIO of another of Bob Kirshner's graduate students, Chris Smith, who was about to start his first postdoc at the observatory.

I arrived in Santiago from the long flight from Miami, and was taken to the bus station for a 6 hour bus trip to La Serena. There I met Pete Challis from the Space Telescope Science Institute who was also on his way to CTIO, but in his case for a long observing run. In the 6 hours to La Serena, Pete and I covered a lot of ground, and we soon established that Pete had been at Michigan as an undergraduate with my PhD supervisor, Bob Kirshner, and was interested in changing jobs. I told Pete that Bob was looking for someone to help him manage his Hubble Space Telescope Observations, and in that way Pete and Bob were reconnected, and they continue to work together to this day.

When we arrived at La Serena, I was met by Mario Hamuy and Mark Phillips, who were, in addition to picking me up, putting a wooden box full of photographic plates from the Curtis Schmidt Telescope onto the bus for its return journey to Santiago. While I slept the photographic plates made their journey south to the University of Chile where Jose Maza and his team would search them the following day for supernovae. The Calan/Tololo Survey used this technique to efficiently discover more than 50 objects from 1990–1993. The survey had regularly scheduled CTIO-4m and CTIO-1.5m time to obtain spectra as they were sufficiently regular at discovering objects that they could plan in advance on their discoveries. For photometry, because they only needed a small amount of time, they borrowed time from cooperative astronomers observing on CTIO telescopes who enjoyed the excitement of observing an astronomical object that changed over the course of a few nights.

A few days after arrival, I asked Mario how his work on SN Ia was going, and he said he was depressed. He showed me his first couple of objects, and one of them, SN 1990af, looked pretty normal with respect to its spectrum, but compared to Bruno's template, it clearly rose and fell more quickly, More significantly, SN 1990af was significantly fainter than the other objects in their sample, despite being at the same redshift. He felt that the Calan/Tololo programme to use SN Ia to measure H_0, and eventually q_0, had run into a snag – the objects they were planning to use to measure distances were not living up to their reputation – they were not standard candles.

1991 was a transformational year for SN Ia. Early in the year, a nearby galaxy hosted SN 1991T. In a paper lead by Mark Phillips that included both the Tololo and Harvard groups [19], as well as a paper lead by Alex Filippenko [20], the object was shown to be highly unusual. Its spectrum had extra features early on but was largely missing the most recognised feature of the class, a strong silicon line at 6130 Angstroms. In addition, the light curve rose and fell significantly more slowly than average and it seemed to be too bright given its host galaxy's distance. Between uncertainties in the amount of dust obscuring the object and the distance to its host galaxy, we could not be absolutely certain that this object was brighter than other SN Ia, although work done by Jason Spyromilio at the Anglo-Australian Observatory indicated that SN 1991T produced more iron than is typical for normal SN Ia [21]. Later in the year, another object, SN 1991bg, occurred in a nearby elliptical galaxy. In papers lead by Leibundgut [22] (Harvard and Tololo groups) and Filippenko [23], this object was shown to have a different spectrum from the norm, and a light curve that faded much more quickly than average. In this case, the object was so much fainter than average, with no evidence of any obscuring dust, that the case was clear.

By 1993, based on the range of objects being studied in the nearby Universe, and consistent with the picture that was emerging from the Calan/

Tololo survey, Mark Phillips wrote his seminal paper which compared the rate that an object faded to its luminosity, finding that faster evolving objects were systematically fainter than their slower evolving siblings [24]. I remained a bit sceptical – while SN 1991bg was clearly different, the objects in Mark's 1993 paper were all in the nearby Universe, and the objects' distances were uncertain. I felt that it was possible that the whole correlation might go away, if only SN 1991bg were thrown out. But this paper got the world thinking, and amongst those were Bob Kirshner, whose new PhD student Adam Riess was looking for a project for his thesis. Bob focused Adam's attention at using the statistical expertise of Bill Press (who was one floor down at the Center for Astrophysics), to develop a technique to model SN Ia light curves and estimate their distances.

I was finishing my thesis on SN II-P during this time, but spent a lot of time talking to Adam about his project. The emerging picture of SN Ia was just so interesting, and despite the need for me to write up, I couldn't stop thinking about how to use SN Ia to measure distances. I submitted my thesis in August 1993, and stayed on at the Center for Astrophysics as a Harvard-Smithsonian Center for Astrophysics Postdoctoral Fellow. I had the benefit of a fellowship in which I could do anything I wanted, but with the opportunity of being embedded in the expertise of Bob Kirshner's group.

Early in 1994, Mario Hamuy from the Calan/Tololo group visited. The Calan/Tololo group had expanded to include Bob Schommer, a CTIO astronomer who had experience in measuring the Hubble Constant using the Tully-Fisher technique, and Chris Smith (another Kirshner student), whose all-round observational and analysis experience was being used to help analyse the SN light curves. Mario was armed with Calan/Tololo's first 13 SN Ia light curves and redshifts and to me what was an astonishing discovery. If they applied Mark Phillips' relationship to this independent set of objects, the scatter about the Hubble law dropped dramatically and demonstrated that the SN Ia provided distances with a precision better than 7 % per object. This was much better than anything I thought could ever be achieved. The Calan/Tololo group allowed Adam to train up his new statistical method with these data – they were at sufficient distance that their relative distances could be inferred with high accuracy from their redshifts – thereby removing one of the principal problems in previous SN Ia distance work.

A month later, during one of the groups observing runs at the MMT, Bob Kirshner, Adam Riess, and Pete Challis received a call from Saul Perlmutter of the Supernova Cosmology Project (SCP) to follow-up a high redshift supernova candidate of theirs. The SCP had struggled to find distant SN Ia over the previous 5 years, but I was excited by the spectrum I saw the following morning from the MMT. Pete had already reduced the data and had eye-balled it as a Type Ia SN at a redshift of z=0.42, something I confirmed during the day from the comfort of my CfA Office. In the ensuing weeks, as we negotiated with Saul's team to publish the spectrum in the International Astronomical Union Circulars, we realised that this event was not alone – the SCP had discovered several such objects in the previous months.

These two events – the development of the ability to measure precise distances with SN Ia, and the capacity to discover these objects in the distant Universe, were the ingredients necessary to finally mount a successful campaign to measure the deceleration parameter. The Supernova Cosmology Project had been working towards this goal since 1988, but it became clear that they had significantly different views on how to approach the problem – especially with respect to measuring precise distances – than my supernova colleagues and I had.

THE HIGH-Z TEAM: MEASURING THE DECELERATION RATE OF THE UNIVERSE

In mid-1994 I went to CTIO for an observing run for a project on clusters that ultimately did not pan out. While I stayed on at CTIO after observing, Nick Suntzeff and I hatched a plan to use the CTIO 4m to mount our own campaign to measure q_0, given that the two essential ingredients were suddenly in place. Measuring q_0 had always been part of the plan of the Calan/Tololo survey, but opportunity knocked a few years earlier than the group had anticipated.

Figure 3. Brian Schmidt, Pete Challis, and Nick Suntzeff discussing the High-Z SN Search at Cerro Tololo.

Type Ia supernovae are not common objects, they occur in a galaxy like the Milky Way a few times per millennium. Since SN Ia take approximately 20 days to rise from nothingness to maximum light, observing the same piece of sky twice with a one month separation (which equates to 20 rest frame days at z=0.5) will yield objects which are typically near maximum light, and therefore young enough to be useful for measuring precise distances. The CTIO 4m telescope was equipped with a state-of-the-art 2048x2048 pixel CCD that covered the widest field of view of a 4m telescope at the time. The weather at CTIO was also impeccable through the Chilean summer – so there would be virtually no chance of being weathered out in a supernova search. This was essential because the experiment required images taken a month apart to be compared, and additional pre-planned telescope time afterwards to follow-up the candidates. Bad weather at any of these times would prove fatal for the experiment, leaving no candidates and lots of telescope scheduled to observe objects that didn't exist. This is a problem I knew that the SCP had faced many times.

Nick and I soon enlisted Mark Phillips, Mario Hamuy, Chris Smith and Bob Schommer (CTIO), and Jose Maza (University of Chile) from the Calan/Tololo SN Search. We also brought on Bruno Leibundgut and Jason Spyromilio, who were now at the European Southern Observatory, as well as Bob Kirshner, Pete Challis, Peter Garnavich, and Adam Riess from Harvard. This provided the observational fire-power for both discovering SN Ia, and for following up our discoveries. The proposal was due as my first child, Kieran, was being born, and Nick Suntzeff and Bob Schommer polished up our team's proposal, and submitted it on September 29, 1994.

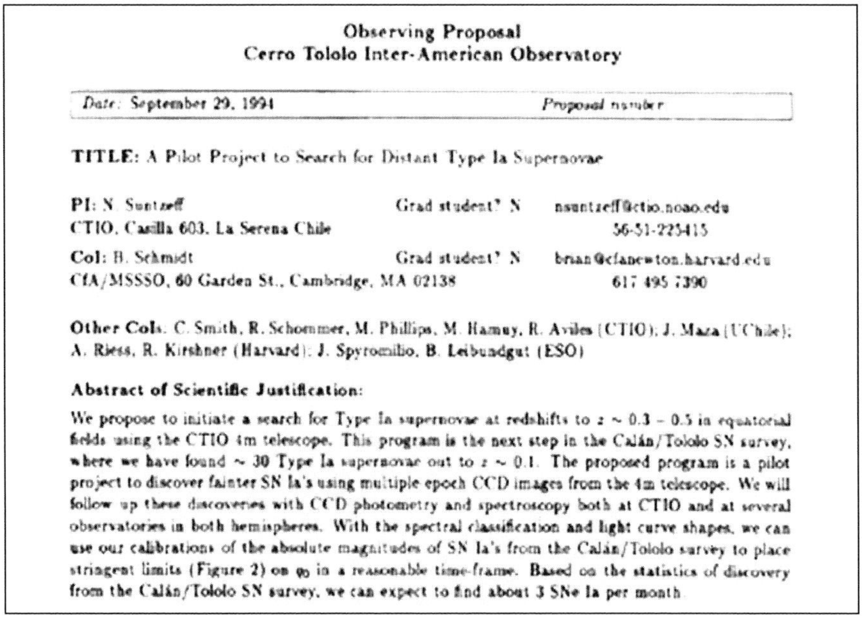

Observing Proposal
Cerro Tololo Inter-American Observatory

Date: September 29, 1994 *Proposal number*

TITLE: A Pilot Project to Search for Distant Type Ia Supernovae

PI: N. Suntzeff Grad student? N nsuntzeff@ctio.noao.edu
CTIO, Casilla 603, La Serena Chile 56-51-225415

CoI: B. Schmidt Grad student? N brian@cfanewton.harvard.edu
CfA/MSSSO, 60 Garden St., Cambridge, MA 02138 617 495 7390

Other CoIs. C. Smith, R. Schommer, M. Phillips, M. Hamuy, R. Aviles (CTIO); J. Maza (UChile); A. Riess, R. Kirshner (Harvard); J. Spyromilio, B. Leibundgut (ESO)

Abstract of Scientific Justification:

We propose to initiate a search for Type Ia supernovae at redshifts to $z \sim 0.3 - 0.5$ in equatorial fields using the CTIO 4m telescope. This program is the next step in the Calán/Tololo SN survey, where we have found ~ 30 Type Ia supernovae out to $z \sim 0.1$. The proposed program is a pilot project to discover fainter SN Ia's using multiple epoch CCD images from the 4m telescope. We will follow up these discoveries with CCD photometry and spectroscopy both at CTIO and at several observatories in both hemispheres. With the spectral classification and light curve shapes, we can use our calibrations of the absolute magnitudes of SN Ia's from the Calán/Tololo survey to place stringent limits (Figure 2) on q_0 in a reasonable time-frame. Based on the statistics of discovery from the Calán/Tololo SN survey, we can expect to find about 3 SNe Ia per month.

Figure 4. The original High-Z SN Search Team proposal.

I had successfully applied for a postdoctoral fellowship in Australia at the Mt Stromlo Observatory, and so in my last few months at the CfA at the end of 1994, I started writing a supernova discovery pipeline. Supernovae are not always easily identified as new stars in galaxies – most of the time they are buried in their hosts, and cannot just be identified without a more sophisticated technique. From colloquia, I knew that the SCP had developed some sort of image subtraction pipeline, and that was the technique they had used to successfully discover their distant objects.

As part of my thesis I had developed techniques of automatically aligning images, but the Earth's atmosphere blurs each image differently, making the shape of a star on each image, known as its point spread function, unique. I had met Drew Phillips at CTIO, and he had developed a technique for convolving images with a kernel to match two images' point spread functions, thereby enabling a clean image subtraction. I used this package as the basis of our pipeline, and set about developing a series of scripts to automatically subtract the massive amounts of data we would get in early 1995. These programmes were meant to take the gigabytes of imaging data that we gathered in a night, align it with the previous epoch, and then match and scale the image point spread functions between the two epochs to make the two images as identical as possible. These two images are subtracted and then the difference image is searched for new objects, which stand out against the static sources that have been largely removed in the differencing process.

During my last months at the CfA, Bob Kirshner's new postdoc Peter Garnavich arrived. Peter was busy principally working on SN 1987A and another nearby object, SN 1993J, during this time, but he was a new colleague with fresh ideas with whom I could discuss the High-Z programme. We instantly became friends, and despite our short overlap at the CfA, Peter is a colleague I have always known I could trust through good times and bad. By the time I left for Australia I felt I had a discovery programme that more or less worked out.

When I arrived in Australia, I had a few weeks to get myself settled before our first observing run started in Chile. I had decided that I would stay put in Australia, rather than travel to Chile, since we were still in the middle of moving, with my wife starting her job and our 4 month old son proving not to be the great sleeper we had hoped for. As we started to implement the pipeline at CTIO it became clear we had a problem or two. The CTIO computing system, which I thought was a lot like my own in Australia, had substantial differences which prevented the software from running. To confound matters, the internet connection between Australia and Chile was about 1 character per second – making it almost impossible for me to do anything remotely. Working with a very patient Mario Hamuy, we slowly marched through the problems. I would email Mario snippets of code to be inserted in the subtraction programme, with Mario reporting back how it worked.

Our first observations were taken on February 25th 1995, and we had another night's data on March 6th. The processing of this data was an unmitigated disaster – nothing seemed to work, and I could not get the data to Australia to diagnose what was going wrong. We used a courier company to express tapes of data to Australia so I could work to fix problems, but that delivery was lost and never arrived. Now, working with the entirety of the CTIO collaboration, we slowly pieced the pipeline together, making it email tiny 16x16 pixel stamps of interesting things to me in Australia. These little mini images, combined with as vivid descriptions as could be mustered by telephone, were all that I had to figure out what was going wrong, or right. We had two nights on March 24th and 29th, and a proposal to write for a continuation of our program, due on the 30th of March. Around March 27th, suddenly the stamps that were being sent to me started producing objects that looked interesting. Several were asteroids – we could tell they were moving – but one was on the outskirts of a galaxy. This object was detected on March 6th, but was not visible on the data of March 24th (The data from this night was poor, so we could not confirm that it was not an asteroid). With these candidates, we submitted the continuation of our program, and set about searching the data from March 29th. Stamp after endless stamp arrived in Australia, and suddenly one, C14 as it was named, looked interesting. It was a new object, buried in a spiral galaxy – it didn't move, and it appeared possibly fainter in our poor data from March 24th. I excitedly called CTIO and the report back from looking at the whole image was positive. Yes, it looked like a supernova!

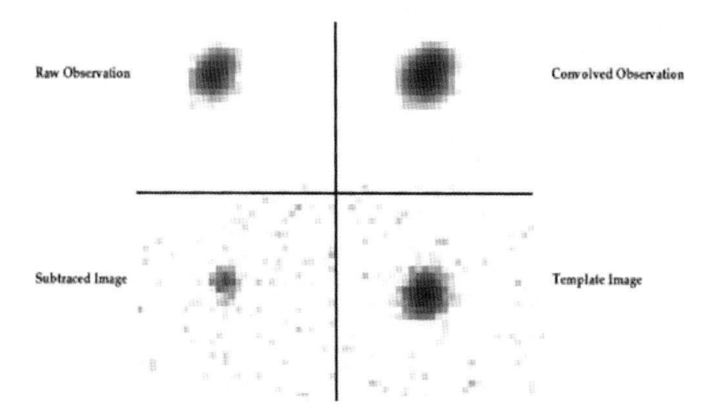

Figure 5. Original stamps for candidate C14 – complete with typos. This object was confirmed as SN 1995K, which at z=0.479, was the most distant SN Ia yet discovered in April 1995. The Observation taken on March 29th, (upper left), was matched (upper right) to the observation taken in February (lower right), and subtracted (lower left).

Using the CTIO-4m spectrograph, Mark Phillips was able to obtain a spectrum of the galaxy. It was at a redshift of z=0.48 – making it potentially the most distant SN yet detected. But this spectrum showed no hint of the supernova, as its light was overwhelmed by its host galaxy's. Bruno and Jason had follow-up time with the ESO NTT at La Silla on April 3rd. Through heroic effort (they observed the object all night) and data reduction (it took a week) the NTT spectrum showed that the object was indeed a SN Ia. In writing the IAU circular, we needed to come up with a name for our team – for lack of anything better, we settled upon the High-Z SN Search team.

In the days that followed, Nick, Mark and Bob Schommer convinced Allan Dressler at Carnegie to take a series of images of SN 1995K with the DuPont telescope at Las Campanas. That data, combined with that taken at ESO and CTIO, provided what is still a very good light curve of a distant SN Ia. We presented the light curve and its place on the Hubble diagram in our Sep 1995 application for telescope time. While 1995K showed q_0=–0.6, the uncertainty was such that we required at least 10 objects to make a statistically significant measurement, and we did not give the actual value much thought.

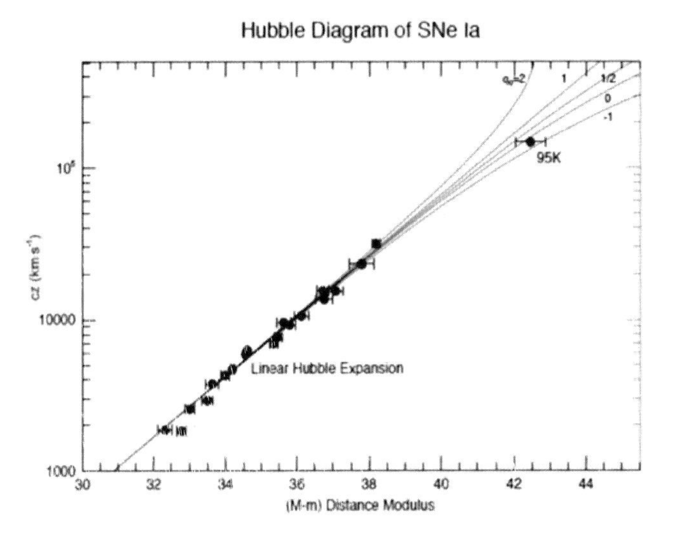

Figure 6. SN 1995K on the Hubble Diagram from our September 1995 telescope proposal.

Supernova aficionados Alejandro Clocchiatti (Catolica University) and Alex Filippenko (Berkeley), along with non-supernova experts John Tonry (Hawaii), Chris Stubbs and Craig Hogan (University of Washington,) were all recruited to the team in 1995 bringing along specific skills and additional telescope time resources.

Alejandro Clocchiatti undertook his PhD thesis at the University of Texas studying Type Ib/c SN – likely contaminants in our experiment which we needed to control using his expertise. Alejandro was also resident in Chile where we could use his physical presence in helping executing observations, as well as providing us additional access to Chilean telescopes.

Alex Filippenko, a member of the community that studied supernovae, had approached me in 1995 to join the High-Z Team. We turned him down on the basis that we didn't want to be seen to poach a member from a competing team. By the end of 1995 it became clear that Alex's expertise and access to Keck were going to be essential for us to successfully undertake our experiment to measure on q_0. So when he asked again in 1996 to join our team, we immediately said yes.

John Tonry, in addition to providing access to telescope time through the University of Hawaii, is widely regarded as one of the most capable observational astronomers of our era. On my trips to Hawaii John and I would discuss the current deficiencies in our experiment, and John would inevitably write new programs to assist with our discovery and analysis of SN Ia. In these bursts of programming John developed our interactive search tool, our spectral analysis tool (SNID, which is still widely used by the community), and the core of our photometric analysis pipeline.

Chris Stubbs was one of the members of the MaCHO gravitational micro-lensing experiment which operated the Mt Stromlo 50 inch telescope, and he brought significant experience in analysing large datasets, which our group sorely lacked.

Craig Hogan was an eminent theorist who had taught me cosmology at the University of Arizona before he moved to the University of Washington. I felt (and still feel) that it was important to have at least one theorist on any large observational programme, and Craig was someone whose theoretical grounding was well matched to the needs of our team.

Given the dispersed nature of our team, we had to gather each year to discuss how the observational programme was progressing, and how we were going to turn all of the data into a definitive measurement of q_0. Our first meeting was in 1996 at Harvard. We had just been awarded Director's Discretionary time with the Hubble Space telescope, and we needed to plan on how to use this great resource effectively. We decided to expand our discovery platform to include the new wide field camera on the Canada France Hawaii Telescope in Mauna Kea, with University of Hawaii astronomer and High-Z team member John Tonry providing access to this unique facility. Running SN searches on two telescopes, twice per year, made me and the team very busy people.

Each observing run was organised chaos. I would arrive a week early, with the latest version of the software. Since we did not have dedicated equipment, the whole pipeline would be re-built at the beginning of each run – and this never proceeded smoothly. Each facility had its own sets of operating systems that, while all UNIX, were sufficiently different so that the code had to be individually compiled for each system. Due to the size of our dataset we needed to operate across multiple machines and disks – hardware that changed each run. This week inevitably ended with the entire team working 20-hour days to ensure that we were able to promptly discover supernovae. This level of effort led to interesting coping strategies – Bob Schommer was famous for playing James Brown at high volume in the telescope control room. It also lead to the occasional mistake. One night in the CTIO-4m control room as Alejandro Clocchiatti watched as I frenetically typed, Alejandro suddenly turned pale and said, "I don't think you wanted to do that". I had just accidentally deleted the night's data. While we pondered how to tell Nick (who was manning the telescope) the news, Nick suddenly screamed, "What happened to all the data?" I saw my career flash before my eyes, but we soon realised the data were stored (in a way that I had previously thought was inane) such that we were able to restore our files and continue on observing.

Spectroscopic follow-up, principally using the Keck 10m telescopes through time allocated to Alex Filippenko (through the University of California), and John Tonry (through the University of Hawaii), was scheduled just a few days after our search runs. Failure to quickly identify candidate supernovae meant our discoveries would be effectively useless. Despite the chaos, through 1995–1997, we did manage to discover, spectroscopically confirm, and photometrically follow 16 distant SN Ia – enough to make a statistically robust measurement of the deceleration parameter q_0.

In early 1997 most of the team assembled in Seattle at the University of Washington, and we agreed that each paper would be led by a student or young postdoc from within the group. I would write the first paper where we laid out our programme and presented our first object, SN 1995K. Peter Garnavich was selected to write the next major paper, one that would include objects observed with HST, and would likely tell us our first statistically significant measurement of q_0. And finally, Adam Riess was selected to write the next paper that would refine the value of q_0 based on several years' data. The data grunts of the group (myself, Adam Riess, Pete Challis, Saurabh Jha, Alejandro Clocchiatti, David Reiss, and Al Diercks) stayed on in Seattle to work together for a week. Initially, the week was supposed to be a working bee where I would tutor the group on how to make photometric measurements of distant SN Ia, and we would as a group analyse our dataset. While the week did not lead to an analysis of our dataset, it instead became an intense workshop where we thought through most of the outstanding issues necessary to complete the experiment. It was one of the most memorable weeks of the High-Z team for me. While Hale-Bopp blazed invisibly above the continual Seattle drizzle, we clocked in 16 hour days from the basement of the University of Washington Physics Department – taking a break to all see the movie, "Swing Blade", at the request of Adam.

Over the course of the next few years, my life was dominated by SN discovery runs, photometric data reduction, and writing the paper on our SN Programme and SN 1995K. The paper was largely complete in 1996, but the ever-increasing data load made it challenging for me to finish. In addition, the complication that SN 1995K was most consistent with negative acceleration made aspects of the analysis challenging. In addition, there were many possible systematic effects that could derail this experiment into giving an incorrect answer, and I was investigating these at this time, one by one.

SYSTEMATIC EFFECTS

In the nearby Universe we see SN Ia in a variety of environments, and about 10% have significant extinction. Since we can correct for extinction by observing the colours of SN Ia, we can remove any first order effects caused by the average extinction properties of SN Ia changing between z=0 and z=0.5. As part of his thesis Adam Riess had developed techniques to correct for dust based on the colours of supernovae [25]. This was essential work to accurately measure the relative distances to SN Ia, and is an essential ingredient in all supernova distance measuring techniques today.

Our supernova discoveries suffer from a variety of selection effects, both in our nearby and distant searches. The most significant effect is the Malmquist bias – a selection effect which leads magnitude limited searches finding brighter than average objects near their brightness limit. This bias is caused by the larger volume in which brighter objects can be discovered compared to their fainter counterparts. Malmquist bias errors are proportional to the square of the intrinsic dispersion of the distance method, and because SN Ia are such accurate distance indicators, these errors are quite small – approximately 2%. In 1995, I developed Monte Carlo simulations to estimate these effects, and remove their effects from our data sets.

As SN are observed at larger and larger redshifts, their light is shifted to longer wavelengths. Since astronomical observations are normally made in fixed bandpasses on Earth, corrections need to be made to account for the differences caused by the spectrum of a SN Ia shifting within these bandpasses. The SCP had showed that these effects can be minimised if one does not stick with a single bandpass for nearby and distant objects, but by instead choosing the closest bandpass to the redshifted rest-frame bandpass [26]. The High-Z SN search took this one step further, designing new bandpasses, specifically made to emulate the z=0 bandpass at several redshifts.

SN Ia are seen to evolve in the nearby Universe. The Calan/Tololo survey plotted the shape of the SN light curves against the type of host galaxy [27]. Early hosts (ones without recent star formation) consistently show light curves which evolve more quickly than those objects which occur in late-type hosts (objects with on-going star formation). This could be a terminal problem for using SN Ia to measure q_0 if were not for the observation that once corrected for light curve shape, the corrected luminosity shows a much smaller correlation as a function of the characteristics of the host.

COSMOLOGY BEYOND NORMAL MATTER

Since 1917, when Einstein first added the cosmological constant to his equations, this fudge factor had been trotted out on several occasions to explain observations of the Universe that didn't conform to the standard model described earlier. The cosmological constant had developed a bad reputation as being incorrectly asserted as the solution to what were ultimately found to be bad observations.

In 1995 I had served as the referee of a paper by Goodbar and Perlmutter [28] exploring if meaningful limits on the value of the cosmological constant could be made by high redshift SN Ia measurements. In my referee report I expressed concern of the relevance of the paper – I felt that the paper failed to demonstrate that a meaningful limit could be made on the cosmological constant. If there was no cosmological constant, then the uncertainty in a SN Ia-based measurement would be sufficiently large as not to be interesting (see their Fig. 2). I had failed to grasp – so strong were my priors against a Cosmological Constant – that if there was a cosmological constant (see their Fig. 3) a meaningful measurement could be made.

The cosmological constant was not new to me. Sean Carroll had written a review on the topic in 1992, while we shared an office during graduate school [29]. I remember that as he worked through hundreds of yellow post-it notes scrawled on his manuscript by his referee, Allan Sandage, I teased him about writing about something as ridiculous as the cosmological constant. This review ended up being extremely useful as I came to grips with how to interpret SN 1995K, and the range of negative q_0 values it implied.

As part of my paper describing the High-Z SN Search [30], the team theorist, Craig Hogan, encouraged me to go beyond the notion of q_0. He was particularly interested in breaking the assumption that the Universe was made up of only normal matter, postulating that it could be composed of other things as well. In our paper we adapted our measurement to the standards of particle astrophysics. That is, we adapted the Friedmann eq (2) to reflect all species of matter

$$H^2 \equiv \left(\frac{\dot{a}}{a}\right)^2 = \frac{8\pi G\rho_{tot}}{3} - \frac{k}{a^2} \tag{9}$$

describing each species of matter by their fraction of the critical density,

$$\Omega_i = \frac{\rho_i}{\rho_{crit}} \equiv \frac{\rho_i}{\left(\frac{3H_0^2}{8\pi G}\right)}, \tag{10}$$

and this matter's equation of state,

$$w_i = \frac{P_i}{\rho_i c^2}. \tag{11}$$

The equation of state for normal matter is w=0, the cosmological constant, w=–1, and photons w=1/3. This formulation made for a less trivial expression for the luminosity distance,

$$D_L H_0 = c(1+z)\Omega_k^{-1/2} S\left\{\Omega_k^{1/2} \int_0^z dz' \left[\Omega_k (1+z')^2 + \sum_i (1+z')^{3+3w_i}\right]^{-1/2}\right\}, \quad (12)$$

where $S(x) = sin(x), x,$ or $sinh(x)$ for closed, flat, and open models respectively, and Ω_k the curvature parameter, is defined as $\Omega_k 1 \equiv -\Sigma_i \Omega_i$. With multiple forms of matter, Mattig's formulation for D_L (Eq. (8)), is no longer valid, but the q_0 expansion (Eq. (6)) is still valid, except that q_0 is given by the expression,

$$q_0 \equiv \frac{-\ddot{a}(t_0)a(t_0)}{\dot{a}^2(t_0)} = \frac{1}{2}\sum_i \Omega_i(1+3w_i). \quad (13)$$

THE DISCOVERY OF ACCELERATION

By the middle of 1997 the High-Z Team had HST observations of 4 objects, and 10 more distant objects to tackle our ultimate goal, measuring q_0. But there were some complications that needed to be sorted out dealing with statistics. In principal, measuring q_0 from several SN distances and redshifts is straightforward. The redshifts have negligible uncertainty, and the distance estimates had distances with uncertainties well described by a normal distribution. A classic χ^2 method seemed entirely appropriate. Except our data was in a part of parameter space where Mattig's exact formula (Eq. (8)) was invalid, and at a redshift where the Taylor expansion solution Eq. (6) was not very accurate. On the other hand, Eq. (12) covered all possibilities, but there were regions of parameter space which were not allowed, like negative matter. In discussions at CTIO with members of the SCP in 1996, it became clear we were both grappling with how to deal with these statistical issues – it wasn't that they hadn't been solved by science, it was just that we were in new territory for us, and we were struggling to figure out a solution. Adam Riess, who had become adept at statistics in his thesis, in discussions with Bill Press, came up with the solution of converting χ^2 to a probability, applying priors to this probability space (e.g. no negative matter), and integrating over this space to find the probability distribution for the parameters of interest. It seems so passé now, but in 1996, none of us had ever seen this technique used before in astronomy. Computationally, this was not trivial, and Adam Riess, Peter Garnavich, and I all wrote our own versions of codes that did these calculations.

The HST data that Peter Garnavich was analyzing was of very high quality, and was consequently the easiest to reduce. By September he had finished his analysis – the data clearly showed $q_0 \neq 0.5$ and a flat Universe composed of normal matter was ruled out. But, this seemed at odds with a paper put out by the SCP at same time [31]. Peter's draft created a range of reactions within the team – what were our control's on systematic errors, and how could we demonstrate the result was robust? This lead us into examining all sorts of possible systematic errors, and while we never quite reached agreement (Chris Stubbs, who had a particle physics background, was particularly critical of our ability to control all errors) it did mean the team had already grappled with this issue when things got substantially more interesting a few months later. My wife and I had just had our second child, and I have to admit to not doing a good job at getting the team to work together constructively around these issues.

In November of 1997, Adam Riess had finished his first pass at measuring his collection of supernovae – a feat that was achieved due to his unique ability to focus on this one thing with all of his might. He sent me a figure with a subject line of "what do you think?" I looked at the figure and it showed that his group of SN Ia were, on average, definitively fainter than even a $q_0=0$ model. The Universe seemed to be accelerating. I remember thinking, "What has Adam done?", and thus opened up an intense exchange between the two of us, checking the result and refining the analysis. At the same time, I was working to submit my paper, which I swore would be submitted in 1997 – just managing to get it in before the New Year's Eve. Finally, on the 8th of January 1998 (Australian Time), Adam and I agreed on all details of the calculation that showed that the Universe was accelerating, and I sent him an email with "Hello Lambda" as the subject line, and a figure of my calculations. Most of the High-Z team had not been shown the analysis at this point. Adam had shown his work to Alex Filippenko, and we told Peter Garnavich, who was presenting his paper (described above) at the American Astronomical Society Meeting, the next day.

The result was perplexing to me – the cosmological constant had a long history of being proposed to explain a set of observations which was later on shown to be fatally flawed. And then there were the results of the other team. The 1997 SCP paper was at such odds to what we were seeing, I felt no one would take us seriously with such a crazy result. What I had not seen was the SCPs new paper which appeared on the 17th of December on the Astrophysics Archive – only learning about it after the AAS press conference on January 8th [32]. This paper indicated that their value of q_0 was much lower than they had previously presented.

On January 9th I came into work to get a report of the AAS press conference from Peter Garnavich. In addition to presenting his HST data from his *Nature* paper showing the Universe was not decelerating quickly, Saul Perlmutter had given the audience a peek at his entire collection of 40 objects – and these objects did seem to be showing the same thing that we were seeing. Saul's object's were systematically fainter than could be explained in a Universe composed only of normal matter. But Saul's team had not yet corrected for dust, a correction that was built into our analysis from the beginning. Adam had chosen this week to get married, and when he returned from a short honeymoon, we had a lot of explaining to do to the team, recounting all of the steps in our analysis. The team's reactions were mixed – some were excited, others were in dis-belief, and still others felt that we had a long way to go to show the result to be robust to errors. While I shared in the scepticism, I also felt that it would be wrong not to publish a result just because we did not like it. I challenged the team to suggest tests that they felt needed to be made before we published. Over the remainder of January and February, under Adam's leadership, the team worked through all of the tests requested, such that by the end of February, the team had agreed to the contents of the paper, and we were ready to announce our result. Alex Filippenko presented our team's work at a meeting in California at the end of February, and it created a media sensation in the United States. Our paper was submitted a week later to the Astronomical Journal, "Observational Evidence for a Cosmological Constant and an Accelerating Universe". Over the next few months, in addition to continuing our punishing programme of SN observations, Peter Garnavich did the first analysis to show that whatever was causing the acceleration, it seemed to have an equation of state a lot like the cosmological constant.

While I felt that we had done all that was possible with our supernovae to understand our uncertainties, I couldn't help worrying that something unexpected would turn up, and nullify our results. In the language of a US Secretary of Defence, we had controlled the known unknowns, but there were always the unknown unknowns – and this was a crazy result. I expected the community to sceptical, and most probably scathing in the assessment of our results.

During this time, the SCP was working frenetically on their own paper and it soon emerged that the conclusions of the two independent experiments were virtually identical [33]. Their experiment had more objects than ours, but less signal per object – in the end the overall significance of the two experiments was about the same. If combined, the two experiments achieved more than 4σ detection of acceleration (Fig. 7 and 8).

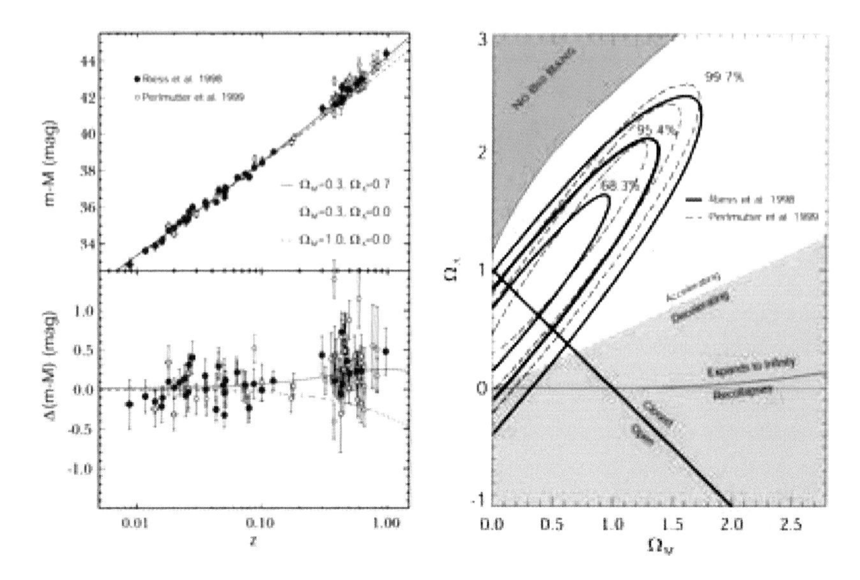

Figure 7a (left). Top Panel: Hubble Diagrams of SN Ia showing High-Z Team and SCP data, with 3 sets of cosmological parameters. Bottom Panel: Data from Top Panel with the model of containing normal matter (30 % of the critical density) subtracted. Figure 7b (right): Probability contours for cosmological fits to the SCP and High-Z teams' data. The results from the two projects show remarkable consistency in their conclusion that Universe has a significant matter component consistent with the equation of state of a cosmological constant.

To my surprise, the accelerating Universe was received with a warmer reception than what I was expecting. The positive reception was due, I believe, partially to the fact that two highly competitive teams arrived independently to the same answer. But, the discovery also provided a solution to some major failings of the prevailing Cold Dark Matter model (CDM) – a model in which initial conditions were set by a period of inflation [34]. This model predicted a geometrically flat Universe with a distribution of initial fluctuations described as a nearly-scale-invariant Gaussian random field. CDM was in conflict with the distribution of galaxies on large scales, as were the prevailing combination of measurements of the Hubble constant, matter density, and age of the Universe. It was realised that the addition of a cosmological constant could fix all of these problems [35] [36] [37].

In 2000 the MAXIMA and Boomerang experiments made measurements of the Cosmic Microwave Background which demonstrated that the Universe was flat to within 10 % – i.e. $\Omega_k \sim 0$ [38] [39]. This measurement was essentially impossible to reconcile with our supernova distances unless the Universe was full of something like a cosmological constant. It was at that moment in 2000 that I finally felt secure that our findings would stand the test of time.

CONCLUDING REMARKS

In the 13 years since the discovery, the accelerating cosmos has received intense scrutiny throughout physics. On the observational side, increasingly large samples of Type Ia supernovae have improved the precision of the measurements of acceleration to the point where they are now systematically, rather than statistically limited [40] [41] [42] [43].

Measurements of the Cosmic Microwave Background have established an increasingly precise measurement of the angular size distance to a redshift of approximately $z \sim 1090$, as well as the physical conditions of the Universe from just after the Big Bang through to the time of recombination [44]. The scale of Baryon Acoustic Oscillations (BAO), which size are understood through modelling of the CMB, have been traced over time through their imprint into the population of galaxies. Astronomy can now connect the scale of the Universe from $z \sim 1080$ to $z=0.2$ [45], $z=0.35$ [46] and $z=0.6$ [47]. Together, the measurements listed above, and most others, remain consistent with a Universe where the acceleration is caused by Einstein's Cosmological Constant ($\Omega_\Lambda \sim 0.73$, $w=-1$), the Universe is geometrically flat, and the remainder of the matter is dominated by pressure-less ($w=0$) matter [48], split between Baryons ($\Omega_B \sim 0.045$) and Cold Dark Matter ($\Omega_{CDM} \sim 0.225$). This basic model is often described as the Flat Λ-CDM Model.

An enormous body of theoretical work has been undertaken in response to the discovery of the accelerating Universe. Unfortunately, no obvious breakthrough in our understanding has yet occurred – cosmic acceleration remains the same mystery that it was in 1998. The future will see bigger and better experiments that will increasingly test consistency of our Universe with the Flat Λ-CDM Model. If a difference were to emerge, thereby disproving a Cosmological Constant as the source of acceleration, it would provide theorists with a new observational signature of the source of the acceleration. Short of seeing an observational difference emerge, we will need to wait for a theoretical revelation that can explain the standard model, perhaps informed by a piece of information from an unexpected source.

REFERENCES

[1] J. Norton, "How Einstein Found His Field Equations: 1912–1915", *Historical Studies in the Physical Sciences, 14* (1984), pp. 253–315. Reprinted in D. Howard and J. Stachel (eds.), *Einstein and the History of General Relativity: Einstein Studies* Vol. I, Boston: Birkhauser, pp101–159 (1989).
[2] W. de Sitter, *MNRAS*, **78**, 3 (1917).
[3] A. Einstein, *Sitzungsberichte Berl. Akad.*, **1**, 142 (1917).
[4] A. Friedmann, *Zeitschrift für Physik*, **10**, 377 (1922).
[5] V. M. Slipher, *Proc. Amer. Phil. Soc.*, **56**, 403 (1917).
[6] G. Lemaître, Annales Société Scientifique de Bruxelles, **A47**, 49 (1927).
[7] Gaither's Dictionary of Scientific Quotations (2008).
[8] H. P. Robertson, *Phil. Mag.*, **5**, 835 (1928).
[9] E. Hubble, *Proc. Natl. Acad. Sciences*, **15**, 168 (1929).
[10] A. R. Sandage, *ApJ*, **133**, 355 (1961).
[11] M. L. Humason, N. U. Mayall & A. R. Sandage, *ApJ*, **61**, 97 (1956).
[12] B. Tinsley, *ApJ*, **178**, 319 (1972).
[13] A.V. Filippenko., *ARAA*, **35**, 309 (1997).
[14] S. Chandrasekhar, *ApJ*, **74**, 81 (1931).
[15] R. P. Kirshner, J. Kwan, *ApJ*, **197**, 415 (1975).
[16] B. P. Schmidt *et al.*, *ApJ*, **432**, 42 (1994).
[17] B. Leibundgut, *Thesis, University of Basel* (1988).
[18] B. Leibundgut *et al.*, *ApJ*, **371**, 23 (1991).
[19] M. M. Phillips *et al.*, *ApJ*, **103**, 1632 (1992).
[20] A. V. Filippenko *et al.*, *ApJL*, **384**, 15 (1992).
[21] J. Spyromilio *et al.*, *MNRAS*, **258**, 53 (1992).
[22] B. Leibundgut *et al.*, *AJ*, **105**, 301 (1993).
[23] A. V. Filippenko, *AJ*, **104**, 1543 (1992).
[24] M. Phillips, *ApJL*, **413L**, 105 (1993).
[25] A. G. Riess, W. H. Press, R. P. Kirshner, *ApJ*, **473**, 88 (RPK96) (1996).
[26] A. Kim, A. Goobar, S. Perlmutter, *PASP*, **108**, 190 (1996).
[27] M. Hamuy *et al.*, *AJ*, **112**, 2391 (1996).
[28] A. Goobar, S. Perlmutter, *ApJ*, **450**, 14 (1995).
[29] S. M. Carroll, W. H. Press, E. L. Turner, *ARAA*, **30**, 499 (1992).
[30] B. P. Schmidt *et al.*, *ApJ*, **507**, 46 (1998).
[31] S. Perlmutter. *et al.*, *ApJ*, **483**, 565 (1997).
[32] S. Perlmutter *et al.*, *Nature*, **391**, 51, (erratum **392**, 311) (1998).
[33] S. Perlmutter *et al.*, *ApJ*, **517**, 565 (1999).
[34] A. Guth, *PhysRev D*, **23 (2)**, 347 (1981).
[35] G. Efstathiou, W. J. Sutherland, S. J. Maddox, *Nature*, **348**, 705 (1990).
[36] L. M. Krauss, M. S. Turner, *General Relativity and Gravitation*, **27**, 1137 (1995).
[37] J. P. Ostriker, P. J. Steinhardt, *Nature*, **377**, 600 (1995).
[38] S. Hanany *et al.*, *ApJL*, **545**, L5 (2000).
[39] P. de Bernardis *et al.*, *Nature*, **404**, 955 (2000).
[40] M. Hicken *et al.*, *ApJ*, **700**, 1097 (2009).
[41] R. Kessler *et al.*, *ApJS*, **185**, 32 (2009).
[42] W. M. Wood-Vasey, *ApJ*, **666**, 694 (2007).
[43] J. Guy *et al.*, *A&A*, **523**, 7 (2010).
[44] E. Komatsu *et al.*, *ApJS*, **192**, 18 (2011).
[45] W. Percival *et al.*, *MNRAS*, **401**, 2148 (2010).
[46] D. J. Eisenstein *et al.*, *ApJ*, **633**, 560 (2005).
[47] C. Blake, *MNRAS*, **415**, 2892 (2011).
[48] M. Sullivan, **737**, 102 (2011).

Adam G. Riess. © The Nobel Foundation. Photo: U. Montan

Adam G. Riess did not submit an autobiography. See https://www.nobelprize.org/prizes/physics/2011/riess/facts/

MY PATH TO THE ACCELERATING UNIVERSE

Nobel Lecture, December 8, 2011

by

ADAM G. RIESS

Johns Hopkins University, Baltimore, MD, USA, Space Telescope Science Institute, Baltimore, MD, USA.

INTRODUCTION

I think one of the most amazing facts about the universe is that it is expanding. I never would have guessed it. Even as an undergraduate, once I'd learned a little physics, I would have thought that the universe was eternal, static, and always in equilibrium. So in graduate school when I found out that the universe was expanding, I was awestruck. Then I learned if we could measure the expanding universe, the way we record the growth of a child with marks on a doorframe (Fig. 1), we could determine the *age* of the universe and predict its ultimate fate. This was staggering! I knew this is what I wanted to do. Since that time, charting the expanding universe to determine its nature has been my passion. Though I have to add: knowing what I know now, that the universe is not only expanding but also *accelerating*, I feel like King Alfonso X of Castile who saw Ptolemy's theory of the Cosmos and reportedly said "If the Lord Almighty had consulted me before embarking on creation thus, I should have recommended something simpler."

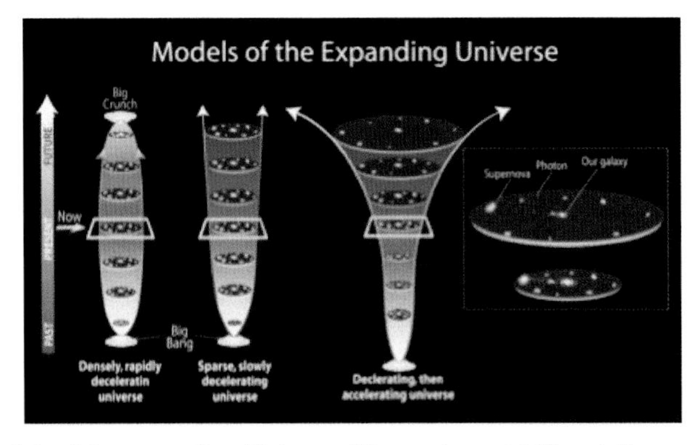

Figure 1. Models of the expanding Universe. We use the redshift and distance (brightness) of supernovae to measure the change in scale with time, respectively. Together these measure the expansion history.

THE EDUCATION OF A COSMOLOGIST

After junior year in high school, I spent a month that summer at the New Jersey Governor's School of Science. I went into the program thinking I was interested in genetic engineering but on a whim, I took a course on Einstein's theory of special relativity. When Dr. Jim Supplee of Drew University explained time dilation, length contraction, and the relativity of simultaneity I was hooked. From 1988 to 1992 I studied physics at the Massachusetts Institute of Technology. "Course 8", as physics was called at MIT, was demanding and MIT stretched my mind and capabilities more than any experience before or since. The hardest but most rewarding course was Junior Lab where we reproduced the great experiments of Nobel Prize winners from the first half of the twentieth century. I learned to love the process by which experimentalists and observers take raw measurements from a piece of equipment and convert them into basic statements of fact and tests of hypotheses. I would remain an ardent "data reductionist" for the rest of my career.

As I was finishing at MIT, Professor Robert Kirshner, then chairman of Harvard's astronomy department, called to tell me I was accepted to the astronomy graduate program and waitlisted for the physics program. He asked if I'd like to come visit and offered, as was standard practice, to pay my expenses. Given my proximity, he said, a subway ("T") token was in the mail to me! I rode two stops up the red line and was brought to the office of a model graduate student named Brian Schmidt. I would later learn critical lessons from him, but at the time we discussed the life of a graduate student. I decided to go to Harvard to earn a doctorate in astronomy. I knew next to nothing about astronomy and astrophysics but had a vague feeling that I wanted to learn more. That summer before grad school I worked on the MACHO Project at Lawrence Livermore National Laboratory where I briefly met Saul Perlmutter. So before I had even started grad school, I had already met many of the people who were going to make an enormous difference to the direction and shape of my career.

The only preliminary exam at Harvard at the time was based on Frank Shu's wonderful book "The Physical Universe", which I happily read from cover to cover. By the time I got to Chapter 14 on the expansion of the universe and Chapter 15 on gravitation and cosmology I knew what I wanted to work on: measuring the rate at which the universe expanded, or rather, the rate at which the expansion was decelerating. Cosmologists were then unable to measure that rate – and correspondingly, the universe's age – to better than a factor of two. Worse yet, their inaccurate measurements of the rate and the age meant, as Chapter 15 pointed out and recent work by Allan Sandage and others confirmed, that they also could not determine the mass density and fate of the universe. To solve both problems, they needed to gauge distances with precision across billions of light years. So far, their best indicator of distance was the brightest galaxies in clusters.

In the spring of 1993, I went to talk to Professor Robert Kirshner about choosing a research project. Kirshner and his student, Brian Schmidt, were finishing work on another distance indicator called core-collapse supernovae. He told me about some new work by Mark Philips of Cerro Tololo Inter-American Observatory on a third distance indicator, another class of supernova called Type Ia supernovae. Type Ia supernovae were thought to arise when carbon-oxygen white dwarf stars accreted mass from a companion star, grew beyond the Chandrasekhar limit, and exploded. After their explosions, their light output rose, reached a peak and then declined. Because it was thought all explosions were the same, the peak brightness could be used to determine their distances: brightness falls off in a regular way with distance. And because Type Ia's are the most luminous of the common supernova types, peaking at 4 billion solar luminosities, the distances at which they can be seen are extremely high. But there was a catch. Some of these supernovae were not identical; some were *intrinsically* more luminous than others, causing astronomers to over- or under-estimate their distances. Mark found that the intrinsic luminosity at the peak appeared to correlate with the rate at which the light output declined after reaching peak. The more slowly their light declines, the more luminous they are. Thus the Type Ia's decline rate could be used to improve the precision of their distance estimates. Unfortunately the problem remaining was that variations in the amount of dust along the line of sight to a supernova changed its brightness and therefore degraded the precision of the distance estimates.

To understand these variables, a team led by Mario Hamuy, with Mark Philips, Nick Suntzeff, Robert Schommer and Jose Maza in Chile were conducting the first large-scale program, the Calan/Tololo Survey, measuring the light curves of Type Ia supernovae. This survey would be groundbreaking, gathering the data that would provide the first proof that Type Ia supernovae would make excellent distance indicators.

Bob Kirshner was an expert in observing supernovae. He also had great common sense, a nose for what was important, and the ability to marshal the resources his students needed to succeed. He suggested we collaborate with another Harvard professor, William Press, an expert in developing algorithms – or "Numerical Recipes", the title of his famous book – for analyzing complex data (Fig. 2). This first summer of work was slow going. Bill Press was away and I tried to glean from his latest paper "Interpolation, Realization, and Reconstruction of Noisy, Irregularly Sampled Data" (Rybicki and Press 1992) and from a preprint of Mark's (Philips 1993) a way to make optimal use of the supernova's light curve – that is, the rise, peak, and decline of its light – to predict its true luminosity. At the time, the Calan/Tololo team had not yet finished its survey, so we could find little data with which to test new algorithms for predicting a supernova's luminosity, and even worse, most of that data was fatally contaminated. Bruno Leibundgut, an important future collaborator, had compiled an atlas of the historical observations going back to the 19th century of Type Ia supernova, most of them observed with photographic plates. But photographs, due to their analog nature, do not allow one to accurately separate out the light of a fading supernova from the background light of its host galaxy. Thus the relative rates of rise and decline measured by my newly developed algorithms were less a clue about the supernova's luminosity than they were a measure of the error made in background subtraction. Before any progress could be made in improving measurements of the expansion rate, Type Ia's would have to be observed with the more precise, digital CCD detectors.

However, this time was well spent learning from Brian the techniques of using CCDs to measure the brightness of stars and I wore a furrow in the carpet between our offices. Later I shared a couple of observing runs at Mount Hopkins with Brian, Bob, and Peter Challis. Bob and I decided that part of my thesis would be to collect a large sample of Type Ia light curves, the first sample using CCDs on the skies over the northern hemisphere (the Cerro Tololo survey had been done in the skies of the southern hemisphere). I began searching the circulars from the International Astronomical Union for reports (most from amateurs) of Type Ia supernovae that had just exploded. Every time I found one, I needed follow-up observations of its light curves, so I would trade our group's future time on the Mt. Hopkins 1.2 meter telescope for present time from other observers, a typical cost of 30 minutes a night. We later formalized this time swap arrangement with the telescope's time allocation committee. This was a practical approach to collecting nearly nightly observations without having to live at the telescope.

After identifying a supernova candidate and confirming from its spectrum whether it was a young Type Ia (it often was) with the help of future collaborators Peter Garnavich and Pete Challis, I would contact the present observer to request the appropriate exposures, filters, and exposure times and I would provide a finder chart. The next day I would transfer the observations and calibration data from the computers at Mt. Hopkins to Harvard. Starting with SN 1993ac in October of 1993 and ending with SN 1996bv in November of 1996 I collected 22 well-observed Type Ia light curves, 1200 observations in all. A few years after the Calan/Tololo survey published their final sample of 29, we published ours, thus doubling the world's sample of high-quality, multi-filter CCD-measured light curves. These two samples would form the cornerstone of our team's discovery of acceleration.

At the end of that summer, when Bill Press returned, we worked together on a small sample of 8 well-observed supernovae, using the Rybicki and Press method to decouple the luminosity of a supernova from its distance in one passband – a narrow range of wavelengths – only. The virtue of this method, which we called the Light Curve Shape (LCS) method, was that it provided a true, empirical model which included the covariance of the data and model. So using the LCS method, we could use data from any point in the life of the supernova, and our error estimates would be more rigorous than prior approaches. When applied to the first ten SNe Ia from the Calan/Tololo survey, the LCS method reduced the scatter in the measurements of the expansion rate by more than half. It thereby reduced the number of supernovae required for a significant measurement of expansion or the expected deceleration of expansion by a factor of 5.5. I was fortunate to be able to test the method on data taken by the Cerro Tololo team in the south, and I want to repeat here the gratitude I expressed to the Calan/Tololo Survey Team then. I quote the acknowledgment from the LCS paper written with Bob and Bill: "We are grateful to Mario Hamuy, Mark Phillips, Nick Suntzeff and the entire Calan/Tololo collaboration for the opportunity to study their outstanding data before publication." This was a favor we were happy to be able to return by providing them early use of our own northern supernova sample a few years later for a paper written by Mark and the rest of the Calan/Tololo members.

Figure 2. Left: Left to Right, Bill Press, Adam Riess and Bob Kirshner at the Harvard Center for Astrophysics. Right: Application of the MLCS method to nearby supernovae to differentiate the 3 effects of distance, dimness and dust on the flux of supernovae, reprinted from Science Magazine in 1995.

Despite the improvements afforded by LCS, I knew we were still ignoring one of the biggest bugbears for accurately measuring distances in the Universe: the presence of dust. Dust in the galaxies hosting supernovae dims their light, fooling you into over-estimating their distances. Dust also reddens a supernova's colors: the dust grains are of a size that scatters blue light more efficiently than red, leaving more red light to pass through and fooling you into thinking the supernova is redder than it is. Yet when previous workers, like Allan Sandage and Gustav Tammann tried to quantify the dimming from dust by measuring this reddening of the supernovae's colors and assuming that they all had the same intrinsic color, the scatter in the distance measurements went up! As Bob said at the time "This is a bad sign that this is the right thing to do!"

At this point I had a new idea, probably the first I had had on my own. I could use the LCS method to disentangle the color intrinsic to the supernova (we learned that dimmer ones were intrinsically redder) from its extrinsic reddening by dust by using several different-colored filters. This new Multicolor Light Curve Shape (MLCS) was a technique which could distinguish between the effects of distance, dust, and dimness on the brightness of Type Ia supernovae. With the MLCS and the new supernova samples, we could measure the expansion rate of the Universe to unprecedented precision, reducing the scatter from the narrow-waveband approach of LCS by 50% (Fig. 2). More importantly, MLCS could remove the presence of interstellar dust as a major source of uncertainty. By distinguishing an intrinsic and extrinsic change in supernova color it was further possible to make use of additional understanding (known in statistics as a "Bayesian Prior") about the properties of dust to further improve the distance measurement. This addition would prove of particular importance in our later Nobel discovery: much of the color data we collected had a low ratio of signal-to-noise, a regime where such an approach excels and we used it for all our subsequent distance measurements.

As an aside, Robert J. Trumpler had first identified the importance of dust in determining the positions of stars in the 1930s, so I felt particularly honored to receive the 1999 PASP Trumpler Award for "the doctoral thesis judged particularly significant to astronomy".

Using MLCS and the vastly improved sample of low-z supernovae we were able to make a number of additional, important measurements: the motions of the galaxies of our Local Group in relation to the universal standard frame of reference, the cosmic microwave background, the reddening properties of the dust in galaxies hosting supernovae and the linearity of the local expansion.

THE BIRTH OF THE HIGH-Z TEAM

In early March of 1994, I was still in graduate school and was on an observing run at the Multiple Mirror Telescope on Mt. Hopkins with Bob and Pete. While we were in the telescope control room, we received an excited call from Saul Perlmutter of Lawrence Berkeley Lab about a possible distant supernova that his team found. We obtained its spectrum for the Berkeley team immediately. The shift in a supernova's spectrum due to the expansion of space gives its redshift (z) and the relation between redshift and distance is then used to determine the expansion rate of the Universe. Supernovae with greater redshifts and distances reveal the past expansion rate because their light was emitted when the Universe was younger. When compared to their nearby brethren they can measure how the expansion rate has changed over time. SN 1994G at $z = 0.425$, or about 5 billion light years away was the most distant Type Ia supernova known! This experience sparked an interest in all of us to fish for supernovae in the higher redshift waters. Later that year my thesis committee asked me to calculate the number of Type Ia supernovae we would need to observe at this record redshift if we added them to our nearby sample and then used MLCS to usefully measure how much the universe was decelerating. I made a table with the answer: a couple dozen should suffice.

About this time Brian began having discussions with Nick Suntzeff about forming a new team to compete with Saul's. Brian's reasoning was that a team combining Calan/Tololo and Harvard groups, plus a few of their past members, would combine a solid understanding of supernovae and the critical low-redshift samples, and would therefore be able to hit the ground running. This newly formed "High-Z Team" (Nick was the principal investigator of the first proposal but Brian took the helm by 1996) quickly bought a set of custom filters to deal with the large redshifts of the targeted supernovae. Brian developed software whose importance is hard to overstate; it let us subtract the galaxy light to discover the supernovae while accounting for variations in the image quality. We were now in the race to measure the deceleration of the expansion rate and to predict the future of the Universe.

By April of 1995 Brian had found our team's first distant supernova, another record breaker at $z = 0.478$ and had measured its light output at various dates. I used MLCS to measure its distance and compared the result to different possible values for the universe's deceleration. I showed the result to the rest of the team and they chuckled nervously: that single supernova lay in the region of the diagram indicating the Universe was accelerating. We consoled ourselves that the error bar was big and one can be unlucky with a single object.

In the summer of 1995 a conference about Type Ia supernovae was held in Aiguablava, Spain, and members of both teams attended (Fig. 3). Saul's team, called the Supernova Cosmology Project (SCP), already had 7 high-redshift supernovae and was getting close to an answer. We had 1 high-redshift supernova and the hope that by the next year, we'd catch up to them.

In 1996, in the 6th cycle of Hubble Space Telescope operations, the director of the Space Telescope Science Institute, Robert Williams, awarded 28 orbits of his director's discretionary time on the Hubble Space Telescope to both teams to follow up the high redshift supernovae they found from the ground. The Supernova Cosmology Project was miffed that we were elevated in the process to equals, but Williams felt the Space Telescope should support competition for such an important problem.

That summer after finishing my PhD I began thinking about how well we could measure the deceleration if we followed different strategies. Would we be better off with, say, 3 measurements of 8 supernovae … or 6 measurements of 4 supernovae? I wrote a Monte Carlo simulation (which allows the computer to repeat the experiment thousands of times) and found we should choose 6 measurements of 4 supernovae, and so choose quality over quantity. That spring we also had good luck with telescopes on the ground, finding 8 new SNe Ia out to $z = 0.62$.

Figure 3. Upper left: High-Z Team in Aspen. (Back row from left: Tonry, Suntzeff, Leibundgut, Filippenko and Hamuy. Front row: Jha, Riess, Schmidt and Kirshner.). Upper right: Riess, Goldhaber and Schmidt in Aiguablava. Lower left: Both Teams in Aiguablava. (From left: Pennypacker, Filippenko, Riess, Schmidt, Nugent and Suntzeff.) Lower right: High-Z Team observing at Keck Summit at fall 1996. (From left: Riess and Filippenko.)

By now, I had finished my PhD and needed to find a post-doc position. Unfortunately, our newly formed High-Z team did not have funds to hire a post-doc so I had to go out onto the job market. I received a firm offer from the Supernova Cosmology Project, communicated by the late Gerson Goldhaber, to work for them – an offer I nearly accepted until I came off the waitlist for a coveted Miller Fellowship at UC Berkeley, a position which offered a small research budget and full autonomy. At Berkeley was Professor Alex Filippenko who was an expert in supernova spectroscopy and an enthusiastic supporter of junior scientists. Moreover, he'd recently switched teams from the SCP to the High-Z Team and brought access to the large Keck Telescope with him. I happily went to Berkeley to work with Alex and continued working in the High-Z Team. Interacting with Alex, an energetic astronomer with great attention to detail, was a highlight of my time at Berkeley. Later that year the SCP began to discuss their initial results from their first seven supernovae. Their data supported a strongly decelerating universe with enough matter so that its mutual gravitation might even cause the future universe to re-collapse. They discussed this result at a conference Bob attended in 1996 in Princeton, and published it in 1997 [4]. The result was close to what was expected at the time, but it was clear that more data were badly needed.

By this time, our team was running into the usual academic management problem of having more chiefs than Indians. The original 9 members of the team had grown to ~15 including the additions of John Tonry, Christopher

Stubbs, Peter Garnavich, Craig Hogan and Peter Challis. And only a few of us were in the "sweet spot" of an academic career – that one fleeting intermediate post-doctoral stage, in which the scientist is fully up-to-speed on the subject but still has maximal time to devote to research. In astrophysicists' collaborations, the rate-limiting step is usually not CPU power or telescope time; it's the lack of human power.

We were having trouble analyzing our data, partly because we needed too much time to find and re-observe the supernovae; and partly because, unlike the Supernova Cosmology Project which was based in Berkeley, we were spread around the globe and couldn't easily work together. By the beginning of 1997 we were already falling behind, collecting new data without yet analyzing the old. Brian suggested and the team agreed on a solution: offer individual junior members the opportunity and burden of pulling the past data together, whipping it into shape, drawing conclusions, and writing up the results. The reward was to be the leader of the study and the lead author on the published result. The junior members would thus gain the badly needed academic credentials needed to land a faculty job in exchange for their sweat equity. To jump start the process, junior members of the team (Brian, myself, Peter Garnavich, Pete Challis, Alejandro Clocchiatti, Al Diercks and David Reiss) got together in Seattle in March 1997 at the invitation of Chris Stubbs for a working summit we called "Reduce-Fest" to work on the software to analyze telescope images. We made some progress but in the end realized that our reference images weren't good enough to let us reliably subtract host galaxy light from the images of the supernovae. Brian and I had vociferous arguments about the best way to measure the brightness of the supernovae when the image statistics were dominated by random fluctuations in the brightness of the night sky – statistics that in hindsight were integral to the future work. Brian and I argued a lot about science, but the arguments were always the good kind – wrestling with how to proceed on a thorny problem. When Brian moved to Australia these arguments occurred at strange hours for both of us and had large phone bills, paid by Alex.

Our team began to fear that the lack of a "first result" paper from the High-Z Team to match that already published by the Supernova Cosmology Project could limit our ability to compete successfully for telescope time. We knew we still had a lot of work to do to analyze the full sample of supernovae in hand, but we saw a way of showing progress. In 1997 we had used our Hubble Space Telescope time to follow 4 new high redshift supernovae, and the quality and homogeneity of the Hubble data allowed for a much faster analysis of those data. Peter Garnavich volunteered to play the "scout", leading a quick march to analyze this Hubble data and to write a short letter about the results [1]. What Peter and our team saw in that first reconnaissance of the distant universe was already revealing. We could see that the universe was not decelerating strongly enough to re-collapse in the future.

Meanwhile, the Supernova Cosmology Project had changed their earlier conclusions based on their first 7 high redshift supernovae [3]. They now included their new highest redshift supernova which, owing to their Hubble observations, was also their best measured. Both teams were now agreeing that the Universe was not decelerating enough to re-collapse. By the end of 1997, and at a dual press conference in January 1998 at the AAS, both the High-Z team and the Supernova Cosmology Project presented this result and hints of more results to come.

THE ACCELERATING UNIVERSE

Later in the year 1997 I began leading the effort to analyze the bulk of our team's data. Most of what I did was technical work. After using the new reference images to subtract the light of the galaxies, plus a software package Brian wrote, I measured the brightness for six of our supernovae relative to stars in the fields. Attendees of "Reduce-Fest" pitched in by re-measuring the single supernova they had worked on in Seattle. Next I calibrated the brightness of the reference stars against three nights of observations of standard stars. Because the supernovae had been monitored with different CCDs at different facilities I also needed to measure differences in CCDs' wavelength responses using the calibrated reference stars seen through thin clouds. A quick consultation with Nick Suntzeff was sufficient to conclude that intervening clouds were grey enough for this approach to work. I then set to work improving my light-curve measuring algorithm, MLCS. I needed to incorporate into it the new supernova light curves from my thesis and from the Calan/Tololo Survey, plus a few new improvements garnered in the prior two years. These improvements included a second order correlation now apparent between light curve shape and the supernova's luminosity, and the use of Monte Carlo simulations of the lines of sight through galaxies to improve upon the measurement of the dust between us and the supernova. This latter idea came from a new paper in 1997 by Hatano, Branch and Deaton. This new, improved version of MLCS ended up as the Appendix to the paper I was already writing [5].

If you are ever in a room full of physicists and astronomers and want to figure out which is which, ask each if they know how to calculate a "K-correction". Both will know how to do it, but only the astronomers will know it by its name. The need for this correction arises when we measure the distance to a supernova from its brightness because the way we see a supernova is also affected by the universe's expansion. Besides causing the redshift, cosmic expansion also dilates (expands) time intervals over which supernova light is collected, changes the size of the increments in brightness, and shifts the portion of the spectrum we observe. (A physicist would mutter about need for "relativistic corrections"). Brian and I iterated back and forth a few times before we were both comfortable applying these corrections. I later heard from members of the Supernova Cosmology Project that despite their own extensive study of these corrections, errors in making the corrections likely limited the accuracy of the analysis of their first seven supernovae. K-corrections are a tricky step.

By the fall I had run all the data through MLCS and later, with Mark

Phillips, through his own algorithm so we could compare results. Then, to reclaim a half dozen other supernovae previously abandoned because our observations didn't cover their full light curves, I used a new "Snapshot Method" I developed that year with Peter Nugent which used the spectrum to determine the supernova's progress along its incomplete light curve. I was keen to make use of every scrap of data we had, since I had begun to hear from Supernova Cosmology Project members (during games of touch football played in the muddy parks of Berkeley) that they now had about 40 supernovae at high redshift.

The fewer numbers of high-redshift supernova that our team had did have compensations. First, our team was able to muster a larger sample at low redshift by adding to the 17 supernovae from the Calan/Tololo Survey that both teams used, another 17 from my thesis and my Snapshot paper. Second, because of our extensive color measurements and use of the Bayesian prior, our high redshift supernovae had half the scatter of the other teams' sample. In all, during that fall of 1997, a lot of new and important developments for our supernova cosmology work came together and critically increased the credibility of our subsequent discovery.

I had measured the supernovae used to determine the universe's expansion rate in the present and in the past, and could then measure how that rate changed over the last few billion years. Knowing how the expansion decelerated – a quantity called q_0 – I could predict the amount of mass the universe must have – a quantity called Omega matter, or Ω_M. The higher the universe's mass the more its gravity pulls against its expansion, and the more the universe decelerates. The equation is surprisingly simple, $q_0 = \Omega_M / 2$. But what I initially measured and wrote in my lab notebook in the fall of 1997 was stunning! The only way to match the change in the expansion rate I was seeing was to allow the universe to have a "negative" mass. In other words, up-ending the equation, the Universe wasn't decelerating at all – it was accelerating (Fig. 4)!

That simple equation assumes that matter is the only important component of the Universe. So initially, I hadn't considered any force besides the gravity of matter, and now my computer programs were telling me that only an imaginary negative mass could match the apparent acceleration and cause the reverse of attractive gravity. The more complete equation for the deceleration parameter, q_0, in Einstein's Theory of General Relativity is $q_0 = \Omega_M / 2 - \Omega_\Lambda$ where Ω_Λ, or Omega lambda, is the energy density of empty space. Einstein called the Λ component the cosmological constant, and in effect it's a repulsive gravity. A contemporary particle physicist would call it vacuum energy, that is "the zero point energy summing all possible particles in the vacuum", and then would complain that calculating it yields a nonsensically enormous answer.

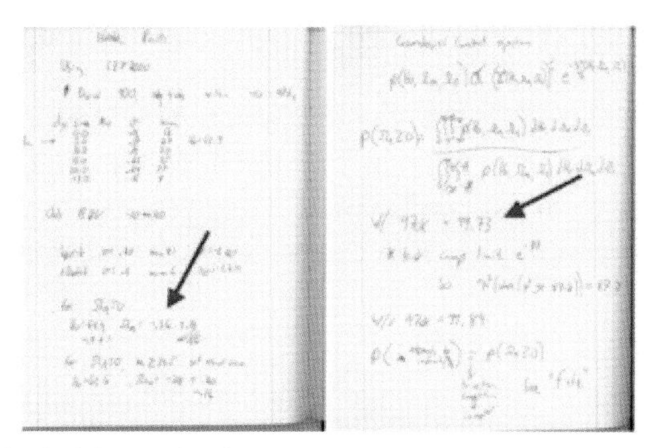

Figure 4. The author's lab notebook. Left: After performing various tests of the data I decided to analyze the expansion rate data in terms of the mass density for the Universe it suggested. The answer I got, –0.36 ± 0.18 made no sense, unless the Universe was accelerating! Right: A few days later I calculated the significance of the cosmological constant to 99.7% – 99.8% confidence no matter what the mass density. If the mass density was even the smallest conceivable amount, ~0.2, the confidence rose to 4–5 sigma.

Since there is no such thing as negative mass, in something like confusion and desperation (in the same spirit Einstein had introduced it long ago), I re-introduced the famous cosmological constant to the equation and immediately found that its repulsive gravity could explain the acceleration I was seeing (Fig. 4). Its presence was significant in a statistical sense as well as an absolute sense – in fact the Universe was 70% in this form alone! This was remarkable and even my modest experience told me that such "discoveries" are usually the result of simple errors. So I spent a couple of weeks double checking my results but could find no errors. Then I thought hard about unexpected astrophysical sources of contamination to the observations. Here my thesis on recognizing and correcting for the effects of interstellar dust was helpful. Although dust in the high redshift galaxies could mimic the effect of acceleration and dim the light of distant supernovae, my use of MLCS and Mark Phillips's use of his own dust-correcting algorithm made this unlikely. With growing confidence in the results, I first told Brian who spot-checked the final calculation. At the beginning of January 1998 he wrote me that he was getting the same answer. Later, the media quoted him saying: "My own reaction is somewhere between amazement and horror."

I was also able to rule out a number of other concerns. One was that high redshift supernovae, born when the universe was younger, might be somehow different. But a comparison of the distances from nearby supernovae in old (elliptical) and young (spiral) galaxies limited the possible size of the difference to less than a third of the size of the acceleration signal. Moreover, my MLCS and Mark's own algorithm gave the same result: the light curves and spectra of the nearby and high redshift supernovae were indistinguishable. Another concern was that a kind of exotic type of dust in the host galaxies could have had unusually large grains that would not redden the supernova light and could therefore go undetected. I calculated that the low scatter in the high redshift supernovae limited that kind of dust to an insignificant amount. A third concern was a well-known bias: astronomers tend to find the brightest objects of a class preferentially. Brian did a simulation showing that this effect was insignificant too. We even calculated possible but unlikely explanations – the presence of a local void in the Universe, contamination of the sample by other supernova types and an effect known as gravitational lensing – and ruled out the significance of all of them.

Coincidentally, another exciting event was occurring in my life. Nancy Joy Schondorf and I were married on January 10, 1998, the best day of my life. We planned a honeymoon to Hawaii, after the next supernova observing run on the Big Island.

Meanwhile, the rest of the team did their own spot-checking of the results and more thinking and they too could not find any mistakes. It is amusing to look back at the emotions expressed about the result in the emails between team members over a couple of days in early January of 1998:

A. Filippenko, Berkeley, CA, 1/10/1998 10:11 am:

"Adam showed me fantastic plots before he left for his wedding. Our data imply a non-zero cosmological constant! Who knows? This might be the right answer."

B. Leibundgut, Garching, Germany, 1/11/1998 4:19 am:

"Concerning a cosmological constant I'd like to ask Adam or anybody else in the group, if they feel prepared enough to defend the answer. There is no point in writing an article, if we are not very sure we are getting the right answer."

B. Schmidt, Australia, 1/11/1998 7:13 pm:

"It is true that the new SNe say that [the cosmological constant] is greater than zero... How confident are we in this result? I find it very perplexing..."

M. Phillips, Chile, 1/12/1998 04:56 am:

"As serious and responsible scientists (ha!), we all know that it is FAR TOO EARLY to be reaching firm conclusions about the value of the cosmological constant..."

R. Kirshner, Santa Barbara, CA, 1/12/1998 10:18 am:

"I am worried. In your heart you know [the cosmological constant] is wrong, though your head tells you that you don't care and you're just reporting the observations... It would be silly to say 'we MUST have a nonzero [cosmological constant]' only to retract it next year."

J. Tonry, Hawaii, 1/12/1998 11:40 am:

"...who remembers the detection of the magnetic monopole and other gaffes?...on the other hand, we should not be shy about getting our results out with appropriate disclaimers..."

A. Filippenko, 1/12/1998 12:02 pm:

"If we are wrong in the end, then so be it. But at least we ran in the race."

A. Riess, Berkeley, CA, 1/12/1998 6:36 pm:

(Sent on the eve of our honeymoon over understandably icy stares from my wife!)

"The results are very surprising, shocking even. I have avoided telling anyone about them because I wanted to do some cross checks (I have) and I wanted to get further into writing the results up before [the other team] got wind of it... ...The data require a nonzero cosmological constant! Approach these results not with your heart or head but with your eyes. We are observers after all!"

A. Clocchiatti, Chile, 1/13/1998 7:30 am:

"If Einstein made a mistake with the cosmological constant... Why couldn't we?"

N. Suntzeff, Chile, 1/13/1998 1:47 pm:

"I really encourage you [Adam] to work your butt off on this. Everyone is right. We need to be careful and publish good stuff with enough discussion to make it believable to ourselves. ... If you are really sure that the [cosmological constant] is not zero – my god, get it out! Take responsibility as organizer and grunt. I mean this seriously – you probably never will have another scientific result that is more exciting come your way in your lifetime."

Nick was absolutely right.

On February 20 we had a team-wide teleconference to decide whether to go forward with the paper I was writing, a draft of which I had earlier sent around. We decided to proceed. Somehow – I'm still not clear how – Jim Glanz of Science Magazine got wind of our result, interviewed many of us, and broke the story. Alex Filippenko on our team discussed the results at the UCLA Dark Matter conference in Los Angeles back to back with the Supernova Cosmology Project team, with both teams now claiming to see acceleration. I finished our paper with long nights and with crucial assistance from team members and on March 13, I submitted it: "Observational Evidence From Supernovae for an Accelerating Universe and a Cosmological Constant" (Fig. 5). It was accepted in May 6 [5]. The Supernova Cosmology Project published the same conclusion 9 months later [2]. Together the two team's conclusion became the "Breakthrough of the Year" in 1998 of Science Magazine (Fig. 5).

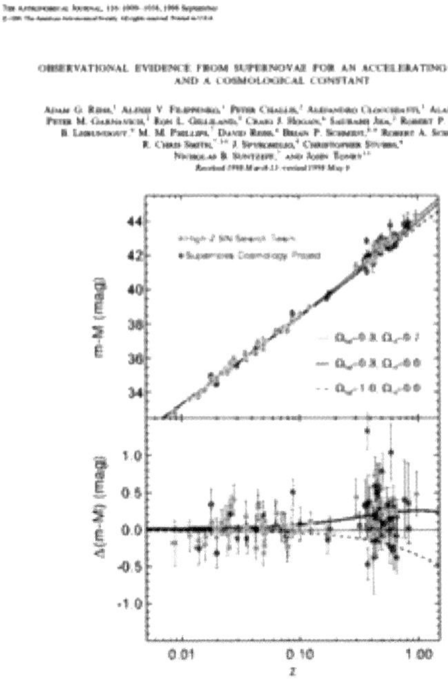

Figure 5. The High-Z Team discovery paper [5] and the Hubble diagrams over-plotting the supernovae measured by both teams.

EXTRAORDINARY CLAIMS REQUIRE EXTRAORDINARY EVIDENCE

Our finding, that the Universe was currently accelerating, immediately suggested a profound conclusion: the universe's cosmic energy budget is dominated by a type of smoothly distributed "dark energy". Dark energy, a new component of the universe with negative pressure, causes the repulsive variety of gravity to dominate over matter's attractive gravity. The cosmological constant would be dark energy's poster child, the one with the pedigree, which came from Einstein. Was cosmic acceleration caused by dark energy the correct interpretation of the supernova data, or did we make a mistake in the interpretation of the supernova data? Past attempts to use distant objects (like the brightest cluster galaxies) to measure the change in the expansion rate had been foiled by evolution of the object's intrinsic luminosity. So we needed a good test of cosmic speedup. As scientists often say, extraordinary claims require extraordinary evidence.

The situation circa 2000 was well summarized in the popular book by Donald Goldsmith "The Runaway Universe", published that year. The essence of what we had learned from high-redshift supernovae in 1998 was that they were 20% fainter than we expected. We had assumed the usual rule that "fainter means farther" and had come to our conclusion of acceleration and dark energy. In 1999, Anthony Aguirre of the Harvard CfA wrote a series of bracing papers pointing out the supernova evidence could be mimicked by gray dust – mythical stuff akin to the Loch Ness Monster or Bigfoot of astronomy. His idea was that, in addition to the familiar, interstellar dust grains that redden and dim supernova light, intergalactic space was full of dust "needles" one-tenth of a millimeter long and one-tenth that length in width. Aguirre calculated rigorously that such needles could absorb light equally well at all visible colors making them look "gray". Gray dust would be nearly impossible to detect. Such a possibility seemed outlandish, but so of course did dark energy. Aguirre showed that at the time no astronomical observation ruled out the possibility of gray dust between galaxies in amounts sufficient to explain away the evidence for acceleration and dark energy. When trying to discriminate between improbable options Occam's Razor becomes a blunt tool.

Fortunately a good test of competing hypotheses had just become available. If distant supernovae appeared faint not because the universe was accelerating but because of some astrophysical cause – like a pervasive screen of gray dust making supernovae appear dim, or past supernovae being born dimmer – then the extra dimming seen in distant supernovae would be expected to steadily increase as the distance to the supernovae increases. That is, if we were to look twice as far through a uniform screen of gray dust, we should see twice as much dimming. But if instead dark energy had begun increasing the size of the universe only recently, then the more distant supernovae should bear witness of the universe at an earlier stage in which dark energy was subordinate to matter. At that earlier stage, matter's attractive gravity would have been decelerating the expansion and forming structures like galaxies and clusters. While acceleration increases distances and dims the light of supernovae, deceleration does the opposite, shortens distances and brightens light. The cumulative effects of both phases would likely look different than the effects of either gray dust or evolution if we measured supernovae beyond a redshift of one. As Goldsmith wrote in 2000:

"Gray dust and systematic differences can mimic the effects of a nonzero cosmological constant with high precision only so long as we examine distant supernovae within a relatively constricted range of distances… [I]f however astronomers observe distant supernovae over a much larger range…then cosmological models allow astronomers to disentangle all other effects from the crucial one: the acceleration produced by a nonzero cosmological constant… astronomers must therefore not rest on their current supernova assets… they must push their frontiers farther into space. Only then can they eliminate the possibility that gray dust has fooled them and show that the runaway universe deserves general acceptance."

This was a call to arms we had already accepted!

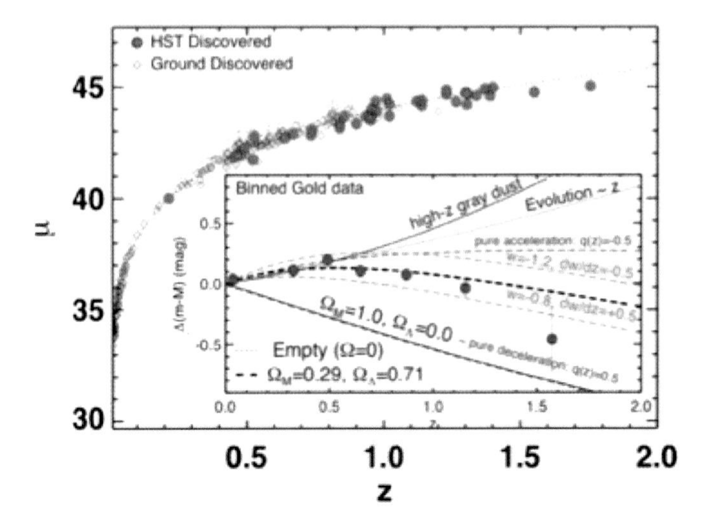

Figure 6. Supernovae observed with Hubble at $z > 1$ confirm the result. By observing the transition from acceleration to deceleration (looking back in time) the High-Z Team papers could rule out simple astrophysical dimming (dust or evolution) as an alternative to acceleration with a mixed dark matter and dark energy Universe.

Unfortunately, finding supernovae so far back and away is difficult because they are so faint – a 60-watt light bulb held at a distance of 400,000 miles, twice the distance of the Moon. Five years of attempts by both teams had demonstrated that such faint supernovae were too hard to find reliably from the ground. So far, each team had found only one supernova each at $z = 1.2$, the Supernova Cosmology Project in 1998 and the High-Z Team in 1999. The Hubble Space Telescope could see more deeply than any ground-based observatory, but the small field of view of its main camera, Wide Field Planetary Camera 2 (WFPC2) made finding supernovae unlikely unless they occurred in the younger universe in dramatically higher numbers. In 1997 High-Z Team members Ron Gilliland and Mark Phillips made a reconnaissance of the high-redshift universe by re-observing the Hubble Deep Field with WFPC2 two years after it was first observed. They discovered two supernovae, one of which, SN 1997ff, appeared to be a high-redshift Type Ia because it was in an old, elliptical host. While interesting, this single observation did not allow measurements of the supernova's light curve or colors to estimate its distance, and thus could not provide the test of the gray dust (or evolution) dimming. However, in 2001, after I had moved to STScI, I began looking for additional observations of SN 1997ff and encountered incredible serendipity!

The PI, Rodger Thompson, of the first near-infrared camera on Hubble, NICMOS, had used his camera to stare deeply into the Hubble Deep Field. By unbelievable good fortune, SN 1997ff had been observed for a couple of weeks shortly after its discovery in the very corner of the NICMOS field. In fact it was so close to the corner that it fell in and out of the frame through the intentional jiggling of the field (a technique for reducing the pixelization of the image). I spent a number of months retrieving the color and light curve data needed to determine the supernova's distance and redshift. SN 1997ff was a Type Ia supernova at z = 1.7, by far the greatest distance ever observed. More importantly, it was about 60% brighter than we would have expected for the gray dust (or evolution) scenario. This was a good sign that the Universe had once been decelerating and that supernovae were faithfully tracing the history of cosmic expansion.

However, we did not want to hang such an important conclusion on a single, serendipitous supernova. We really wanted to find more. This became a possibility in 2001 when a new camera was due to be placed on Hubble, the Advanced Camera for Surveys (ACS), built by Holland Ford of Johns Hopkins University. ACS would improve Hubble's "discovery space" (area times depth) by an order of magnitude. I led a proposal in 2001 to use ACS to find and measure half-a-dozen Type Ia supernovae at z's greater than 1 by "piggybacking" on a survey to collect images of distant galaxies, the GOODS Survey led by Mauro Giavalisco. The idea was that the GOODS Team would reimage a field composed of 15 ACS pointings every 45 days. Our new team, the Higher-Z Team (with members Tonry, Filippenko, Kirshner Leibundgut, Challis and Jha from the old High-Z Team and new members Casertano, Strolger, Ferguson, Giavalisco, Mobasher and Dickinson) would subtract these images to find fresh supernovae. Unlike previous ground-based surveys, we would already have a good idea of the type and redshift of the supernovae from color measurements obtained by the GOODS Team. If we found a supernova at z greater than 1 with the right colors to be a Type Ia supernova, we could interrupt Hubble's schedule to re-observe it with ACS and NICMOS. Our proposal was accepted, but that did not guarantee our plan. The NASA crew of the Space Shuttle flight STS 109 first had to install ACS! One of the greatest privileges I have ever had as a scientist was to present the science case for the observations to the astronauts before their trip to Hubble. These folks (John Grunsfeld, Mike Massimino, Scott Altman, Jim Newman, Duane Carey, Nancy Curry, Richard Linnehan) were real life heroes who risked their lives to refurbish Hubble. Their successful mission in March 2002 brought Hubble again to the peak of its capabilities and allowed us to undertake the proposed investigation.

Compared to our searches using ground-based telescopes, using Hubble to find and follow supernovae had advantages and disadvantages. On the positive side: the weather was always good in space, the image quality always razor sharp and moonlight never got in the way of our observations. On the negative side: the Hubble doesn't have the protection of Earth's atmosphere and magnetic fields that ground-based telescopes do, so cosmic rays strike Hubble about 100 times more frequently. A cosmic ray strike can look quite similar to a supernova: they both appear as a new source not in the prior image. Because cosmic rays affect only about 2% of the pixels in a Hubble image, and because the odds of consecutive strikes is small, most astronomers can distinguish a real astronomical source from a cosmic ray strike by obtaining a second image. However, because a supernova could appear on any pixel of the ACS's 16 million pixels, we determined we needed 4 consecutive images to rule out the fluke of even 3 consecutive strikes. Another challenge was imposed by the way Hubble is scheduled: observing schedules are uploaded to the telescope weekly. We could alter Hubble's schedule only on a Tuesday by noon, in advance of the next schedule upload. It was ironic that a supernova's light could travel for 9 billion years but needed to reach Hubble shortly before Tuesday to be of any use! We found that if you look for a supernova only on the weekends, you could naturally meet this requirement.

Our year-long program from 2002 to 2003 to measure supernovae on Hubble was highly successful. We found 6 Type Ia supernovae at redshifts over 1.25 [6]. They allowed us to rule out gray dust and evolution and to clearly determine that the universe was decelerating before it began accelerating (Fig. 6). In physics, a change in the value or sign of deceleration (which results from a change in force) is called a jerk. So when we announced this result in a conference in 2003 we described the change from the universe's past deceleration to its present acceleration as evidence of recent "cosmic jerk". I saw Dennis Overbye from the New York Times in the front row as I discussed the result and asked him to please not run a picture of me next to a headline "Cosmic Jerk Discovered" – to no avail.

Over the next two years we continued using Hubble to collect Type Ia supernovae at redshifts greater than 1 and by 2007 we published a sample of 23. Not only did this data appear to confirm that supernovae were faithfully communicating the history of cosmic expansion; they could also be used to help determine if the properties of dark energy had been changing over the last 10 billion years. So far those properties seem unchanged, adding credence to Einstein's cosmological constant. But to quote Edwin Hubble from "The Realm of the Nebulae":

"With increasing distance, our knowledge fades, and fades rapidly. Eventually, we reach the dim boundary – the utmost limits of our telescopes. There, we measure shadows, and we search among ghostly errors of measurement for landmarks that are scarcely more substantial. The search will continue. Not until the empirical resources are exhausted, need we pass on to the dreamy realms of speculation."

Although Hubble did not know about acceleration, his description fits our work. Today we seek to learn more about the cause of cosmic acceleration by refining a wide range of cosmological measurements. Many feel this challenge is one of the greatest for cosmology and fundamental physics and I agree. Since about 2003, WMAP data, measurements of Baryon Acoustic Oscillations (BAO), large-scale structure, weak lensing and the integrated Sachs Wolfe effect also give strong evidence for dark energy in ways independent from supernovae. The BAO technique alone now provides independent confirmation of the recent acceleration. For my own work I have been focusing on improving the measurements of the present expansion rate, also known as the Hubble constant, because knowing this to percent level precision would significantly aid the present quest. Already we have improved the determination of the Hubble constant by a factor of

3 to 3.5%. This, combined with the WMAP measurements of the cosmic microwave background are good enough to measure the properties of dark energy about as well as high redshift supernovae – to around 10% – and to provide another independent check of the results. An ambitious goal will be to achieve about a 1% measurement. I don't expect to win another Nobel Prize from this new work, but it should keep me out of trouble for a while.

I want to close by expressing my deep gratitude to the amazing people and facilities I have been honored to work with. To my colleagues on the High-Z and Higher-Z Teams, it has been a pleasure to share this scientific adventure with you. I thank those who built the instruments and facilities at CTIO to allow us to find supernovae (Bernstein and Tyson) and the Calan/Tololo Survey for helping ground and inspire the subsequent work. I thank the men and women who have helped make the Hubble Space Telescope the premier scientific instrument of our time and the astronauts who risked their lives to maintain it. Most of all I thank my family, my wife Nancy and my children for help keeping me sane and for reminding me that the world down on Earth is at least as interesting as the universe around it.

REFERENCES

1. P. M. Garnavich *et al., Astrophysical Journal,* **493**, L53 (1998).
2. S. Perlmutter *et al., Astrophysical Journal,* **517**, 565 (1999).
3. S. Perlmutter *et al., Nature,* **392**, 311 (1998).
4. S. Perlmutter *et al., Astrophysical Journal,* **483**, 565 (1997).
5. A. Riess *et al., Astronomical Journal,* **116**, 1009 (1998).
6. A. Riess *et al., Astrophysical Journal,* **607**, 665 (2004).

Physics 2012

Serge Haroche and David J. Wineland

"for ground-breaking experimental methods that enable measuring and manipulation of individual quantum systems"

The Nobel Prize in Physics

Speech by Professor Björn Jonson of the Royal Swedish Academy of Sciences.

Your Majesties, Your Royal Highnesses, Ladies and Gentlemen,

In his epilogue at the Master's Degree Celebration in Lund 1820, Professor Esaias Tegnér – also a famous poet and author – said

This is Man's wonderful ability:
to be able to grasp the inner essence of phenomena,
not what they appear to be, but what they mean,
and the reality that we see with our eyes
is a symbol only of something higher.

What is it that our eyes see? It is light. Everything we see around us – colours, shapes, and objects – comes from light that strikes our eyes, which forwards the information to be analysed by our brain. What we see can be described and understood with what we today call classical physics. But if we try to describe the heart of matter, classical physics is not enough!

Because of this, physicists in the 1920s developed a theory for the innermost parts of matter. It is an extremely beautiful theory called quantum mechanics. In the quantum world, light, atoms, nuclei and elementary particles are all described as being both particles and waves. Quantum physics has been extremely successful in modern physics. But to really observe matter at the quantum level one encounters major problems since almost any attempt to make a measurement causes the quantum properties to simply disappear. It has therefore been a dream since the dawn of quantum physics to be able to study individual quantum systems. This dream has now become a reality.

This year's Nobel Laureates, together with their research groups, have used cleverly designed experiments to successfully perform the trick of controlling and measuring single isolated quantum systems without destroying their quantum properties. Serge Haroche and David Wineland have opened the door to the observation of the most fundamental quantum world. In the laboratory, they have shown that it is possible to perform tricks according to the rules of quantum physics that were previously not thought possible. They have shown how to study and control and count quantum systems. One may call this mastering without touching – since the enchantment of the micro-world would otherwise disappear.

Their experimental methods have many things in common. David Wineland captures individual electrically charged atoms, ions, in an electrostatic trap, and cools, measures and controls them with photons of light. If the temperature is close to absolute zero, both the energy and the frequency of oscillation of the ion can only obtain certain distinct values; they are what we call quantised. By shining laser light on the ion it can then be put in a state called superposition, a state simultaneously occupying two different energy levels! The superposition of energy can then be transferred into a corresponding superposition of vibrational states, also with laser light. A superposition has been transferred without destroying it.

Serge Haroche takes the opposite approach – he captures single photons between two perfectly reflecting mirrors and keeps them there long enough to be able to measure and control them. This is done with the aid of very highly excited atoms, which pass the trap and tickle the photons just a bit. These atoms act like antennas that gather information about the number of photons in the trap and about their condition. This is simply spying at the quantum level. It is also remarkable to note that Haroche can study single photons, given that a regular 9 W bulb emits more photons in a billionth of a second than there are people on earth.

Both Laureates represent a research field called quantum optics, a field where the interaction between light and matter is studied at a fundamental level. This field of research has progressed considerably since the mid-1980s. Their research has given us new tools for future experiments and is therefore of the highest scientific interest. Their research also provides hope for the construction of future super-fast computers based on quantum physics. Perhaps the quantum computer might change our way of living during this century in the same radical way that the IT revolution did in the last. Their experiments have also led to the development of extremely accurate optical clocks that will eventually become the new time standard. Today they have already achieved an accuracy of about a hundred times more than today's clocks based on caesium. This accuracy is so unimaginable that if you could let such a clock run for a period of time corresponding to the age of the universe, it would only lag by a few seconds.

Professeur Haroche, au nom de l'Académie Royale des Sciences de Suède, j'ai le privilège de vous transmettre mes plus chaleureuses félicitations pour votre travail exceptionnel. Je vous demande maintenant de bien vouloir vous avancer pour recevoir votre prix Nobel des mains de Sa Majesté le Roi.

Professor Wineland, on behalf of the Royal Swedish Academy of Sciences it is my privilege to convey to you my warmest congratulations for your outstanding work. I now ask you to step forward to receive your Nobel Prize from the hands of His Majesty the King.

Serge Haroche. © The Nobel Foundation. Photo: U. Montan

Serge Haroche

I was born on September 11th 1944 in Casablanca, Morocco, in a Jewish family with mixed Sephardic and Ashkenazi origins. My father's parents were teachers at the Alliance Israëlite universelle (AIU), which operated a network of schools dedicated to the dissemination of French language and culture throughout Jewish communities in North African and Mid-Eastern countries. My paternal grandfather had been one of the first students of the AIU school in the town of Marrakesh, in the last years of the nineteenth century. My paternal grandmother had studied at the same time in Tetouan, in the part of northern Morocco under Spanish influence, in the first AIU school, which had been founded there in 1863. After completing their school years, they both decided to become teachers and, unbeknownst to each other, came to France at the very beginning of the 20th century, to get their degree from the AIU teacher training school in Paris. Although they were both there at the same time, they did not meet then since the boys' and girls' schools were separated.

I found a few years ago in one of my grandfather's notebooks a handwritten recollection of his souvenirs as a young teenager suddenly immersed in 1900 Belle Epoque Paris, seeing for the first time automobiles and trains, enjoying theater shows and silent movies, experiencing with bewilderment a modernity contrasting so much with the medieval atmosphere of Marrakesh at that time. My paternal grandparents met shortly after graduating and returning to Morocco, at the time this country became a French protectorate. They soon married and had seven children, while moving from school to school on successive assignments, ending as headmaster and headmistress of the boys and girl's AIU schools in the town of Salé, near Rabat. My father, born in 1920, was raised there and studied law in Rabat, becoming a lawyer at the beginning of the Second World War.

My mother's family emigrated from Russia in the 1920s, in the years following the Bolshevik revolution. Her parents were physicians and chose to settle in Morocco, a country where their Russian medical degrees were recognized. My mother, born in Odessa in 1921, was raised in Casablanca and met my father at the university in Rabat, just before the war, while she was studying to become a teacher in French and German literature. In my early childhood, I was raised in Russian as well as in French, and I remained bilingual until I went to elementary school at the age of six, unfortunately then very quickly losing my fluency in Russian. The community of Russian Jewish immigrants in Casablanca had kept strong links and I remember meeting many of them in my grandparents' house. Among them was an architect, Casimir Zeligson, who has built several of the Art Nouveau building and villas which still give a special character to the town of Casablanca. The Zeligsons had a daughter, Claudine, with whom my younger brother Joël and I used to play. After a long separation, I met her again in Paris fifteen years later, but here I am anticipating.

I have kept strong memories of this time in Casablanca, of the warm Mediterranean weather, the long rainless summers and mild winters with their flowering bougainvillea and hibiscuses, bathing in the cold Atlantic Ocean whose waves can be so strong and treacherous, especially to a young child. Some remembrances are particularly vivid, like the recurring sirocco storms covering everything with a blanket of sand coming from the Sahara, or the locust invasions, which on several occasions have plagued the fields in the countryside around the city, leaving them barren after they had gone. During the last years of my life there, I witnessed the events leading to the independence of Morocco from the French, understanding that, amid the convulsions and sometime violent events, history was in the making.

When independence finally came in 1956, my parents decided, like many Jews, to leave Morocco and settle in France with their sons Joël, Gilles and me (a fourth son, Michel, was born in Paris in 1959). This was essentially a cultural choice. They had received and given to their children a French education and thought that France was the natural place for the family to go. The first years in Paris were difficult, requiring a painful adjustment, especially during the damp and dark winters, contrasting so much with the vivid colors of the Moroccan weather. Learning at school was a consolation, though. I was a very good student, immediately at the head of my class in the Lycée Carnot where I studied until the "Baccalauréat," the final degree of the French high school education.

I was indiscriminately interested in literature, history, mathematics and physical sciences. In humanities, I could share my tastes with my parents, who raised my three brothers and me to love reading, going to museums and discussing all kind of issues ranging from ancient history to modern politics. In scientific matters, I was on my own, however, the first in the family to wander in a domain which required a mathematics background. I remember how, early on, I was fascinated by astronomy and by calculus, the notion of derivatives and simple differential equations which describe so directly and so well the laws of dynamics obeyed by moving bodies. This was the time of the first artificial satellites, the sputniks which orbited the earth and launched the American-Soviet race to the moon.

I marveled at the fact that I was able, with the elementary calculus I knew, to compute the escape velocity of rockets, the periods of satellites on their orbits and the gravitational field at the surface of all the planets … I understood then that nature obeys mathematical laws, a fact that did not cease to astonish me. I knew, from that time on, that I wanted to be a scientist. For that, I embarked in the strenuous and demanding "*classes préparatoires*" of the famed Lycée Louis-Le-Grand, one of the preparatory schools which train the best French students for the contest examinations leading to the "*Grandes Ecoles.*" They are the engineering and academic schools, which since the French Revolution, have formed the scientific elite of France. These were two years of intensive study where I learned a lot of math and of classical physics. I eventually was admitted in 1963 to the Ecole Polytechnique (ranking first in the national examination, to the great pride of my parents) and at the Ecole Normale Supérieure (ENS). I chose to enter the latter because, at that time, it offered a much better opportunity to embark in a scientist career.

The years as a student at ENS (1963–1967) have left me wonderful memories, contrasting sharply with the strenuous training of the preparatory school. Here, in the middle of the Latin Quarter, I was free to organize my time as I wished, to meet and discuss with students working in all kinds of fields in science or humanities and to enjoy all the distractions and cultural activities Paris has to offer. And I was paid for that, since the "Normaliens" as the ENS students are called, are considered civil servants and receive a generous stipend! These were my formative years as a scientist. Coming so to speak from the physics of the 19th century which was taught in the classes préparatoires, I was immediately thrown into modern physics and the quantum world by the classes of exceptional teachers. Alfred Kastler gave us a lyrical description of the dance of atomic kinetic moments and gave atoms and photons a near poetic existence. Jean Brossel brought us back to Earth by describing the great experiments thanks to which quantum concepts were established, instilling in us the austere passion for precision. And Claude Cohen-Tannoudji revealed the theory's formalism to us with extraordinary depth and clarity. I still remember three books I read avidly at the time: *Quantum Mechanics* by Albert Messiah, where I truly understood the depth and beauty of the quantum theory; *Principles of Nuclear Magnetism* by Anatole Abragam, who introduced me to the subtle world of atomic magnetic moments; and Feynman's *Lectures on Physics*, which was a revelation.

But physics was only one side of that time's story. In the spring of 1964, I met Claudine Zeligson again by chance in a Latin Quarter café. She had followed her own path from Casablanca to Paris, where she was studying English, psychology and sociology at La Sorbonne. We resumed a relationship which had been interrupted by a fifteen year latency period and married in 1965, at the age of 21. She later embarked in her own career as a scholar, doing research in sociology and anthropology at the French National Center for Scientific Research (CNRS). We have been together since then, sharing all aspects of life, including a common love for music and painting, movies and travels. Without her love, intellectual stimulation and constant support, nothing would have been possible.

Enthralled by the mysterious beauty of the quantum world, it did not take me long to decide that I wanted to become a quantum physicist. That was the time when various optical methods for the manipulation of atoms were being invented in the *Laboratoire de Spectroscopy Hertzienne* of ENS, which was to be called later the *Kastler Brossel* laboratory. I remember the day in the fall of 1966 when, as a young student, I witnessed the joyous turmoil following the announcement of Kastler's Nobel Prize, attributed to the invention of optical pumping methods. I measured my luck to have just started working in a field which was getting such a worldwide recognition. The Kastler lectures I had attended were immediately illustrated by everyday reality at the ENS laboratory, where in 1967 I started my thesis under the enthusiastic supervision of Claude Cohen-Tannoudji.

I have described in my Nobel Lecture how my interests in physics have naturally evolved from that time on, motivated by the challenge to probe and control the atom-photon interactions processes at the most fundamental level. This adventure started with my PhD work with Claude on the "dressed atom" formalism, analyzing atoms irradiated by radiofrequency fields as being dressed by a cloud of photons surrounding them. We learned how to describe the atoms coupled to photons as a combined entity, whose energy level structure revealed, in a synthetic way, all the properties of the system. At that time, it was unusual to describe the field, especially in the radiofrequency or microwave spectral range, in terms of photons. Some famous physicists even believed that all phenomena in atomic physics, including spontaneous emission, could be explained classically and that the photon was a superfluous concept. They were wrong, as we now know, but it is true that the classical picture viewing the field as a time-varying electromagnetic wave is sufficient to account for all effects involving huge amounts of light quanta.

Claude and I, though, were persuaded that the quantum description of the dressed atom formalism went deeper, giving a more satisfactory interpretation of the phenomena and a better insight into novel effects which were more difficult to predict within the classical approach. Since that time, Claude and I have kept using the dressed atom point of view, along with many other scientists who have adopted it. My main interest has been the interaction of atoms with invisible microwave photons, the kind of photons with which I became familiar during my PhD studies. Claude, with the students he trained after me, extended the formalism to deal with the interaction of atoms with optical fields and used it to explain in a particularly illuminating way the subtle effects involved in the cooling and trapping of atoms with laser light. This eventually led him to the famous studies which were recognized by his 1997 Nobel prize.

The experiments I performed to illustrate the "dressed atom" formalism during my thesis work were done with classical spectral lamps. The laser sources which underwent a spectacular development during the 1960s promised to open new perspectives in atomic physics and I realized that I needed to learn how to use these new tools. After completing my PhD, I chose to become a postdoctoral fellow at Stanford University, in the laboratory of Arthur Schawlow, one of the inventors of the laser. Our son Julien, born in 1970, was a young toddler and our daughter Judith a new born baby when Claudine and I arrived with them in the San Francisco Bay Area in September 1972. The year we spent there has left us wonderful memories. The California weather with its balmy winters reminded us of our childhood in Morocco. We enjoyed a way of life which was very different from the one in Paris, living in a large house on the Stanford campus, close to the natural wilderness of the Pacific coastal mountain range and the ocean beaches, as well as to the sophistication of San Francisco and Los Angeles with their theatres, opera houses and gourmet restaurants. It was an ideal place to raise young children.

In Art Schawlow's lab, I had a lot of fun with the marvelous toys that were the first tuneable lasers sent as prototypes to California laboratories by the commercial companies of what was to become Silicon Valley. Art's enthusiasm was contagious. Every day, a new idea would spring up, sometimes wacky, sometimes brilliant. There came the first "edible" laser the day he had the idea of turning those ghastly food jellies of garish colors that he loved to eat into laser amplifying mediums, but also many demonstrations of clever spectroscopic methods, every time pushing the limits of the precision and sensitivity of measurements further. These new spectroscopic methods led to the Nobel Prize awarded to Schawlow in 1981. A young associate professor, Theodor Hänsch, had joined Art's lab a few months before my arrival there and he was the driving force in the group, always finding new ways to exploit these marvelous laser sources to probe deeper and deeper into atomic spectra.

Art had a great sense of humor, which I believe is essential to maintaining a healthy atmosphere in a laboratory. "To succeed in research," he often said, "one doesn't need to know everything about everything, it's enough to just know a few things that others don't." This sentence, pronounced with his contagious kindness and laughter, went a long way in relieving us of the intimidating weight of universal knowledge, which so often inhibits one, whether it is discouraging, or leads to an overly skeptical attitude about the world and the discoveries still to make. The hospitality of Art and Aurelia Schawlow in their campus home was memorable and we have kept strong connections with them long after Claudine and I had left California. It is during one party at the Schawlows that Claudine and I had a chance to meet Felix Bloch, who had been the first to describe the quantum behavior of electrons in solids and had invented magnetic resonance, the phenomenon I had exploited during my PhD work a couple of years earlier. For the young postdoc I was then, it was an awesome experience to interact with a scientist of this stature, a man who had worked with the founding fathers of the quantum theory. This kind of encounter contributed greatly to the excitement of our life in California.

After a few weeks in Stanford, Art gave me a lab room and a pulsed dye laser and told me it was up to me to find something interesting to do with it. With one of his graduate students, Jeffrey Paisner, I decided to try to study atomic quantum beats. They manifest themselves as time modulations in the fluorescence of atoms following a pulsed optical excitation. Such modulations had been observed previously in experiments realized with spectral lamps, but I suspected that with the increased power and sharper pulses produced by lasers, much faster and stronger beat signals should be obtained. When the experiment worked, almost on the first trial, I was exhilarated. For the first time, a project I had conceived and pursued without supervision was working and yielding interesting results. I presented them in the first International Conference on Laser Spectroscopy which took place in Vail, Colorado, in June 1973. The year spent in California was also an opportunity to meet scientists who have left a strong impression on me, and to start a longstanding friendship with some. This has been the case with Ted Hänsch, who has been a friend since then and whom Claudine and I have met on many occasions in the US first, then in Germany when he settled back there in the 1980s. It is also during the

Vail meeting that Claudine and I got to know Daniel Kleppner of MIT and his wife Beatrice, who have become lifelong friends.

During my postdoctoral year, I had been thinking about a project to study Rydberg states, very excited atomic levels whose energies are close to the atom's ionization limit. These states, of gigantic dimensions on the atomic scale, had been already observed in outer space in radio astronomy, but their preparation and study in the laboratory have had to await the development of lasers. My quantum beat experiments in Stanford had persuaded me that these laser sources were indeed very promising for investigating atomic states previously inaccessible to classical spectroscopy. I was fascinated by the prospect of being able to prepare and manipulate atomic matter under such exotic conditions, which should have properties so different from ordinary atoms. I wrote this proposal in an informal letter to Jean Brossel, who was the director of the ENS physics department. By return mail, he immediately offered me a position at ENS and, when I returned to Paris in the fall of 1973, he gave me a laboratory space and the start-up money required to buy the laser equipment for these experiments.

A bright student from ENS, Michel Gross, joined me and, within a few months, we built a new laboratory and started the investigation of these Rydberg atoms, which have been the workhorses of nearly all the experiments I have done since. Brossel had also secured me a research position of "maître de recherche" in the Centre National de la Recherche Scientifique (CNRS), which allowed me to spend all my time doing research, without any mandatory teaching duties. I had always liked the idea of teaching, though, finding it challenging and stimulating to explain scientific ideas and to communicate about science and research. So when in 1975 the opportunity arose, I applied to a position at Paris VI University and was appointed there as a professor. I started teaching basic physics to premedical students and a course about energy to physics undergraduates. At that time, the teaching load of professors was not as heavy as it has since become and I enjoyed splitting my activities between research and teaching.

Brossel had maintained in his laboratory an atmosphere of freedom and trust which seemed to me natural. In retrospect, I now measure how lucky I have been to work in such a favorable environment, without having to write incessant proposals and to submit to constant evaluations. In this context, I have been able to develop my research in the long term, justifying it only by the papers I published and the recognition they received. Time and trust are, I think, worth even more than money. They are two essential ingredients required for good basic science to flourish. I have been lucky to enjoy plenty of them in my early career as an independent scientist. When the conditions changed in the 1980s and research in France became more and more subject to the rules that the global market started to impose, I had to learn how to write proposals and how to answer repeated evaluation requests. My reputation as a scientist was made, though, and I did not have too much trouble adjusting and receiving the grants required for my group to operate in good conditions.

Time, trust and money are certainly necessary, but the most important factor for the success of my research has been the quality of the people I have had the luck to work with. In 1976, Michel Gross was joined by Claude Fabre who started his PhD work with me. Soon thereafter Jean-Michel Raimond completed the group of bright and dedicated "normalien" students working with me, inventing new ways to explore the radiative properties of Rydberg states.

For that, we needed good microwave sources, operating at frequencies in the tens of GHz range. In another laboratory of the ENS physics department, a solid state physicist, Philippe Goy, was precisely developing and using such sources for his electron cyclotron resonance experiments. I got him interested in our project and he soon joined us for the exploration of the microwave spectra of alkali atom Rydberg states.

This was a very exciting time in the laboratory, when we started to develop the techniques which have allowed us to control the Rydberg atoms and perform the Cavity Quantum Electrodynamics (Cavity QED) experiments described in my Nobel lecture. Michel Gross and Claude Fabre went on to their own research careers after graduating, while Jean-Michel decided to stay in my group. With Philippe Goy, we supervised several PhD students during the 1980s and received many foreign visitors in our laboratory, including Daniel Kleppner and Luigi Moi, a former student of Adriano Gozzini, a friend and colleague of Kastler, working at the Scuole Normale of Pisa, the Italian sister school of the ENS. Moi became a close friend who pursued later his career as a professor at the University of Siena.

In the mid 1980s we were joined by another exceptionally gifted student, Michel Brune, who also stayed with us after his PhD thesis. At that time, Philippe Goy started a small microwave equipment company, AB Millimetre, which has built custom-made microwave sources spectrum network analyzers for customers all over the world. He has no longer worked directly with us since, although we still consult him when we have to solve an arduous technical problem. Jean-Michel, Michel and I have worked together up to now, training generations of students and postdocs in our lab. All the achievements which have led to the Nobel recognition are theirs as well as mine. Beyond the exhilaration of obtaining interesting results, we have shared the pleasure of exchanging ideas in an atmosphere of trust and friendship. We do take the research seriously, but we like to joke about ourselves. Claudine, who visits us often in the lab, likes to define the spirit in our group as a special mixture of self-confidence and derision. Of course we recognize the value of what we are doing, but at the same time, we try not to take ourselves too seriously.

Teaching was an important part of my activities. Since the early 1980s, Paris VI University and ENS had an agreement under which I could do my teaching at ENS to "normaliens"and to a group of selected students accepted in the ENS curriculum. To graduate students, I taught courses in atomic physics and quantum optics, and to undergraduates, I lectured about electromagnetism and quantum mechanics. At the same time, I had a part-time position as a lecturer at the Ecole Polytechnique, where I had also the opportunity to teach very bright students. I found all this teaching very rewarding, especially since I could always do it on topics which had a more or less direct connection with my own research, finding ways to illustrate the lectures with the description of modern atomic physics or laser experiments.

The 1980s were also years when America called. Harvard had offered me a full time professor position in 1981, but I was not ready to leave Paris and the group I was working so well with. Having loved our year at Stanford, I was however tempted to try another experience as a scientist in America. So when Yale, at the initiative of Vernon Hughes, allowed me to come and do research in New Haven for one term each year, while retaining my Paris position, I accepted the challenge to carry out experiments on both sides of the Atlantic. For a few years, I was able to successfully perform atomic physics experiments at both places. In Paris, it was relatively easy since my group was well organized there and Jean-Michel had the maturity to lead the group activities during the time I was away. In New Haven, I collaborated with Edward Hinds who was a professor at Yale, and with Dieter Meschede, a former student of a Munich colleague, Herbert Walther. Dieter had accepted the postdoc position I had offered him to help me start a research program at Yale.

We did some very good work during that time and we had some bright students who have had very successful careers thereafter. I also loved the opportunity to teach physics in an environment very different from the one I had at ENS or at the Ecole Polytechnique. American undergraduate students are somewhat younger and less mature than the "normaliens" I was used to training in Paris. They know less mathematics. Paradoxically, that makes them less inhibited than the French students and more prone to ask questions, sometimes naïve and sometimes deep. I enjoyed the interaction with them very much. I stayed there during the fall term, with frequent round trips to Paris. Claudine and the children visited often, while staying most of the time in France where they worked and studied. Julien and Judith had turned into teenagers who loved to visit the US for a while, but wanted to attend school in France, where they had their friends and their habits. Working in both places was thus a kind of state superposition that was hard to maintain coherently over a long period of time. In the early 1990s, I decided it was time to stop this experiment and to come back full time to my group in Paris, which had an ambitious research project to carry out. It is indeed at that time that the ideas about manipulating and observing photons non-destructively were developed, as well as those about preparing Schrödinger cat states of light.

These ideas sprang from long-term collaboration we had with Brazilian colleagues. I had had the opportunity to visit Brazil for the first time in 1983 when I participated in one of the first French-Brazilian workshop in quantum optics, held in Rio de Janeiro. Claudine and I were immediately seduced by the beauty of the country, its relaxed atmosphere and the gentleness of the Brazilian people. Since then, we have been to Brazil at least once a year and often more, visiting many parts of this huge country, with such diverse people and natural beauty. The physicists I met there in 1983 were enthusiastic about the possibilities opened for science in a country which was emerging from long years of dictatorship. Most of them had studied and obtained their doctorate in the US, where they had fled during the dark times. Some became good personal friends whom we enjoy to visit in Brazil or to welcome in Paris. Luiz Davidovich and Nicim Zagury, two quantum optics theorists from Rio, became soon familiar with the Paris Cavity QED setup and worked on the theoretical aspects of our experiments. Luiz was with us in Paris in 1987 when we operated our two photon Rydberg atom maser, the topic of Michel Brune's doctoral thesis.

Jean-Michel Raimond and Michel Brune were naturally involved in this friendly collaboration. It was during a visit that Jean-Michel made to Rio in 1989 that the ideas about the non-destructive counting of photons first emerged. I remember the excitement we shared while communicating by phone calls and mail between Rio, Paris and New Haven, exchanging ideas and discussing the results of the first computer simulations of photon counting that Jean-Michel was doing with Nicim in Rio. A couple of years later, while Luiz was again in Paris for a sabbatical, we realized that the setup we had in mind to count photons non-destructively could be used to generate Schrödinger cat states of light and to study the phenomenon of decoherence. A 1991 paper in *Physics Today* by Wojciech Zurek, which very clearly described superpositions of harmonic oscillator states and their decoherence, played an important role in this context. When reading this paper, we realized that we were in a position to observe these effects in the laboratory. With our Brazilian colleagues, we wrote a long article in *Physical Review* describing in detail the experiments we were planning. For this, we needed to develop new experimental methods for preparing and manipulating circular Rydberg states and new cavities able to trap photons for a very long time. It took us fifteen years to get a setup allowing us to observe the effects we had predicted in the early 1990s. I have described this adventure in my Nobel Lecture.

The years after my return from Yale were very busy. I resumed my teaching at ENS, under an arrangement which was made possible by the creation of a new structure in the French university system, called the "Institut Universitaire de France" (IUF). Being appointed in the IUF means that, while staying at the University which employs you, for a period of time ranging from 5 to 10 years you get a reduced teaching load and some money to travel or to spend on your research. In 1991 I was lucky to secure one of the first IUF positions and could thus spend more time on research at a time when, in the laboratory, we were struggling to build the setup we were dreaming about in order to manipulate photons "in vivo." Soon after the appointment to IUF, though, the time I could devote to science was reduced by administrative duties. In 1994 I accepted an appointment as chairman of the ENS physics department. This turned out to be a very demanding task, which I assumed for six years. It is also around that time that I was elected a member of the French Academy of Sciences. With great pride, my parents attended my induction in the Academy, under the famed dome of the Institut de France. My father was already very sick at that time. They lived a few more years and passed away in 1998, a few months apart, after having shared the good and the bad times for more than sixty years.

The 1990s were also the years when ideas about quantum computing and quantum information processing with isolated quantum systems started to become popular and competition with other groups around the world became very strong. With Luiz Davidovich and Nicim Zagury we studied various ways to exploit our Cavity QED setup to perform demonstrations of simple quantum information steps. Some of these ideas have remained theoretical. Others have led to actual experiments. We recognized that our system, in which we were trapping photons was complementary to the one that David Wineland and his Boulder group were working on, in which they were trapping atoms. Some of the experiments we were performing in Paris and in Boulder were very close in spirit. We published back to back two physical Review Letters describing the observation of similar Rabi oscillations in a cavity QED and an ion trap. In 1996 we also prepared "Schrödinger cat states" of harmonic oscillators whose features looked alike and we studied their decoherence in experiments bearing strong similarities. It is at that time that David and I became good friends. Claudine and I convinced David and his wife Sedna to join us on a short vacation trip to Italy in the summer of 1996. Since then we have often met in Boulder or in Paris. In 1999, Claude Cohen-Tannoudji and Pierre-Gilles de Gennes approached me to find out whether I would be interested in a position at the Collège de France, a very famous institute to which they both belonged. The Collège de France, founded in the sixteenth century, is a unique institution in the French academic system. I had attended many of Claude's lectures there and I was impressed by the spirit and the values of this institution, in which the professors give public lectures which have to be renewed each year on the topic of their research. No enrollment is required to attend the lectures and there is no final exam. In other words, to prepare a course at the College de France, there is neither a recipe nor a captive audience whose assiduity might be encouraged by the pursuit of a degree. There are only free listeners, who share the same interest and expectation. Each professor tackles this task with their own personality, shaped by their encounters and singular experiences. It makes for a very diverse institution, gathering a broad range of specialists in physical, natural sciences and humanities. The faculty meets three times a year to decide the opening of new positions and to share their views on matters of learning and culture in a unique atmosphere.

The list of former professors at the Collège de France throughout history is impressive. In physics alone, it includes such celebrities as André-Marie Ampère, Leon Brillouin, Paul Langevin, Frederic Joliot, Anatole Abragam, Pierre-Gilles de Gennes and, of course, Claude Cohen-Tannoudji. To be asked to join this club was overwhelming, but I accepted the challenge and paid the traditional pre-election visit to all the professors, fifty in total, to explain them why it was timely to create a chair in the Collège de France on quantum physics. These visits were a unique opportunity to meet colleagues working in fields very far from my own, which turned out to be an enriching experience. I was elected a Professor at the Collège de France in June 2000 and gave the inaugural lecture of my chair in December 2001. I have since given a new course each year, on various topics dealing with quantum information science. Preparing these lectures has been very challenging, and also stimulating for my research. Having to present subtle phenomena as clearly as possible has more than once led me to conceive new experiments to illustrate some physics concepts.

After a few years of teaching, it seemed timely to collect the material from these courses in a book. With Jean-Michel Raimond, who at that time was teaching quantum information at ENS, we embarked in the task of writing a comprehensive volume describing the physics of atoms in cavities, making connections with related problems in quantum optics. This turned out to be a multi-year project, which resulted in the publication of the book "*Exploring the quantum: Atoms, cavities and photons*" in the summer of 2006. It included a detailed theoretical analysis of the coupling of atoms with quantized fields in cavities, along with the experiments we had performed up to then. It also described the related experiments that ion trappers, including David Wineland and his team, had realized and also made comparisons with the physics of cold atoms in optical lattices.

The book appeared exactly at the time when our laboratory got an exceptionally good cavity which finally made possible the experiments we had been dreaming about over the previous fifteen years. In the months and years which followed, with Stefan Kuhr and Igor Dotsenko – two exceptionally gifted postdoctoral fellows both coming from the laboratory of Dieter Meschede, now working in Bonn, Germany – we were able to observe many of the effects we had predicted in theoretical papers and announced in the book. The no-destructive counting of photons trapped in the cavity, the observation of field quantum jumps, the preparation and reconstruction of Schrödinger cat states of the field in the cavity and the direct observation of their decoherence were published in a fast succession of papers. These results were the topics of the PhD work of a succession of very bright and dedicated graduate students, one of whom, Sébastien Gleyzes, joined us a few years later to become a permanent member of our team.

Jean-Michel and I have often reflected upon the fact that our book was completed just before we could include in it the description of the most demonstrative experiments we have performed, which are only analyzed there as proposals. This calls for writing a revised edition incorporating all these results. But the events of the last few years are, for the time being, distracting us from this task. One of these events has been the CNRS Gold Medal, which I received in June 2009. This distinction is accompanied in France by a lot of media attention, which gave me a foretaste of what was going to happen in 2012. During that year I reached my 65th birthday, which was celebrated by a symposium at Collège de France during which I had the pleasure to meet many friends and colleagues coming from all the world for the event. It culminated with a dinner at the Musée Jacquemard-André in Paris, before which we could visit a wonderful exhibition of Renaissance European painting. Reaching this symbolic milestone has been preceded by a sad family event. My youngest brother Michel had died in January 2009, before reaching his fiftieth birthday.

During the last three years, I have been busy with the project of starting experimental physics research at the Collège de France. The Collège laboratory buildings dating from the 1930s have been fully remodeled and we intend to move our research labs there from ENS in the coming months, along with those of Jean Dalibard, a colleague, former student of Claude Cohen-Tannoudji who has just been appointed to the Collège de France. Another professor, Antoine Georges, who is a condensed matter theorist, will join us to make up the Physics Institute of the Collège de France. We hope to attract research teams of junior scientists who will be able to start their independent research career in a favorable environment. It will not be an easy task to build, in the prevailing economy, a laboratory nurturing the kind of values that have allowed my research to develop and thrive when I was young. To succeed, the new institute will maintain a strong link with ENS and the Laboratoire Kastler Brossel, which have kept alive the "time and trust" legacy of Jean Brossel. I find the challenge of building a new research unit on these values to be especially stimulating.

In the fall of 2012, two events occurring a month apart have suddenly made my life, and that of Claudine, more hectic. First, I accepted the post of administrator of the Collège de France. Having received so much from this institution, I considered that it was my turn to take charge of its administration and management, at a time when it was expanding into new directions, especially with the opening of its buildings to new labs, in physics, but also in biology and chemistry. And then, a month later, the announcement of the Nobel Prize in Physics has exposed me to a worldwide media attention that the CNRS Gold medal had hardly prepared me for. I have learned during the last few months that it is essential to be able to say "no" to a lot of solicitations in order to keep some coherence in my life as a scientist and as a person.

Reflecting over the events of the last fifty years, I feel very privileged. On the professional side, I have had the luck to embark in a field – atomic physics and quantum optics – which has undergone fantastic developments over this period of time, improving by many orders of magnitudes the sensitivity of experiments and the precision of measurements. Thanks to advances in laser technology, new domains have been explored, in ultra-low temperature physics or in the study of ultrafast phenomena for instance, that we could not even imagine at the time I was working for my PhD. I did not work myself in many of these fields, but I witnessed these developments as a member of a very active and imaginative community of physicists, sharing the excitement and the bewilderment brought about by all these spectacular advances. And in my own research area of Cavity Quantum Electrodynamics, new developments extending the studies to artificial atoms and to a variety of electromagnetic resonators have kept the subject alive and thriving, with many promising developments to expect in the near future.

But above all, I feel privileged in my personal life. Thanks to Claudine and our children, my interests have extended to many areas beyond physics. By pursuing her own intellectual interests and keeping a very active professional career of her own, Claudine has maintained in our lives a balance between science, arts and humanities which has been very enriching to both of us. Since the beginning of my career as a scientist, as often as possible we have traveled together to conferences and meetings, discovering new places and new people and sharing our impressions about them. We have also taken time for vacation away from physics, visiting the natural and man-made wonders of the world, from Egypt to South Africa, from Patagonia to the Gulf of Saint Laurent, from Angkor to the Maya country, from Machu Picchu to the Galapagos islands. Julien and Judith joined us on many of these trips as children or teenagers and we keep wonderful memories of these special times with them.

They are now pursuing their own full and interesting lives in Paris, Julien as a medical doctor doing clinical research in internal medicine and Judith as a lawyer. We are enjoying the pleasure and wonderment of witnessing the development and awakening to the world of Judith's three children, her twin girls Elsa and Rachel (born in 2005) and her boy, Samuel (born in 2009). I know that Claudine, with her love of privacy and her keen judgment, will help keeping our life in the future not so much different from what it has been up to now.

ADDENDUM, JANUARY 2018

Serge Haroche was born in 1944 in Casablanca. He graduated from Ecole Normale Supérieure (ENS), receiving his doctorate from Paris VI University in 1971. After a post-doctoral visit to Stanford University, he became full professor at Paris VI University in 1975, a position he held until 2001, when he was appointed Professor at Collège de France (in the chair of quantum physics).

He has been part time professor at Yale University (1984-1993), member of Institut Universitaire de France (1991-2000) and chairman of the ENS Department of Physics (1994-2000). In September 2012, he was appointed "Administrateur du Collège de France" (equivalent to President of this institution), a position he held until September 2015. Since then, he has been Professor Emeritus at Collège de France.

Serge's research has mostly taken place in the laboratory Kastler Brossel at ENS. His main research activities have been in quantum optics and quantum information science. He has made important contributions to Cavity Quantum Electrodynamics (Cavity QED), the domain of quantum optics which studies the behaviour of atoms interacting strongly with the field confined in a high-Q cavity, a box made of highly reflecting mirrors.

Serge has received many prizes and awards, culminating in the 2012 Nobel Prize in Physics, shared with David Wineland.

Controlling Photons in a Box and Exploring the Quantum to Classical Boundary

Nobel Lecture, December 8, 2012

by Serge Haroche
Laboratoire Kastler Brossel de l'Ecole Normale Supérieure & Collège de France, Paris

ABSTRACT

Microwave photons trapped in a superconducting cavity constitute an ideal system to realize some of the thought experiments imagined by the founding fathers of quantum physics. The interaction of these trapped photons with Rydberg atoms crossing the cavity illustrates fundamental aspects of measurement theory. The experiments performed with this "photon box" at the Ecole Normale Supérieure (ENS) belong to the domain of quantum optics called "Cavity Quantum Electrodynamics." We have realized the non-destructive counting of photons, the recording of field quantum jumps, the preparation and reconstruction of "Schrödinger cat" states of radiation and the study of their decoherence, which provides a striking illustration of the transition from the quantum to the classical world. These experiments have also led to the demonstration of basic steps in quantum information processing, including the deterministic entanglement of atoms and the realization of quantum gates using atoms and photons as quantum bits. This lecture starts with an introduction stressing the connection between the ENS photon box and the ion trap experiments of David Wineland, whose accompanying lecture recalls his own contribution to the field of single particle control. I give then a personal account of the early days of Cavity Quantum Electrodynamics before describing the main experiments performed at ENS during the last twenty years and concluding with a discussion comparing our work to other research dealing with the control of single quantum particles.

CONTROL OF SINGLE PARTICLES IN A QUANTUM WORLD

Quantum theory told us almost a hundred years ago that matter as well as light have a particle *and* a wave-like character. Atoms and subatomic particles, usually described as discrete entities, can behave as waves while light, generally described as an electromagnetic wave, is also made of discrete quanta, the photons. This wave-particle duality leads to a strange quantum world in which atoms and light fields can exist in superposition states, so to speak suspended between different classical realities. This situation is impossible to comprehend intuitively with our classical minds, which have evolved to understand the macroscopic world surrounding us.

The fathers of quantum physics have been led to this strange description of the world by deductive reasoning based on the discovery of properties of matter and radiation which could not be explained by classical physics. The direct observation of quantum strangeness, though, has been elusive for a long time. Bohr, Einstein and Schrödinger described *thought experiments* in which they imagined that they manipulated and observed single quantum particles in

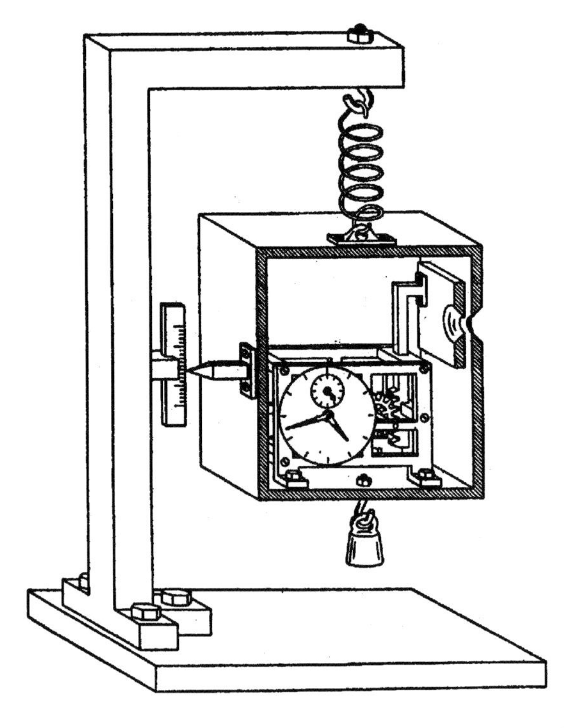

FIGURE 1. The Einstein-Bohr photon box (with permission from the Niels Bohr Archive).

order to reveal the counterintuitive behavior of Nature at the microscopic level. For instance, Figure 1 shows the famous photon box that is supposed to keep photons for a while and to release them on demand, imagined by Einstein and Bohr to illustrate their discussions [1]. This virtual experimental set-up had many parts, among which was a clock to time a shutter which releases the photon, which was drawn on Bohr's instructions with exquisite realism.

While imagining this and other such experiments, the fathers of the theory could not envision that they would one day be feasible. From the 1930s, for sure, single particles could be detected and studied, for instance in bubble or cloud chambers or in accelerator experiments, but they were then observed by their traces or by the debris they produced after being smashed against each other in fiery collisions. In these experiments, their existence and properties were deduced so to speak *post-mortem*. As Schrödinger wrote in 1952 about these studies: *"it is fair to state that we are not experimenting with single particles, any more than we can raise Ichthyosauria in the zoo. We are scrutinising records of events long after they have happened"* [2].

In order to observe quantum strangeness *in the zoo*, one had to manipulate particles of matter or light in a much more gentle way, without destroying them by the mere act of observation. Methods to achieve this subtle manipulation had to await the development of tunable narrow band lasers, of fast computers and of superconducting materials, which are all, in one way or the other, technologies emergent from quantum theory. Using these methods and working in the field of quantum optics, many groups in the world are now able to detect and manipulate single particles, while preserving—even exploiting—their quantum properties. The studies carried on by David Wineland and his team at the National Institute of Standards and Technology (NIST) in Boulder and by my group at the Laboratoire Kastler Brossel of the Ecole Normale Supérieure (ENS) in Paris belong to this very active field of research.

In fact, the Boulder and Paris experiments represent two sides of the same coin: the NIST group is trapping single charged atoms, called ions, in a configuration of fields produced by electrodes and uses laser beams to manipulate and detect the behavior of these particles of matter. In the ENS team we do the opposite, trapping photons in a cavity made of highly reflecting mirrors and using beams of atoms to manipulate, detect these particles of light and study their evolution. The principle of these complementary experiments is sketched in Figures 2a and 2b. In both cases, we perform *"in vivo"* studies, trying to preserve the quantum properties of the manipulated systems.

A simple theoretical model, introduced by Jaynes and Cummings [3] in the early days of laser physics, applies to both situations and contributes to unify

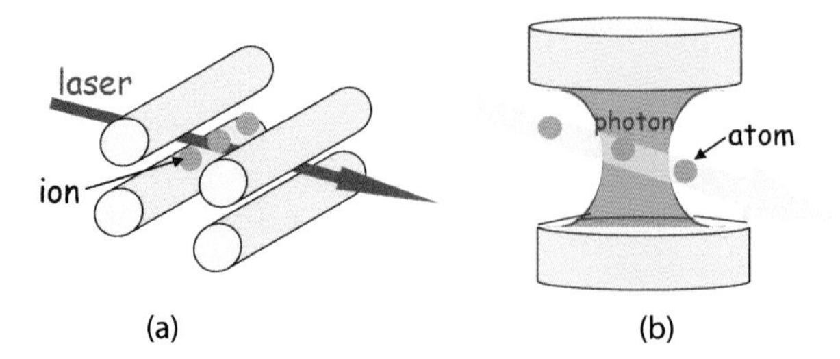

FIGURE 2. Scheme of the NIST ion trap (a) and of the ENS photon box (b).

these two lines of research. It describes the atoms as two-level spin-like systems interacting with a quantum oscillator. In the Boulder experiments, the ions' internal degrees of freedom (making up the "spin") are coupled to the external motion of the particles in the trap undergoing a mechanical oscillation, whose quanta are phonons. The coupling is achieved by irradiating the ions with properly tuned laser beams producing transitions between the ion internal energy levels accompanied by the absorption or emission of phonons. In the Paris experiments, the two-level atoms interact directly with a microwave field mode in the cavity, again a quantum oscillator whose quanta are photons. In this case, the coupling simply describes the processes of photon absorption and emission by atoms crossing the cavity. A simple description of the Jaynes-Cummings model and its application to the ENS and NIST experiments can be found in the book "Exploring the quantum: atoms, cavities and photons" [4].

Figures 3a and 3b illustrate the closeness of the NIST and ENS approaches. They show experimental signals published in two papers which appeared back to back in 1996, in the same issue of Physical Review Letters. We had worked

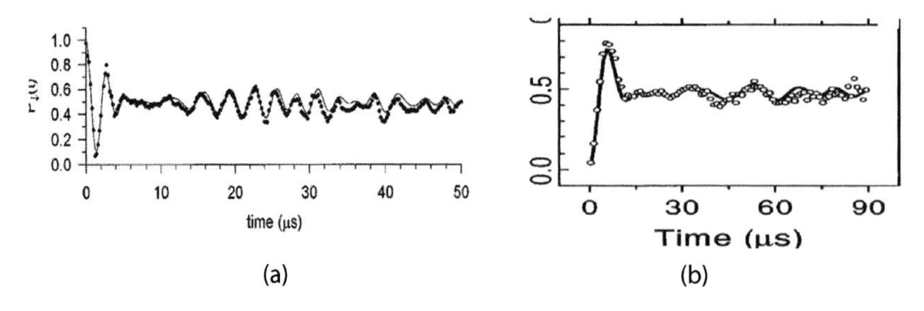

FIGURE 3. Rabi oscillations (a) in the NIST and b) in the ENS experiment (Reprinted with permission from [5] and [6], © American Physical Society; online abstracts at http://prl.aps.org/abstract/PRL/v76/i11/p1796_1 and http://prl.aps.org/abstract/PRL/v76/i11/p1800_1).

independently in Paris and Boulder on quite different set-ups and yet the recordings appear very similar. Fig 3a exhibits the "signature" of an ion oscillating in a state superposition of different vibration quanta in the Boulder trap [5] and Fig 3b shows the corresponding "signature" of atoms having resonantly interacted with a small oscillating electromagnetic field containing a superposition of different photon numbers in the Paris cavity [6]. These signals, which represent transition probabilities between two ionic or atomic states versus time, are called "Rabi oscillations". Their shapes, exhibiting beating between sinusoidal oscillations corresponding to different numbers of quanta, reveal the graininess of the mechanical or field oscillator to which the atoms or the ions are coupled. As will be recalled below, exploiting such Rabi oscillations plays an important role in the manipulation of atomic and field states for quantum information processing.

A PERSONAL ACCOUNT OF THE EARLY DAYS OF CAVITY QUANTUM ELECTRODYNAMICS

The adventure of trapping and controlling single particles had in fact begun well before these Rabi oscillation experiments. David Wineland reminisces in his Nobel lecture about the early history of ion trap physics. Here, I recall how my "photon box" experiments started, at a time when the similarity between these two lines of emerging research was not yet obvious.

I must first recall my thesis work [7,8] under the supervision of Claude Cohen-Tannoudji in the *Laboratoire de Spectroscopie Hertzienne de l'Ecole Normale Supérieure* which later became the *Laboratoire Kastler-Brossel*. It is indeed during this formative period that I learned that matter could be manipulated with light. I got acquainted with the optical pumping method invented by Kastler and Brossel which uses light beams to orient the magnetic moments of atoms and to detect their dance in radiofrequency fields. At that time, we were manipulating and observing large ensembles made of billions of atoms contained in a resonance cell. I was marveling at the fact that all our experiments were explained by the laws of quantum mechanics which had just been exposed to me in the illuminating lectures of my thesis advisor. We only had to trust that there were atoms in our cell, collectively obeying these laws, even if we could not observe them individually.

At the same time, Claude and I developed the *dressed atom* formalism which explains the behavior of atoms exposed to radiofrequency fields described in terms of photons [8,9]. Quantizing the field in this context was a bold move since our fields contained huge numbers of light quanta and could well be described as classical continuous waves. The quantum picture, envisioning our atoms as being surrounded or "dressed" by clouds of photons, was directly inspired by the theory of Quantum Electrodynamics, which was alien to most atomic physicists. However, this approach turned out to be very powerful and allowed us to discover many interesting phenomena in magnetic resonance which were not as clearly apparent in the classical approach [10].

It is around that time that the dream of performing experiments involving small numbers of photons started to take shape in my mind. Thinking about a question of Anatole Abragam, who was a member of my thesis defense committee, I wondered whether it would be possible to design an experiment in which the dressed atom formalism would not be merely a convenient description of the physics, but a necessary approach to explain truly quantum phenomena. I had no idea of how to achieve this when I left Paris for Stanford and my postdoctoral visit with Arthur Schawlow in 1972.

During this very exciting time, I witnessed the development of the first commercial tunable dye lasers, which were going to revolutionize atomic and molecular spectroscopy. These lasers, produced by California start-up companies, were delivered as prototypes to neighboring universities, among which Stanford was probably served first. Having worked for my thesis research with classical lamps, I had to familiarize myself with these new sources of light. Working with Jeffrey Paisner, a graduate student of Art Schawlow, I used a pulsed dye laser to excite Cesium atoms in a superposition of excited states and I observed the quantum beats in the fluorescence light subsequently radiated by the excited atoms [11]. These quantum beat studies drew again my attention to the important concept of superposition state which I had already encountered in a different context during my PhD thesis work when studying resonances related to level crossings in the dressed atom energy diagram [12].

Although the modulations of the fluorescence light characterizing quantum beats were observed on large ensemble of atoms, the basic phenomenon was a single atom interference effect. During the pulsed excitation and subsequent photon emission process, each atom was following several histories at once, leaving the atomic ground state and coming back to it while being transiently brought into different excited states. It was the indistinguishable character of these different paths which led to the final observation of the beats. Even if I was not able to do the experiment at that time, I knew the beats would still be there if I operated on a single atom, in the same way that a Young double slit experiment still works when particles cross the interferometer one at a time. Here again, I was intrigued by the challenge of observing quantum effects at the single particle level.

In 1973 I imported the tunable pulsed dye laser technology to Paris when I went back to ENS to take a permanent position in the National Center of Scientific Research (CNRS), before being appointed as a Professor at Paris VI University in 1975. After my first quantum beat studies, which were performed on states with small principal quantum numbers, it was natural to try to explore more excited atomic levels and I started to record quantum beats emitted by these levels in sodium [13]. Close to the atomic ionization limit, there was a very large number of levels forming a spectroscopic *Terra Incognita* of so called *Rydberg states* with huge electronic orbits. The lasers offered us the opportunity to prepare and study these states and I, like Daniel Kleppner at MIT and many other colleagues in Europe and in the US, was fascinated by the promising properties of these giant atoms, about which I give a few more details below. In particular, I became interested in their extreme sensitivity to microwave radiation.

I started performing microwave spectroscopy experiments on these atoms [14,15], working with a small group of dedicated and bright graduate students. Michel Gross and Claude Fabre, who embarked later on their own independent and successful careers, were followed shortly by Jean-Michel Raimond, who has stayed with me and accompanied me since then on this long research adventure. We were also lucky to get Philippe Goy, a condensed matter physicist wizard in millimeter wave technology, interested in our Rydberg atom microwave spectroscopy experiments. The millimeter wave sources and analyzers that he developed for our research have led him to start a small company which to this day sells custom-made millimeter wave devices to laboratories throughout the world.

For our experiments, we were using the simple set-up sketched in Fig 4: atoms flying in an atomic beam are excited into Rydberg states by lasers and, in the same region of space, they are irradiated by a microwave field. This field is confined in an open cavity made of copper mirrors facing each other, so as to define precisely the zone in which the atoms are irradiated. After having left the cavity, the atoms are ionized between two condenser plates in a ramp of rising electric field which at different times reaches the ionization threshold for the two energy states e and g connected by the microwave transition. In this way, the different ionization signals allow us to discriminate between the two states. This very efficient and selective detection method, first introduced in Rydberg atom physics by Daniel Kleppner and his students [16], is the one we still use today.

I realized in 1979 that if the atoms were initially in the upper state of the transition and the cavity tuned into exact resonance, it was not necessary to inject microwaves in order to observe a fast transfer between the two states. I immediately recognized that the ensemble of excited Rydberg atoms was spontaneously emitting in the cavity, realizing a pulsed maser [17]. The surprising and striking feature was the very low threshold of this maser action, which required only a few hundred atoms as opposed to the billions of atoms involved in ordinary maser or lasers. This unusual order of magnitude was obviously due to the very strong coupling of Rydberg atoms to microwaves. In the conclusion of the paper reporting the result, we noted that this experiment opened the way to the study of much smaller samples, provided we could use better cavities with higher Q factor and longer field damping time.

In fact, we had just started our journey towards single atom and photon detection and manipulation. The field in which we had embarked, dealing with atoms and photons interacting in a space confined by mirrors, was called "Cavity Quantum Electrodynamics" (Cavity QED in short), a name coined by Daniel Kleppner around 1980. I immediately liked this denomination because it once more connected atomic physics to quantum electrodynamics, in the spirit of what Claude Cohen-Tannoudji and I had already attempted ten years earlier with the introduction of the "dressed atom" formalism. Experiments remained to be done to demonstrate that the term "quantum" was legitimate but now at least, we knew in which direction we had to go. The cavity containing the field became the critical ingredient to be improved and the Rydberg atoms the workhorses allowing us to explore the quantum properties of these fields.

We performed several precise quantitative studies of pulsed Rydberg masers, focusing in particular on the evolution of the atomic ensemble during the emission process. Our experiment geometry, with all the atoms being symmetrically coupled to the same field, realized an ideal superradiant sample, a system

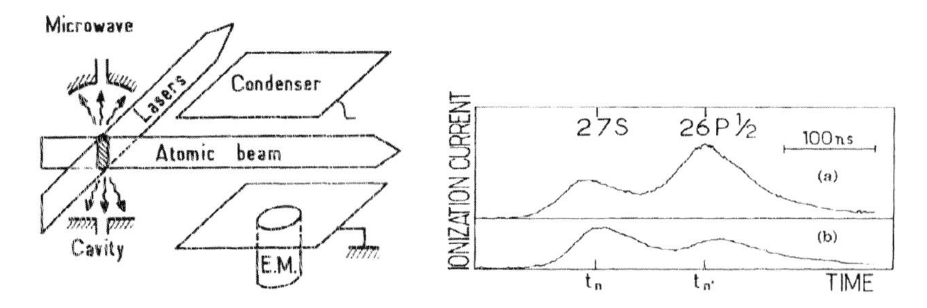

FIGURE 4. The 1979 Rydberg atom maser experiment: Left: sketch of the set-up. Right: time resolved atomic ionization signals discriminating the upper (27S) and lower (26P) states of the atomic transition. The bottom and top traces correspond to off-resonance and on-resonance cavity respectively (reprinted with permission from [17], © American Physical Society; online abstract at http://prl.aps.org/abstract/PRL/v43/i5/p343_1).

that R. Dicke had theoretically studied back in the 1950s [18]: the atoms initially in a fully excited symmetric state by an exchange of atoms remained in such a state throughout the emission process, ending in the final state where all the atoms were in the lower state of the transition. This symmetry entailed a strong collective coupling of the atoms to the field and an emission time much shorter than the spontaneous emission time of a single atom. Our experiment allowed us to measure this shortened emission time and to study the whole dynamics of the emission process. By accumulating statistics over many realizations of an experiment performed with N initially excited atoms, we recorded the probability versus time that n of these atoms had been de-excited and thus n photons emitted (see Figure 5). This realized a first quantitative experimental demonstration of Dicke superradiance [19,20].

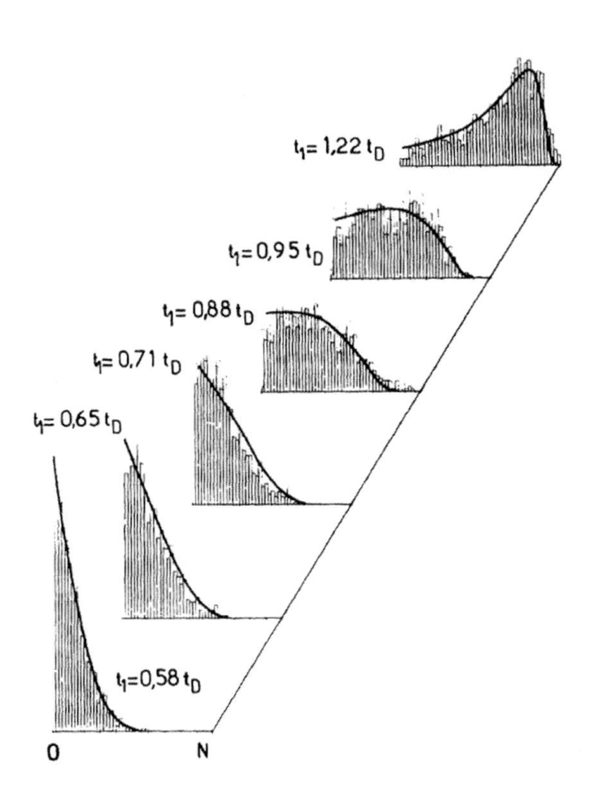

FIGURE 5. Experimental demonstration of Dicke superradiance: histograms showing the probability that n photons have been emitted by a sample of N = 3200 initially excited atoms at times increasing from bottom to top. Times are expressed in unit of t_D, the average emission delay. The curves are theoretical (reprinted with permission from [19], © American Physical Society; online abstract at http://prl.aps.org/abstract/PRL/v49/i26/p1924_1).

We also showed that if the number of decaying atoms was increased beyond a certain limit, the emission started to exhibit a ringing feature, the atomic ensemble getting partially re-excited and undergoing damped oscillations while exchanging its energy with the field in the cavity [21]. This ringing regime of superradiance is described by an equation similar to that of a damped pendulum [22]. In another series of experiments, we also studied the opposite effect, namely the collective absorption of blackbody radiation by Rydberg atoms initially prepared in the lower state of the atomic transition resonant with the cavity field. In this process, the atoms behaved collectively as an ensemble of bosons remaining in a symmetric state under particle exchange, leading to a heat capacity for our atomic system, much smaller than that of a classical gas [23]. Incidentally, the symmetric states along which the atomic ensemble was evolving in these experiments were highly entangled, i.e. non describable as a product of single atom wave functions. We were well aware of this feature, even if the concept of entanglement and all its implications were not of central interest at that time.

During these studies, we did not lose sight of our primarily goal, which was to decrease the threshold of our masers to a single atom. In order to increase the reflectivity of our mirrors, we replaced the copper with superconducting niobium, cooled to a few degrees Kelvin. By strongly reducing the mirror absorption losses in this way, we were able in 1983 to reach a situation where a single atom had its spontaneous emission rate enhanced by the presence of the cavity (Figure 6). This effect, predicted in 1946 by E. Purcell [24], corresponded as we noted to the operation of a single atom transient maser [25]. At about the same time, Daniel Kleppner and his students realized a complementary experiment, which he had proposed a few years earlier [26]. They showed that the spontaneous emission of Rydberg atoms was inhibited, leading to a lengthening of the excited Rydberg state life-time, when the atoms were confined between conducting plates excluding the radiation modes resonant with the atomic transition [27]. Another demonstration of spontaneous emission inhibition had been performed a few months earlier by G. Gabrielse and H. Dehmelt, who had shown that the decay of the cyclotron radiation of a single trapped electron was slowed down when it was located in the microwave cavity formed by the electrodes of the trap which was confining it [28].

One feature of our spontaneous emission enhancement experiment was intriguing me. What would happen if the cavity Q factor was so large as to keep the emitted photon long enough for it to be absorbed by the atoms? As I noted in the lecture notes of a course I gave in Les Houches in 1982 [29], this should lead to a regime of reversible exchange of energy between matter and radiation,

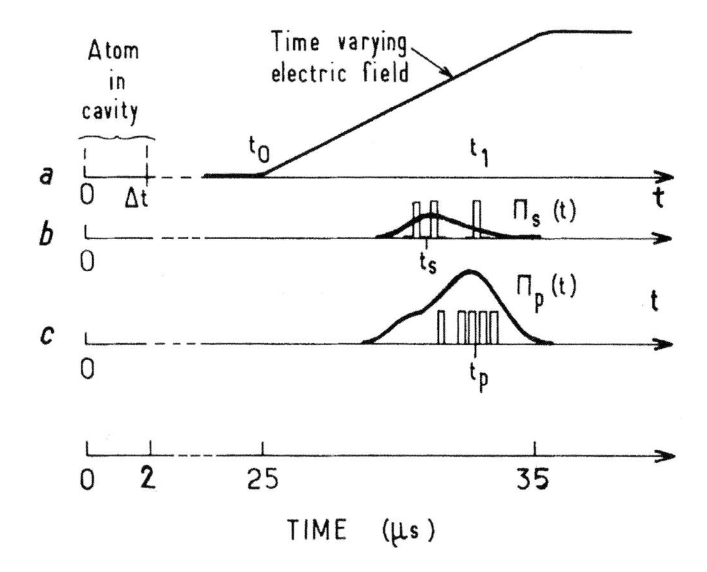

FIGURE 6. Enhancement of spontaneous emission in a cavity: sketch illustrating the experimental time sequence a) ionization field ramp; b) detection of the 23 s initial level; c) detection of the 22P final level (reprinted with permission from [25], © American Physical Society; online abstract at http://prl.aps.org/abstract/PRL/v50/i24/p1903_1).

at the single particle level. If the atom was initially (at time t = 0) in level e and the cavity in vacuum, we expected, according to the Jaynes-Cummings model, that the atom + field system would evolve at time t into the state $\cos(\Omega_0 t/2)|e,0\rangle + \sin(\Omega_0 t/2)|\ |g,1\rangle$, Ω_0 being the so-called vacuum Rabi frequency, and the two symbols in each ket representing the atom's state and the number of photons in the cavity respectively. If the cavity contained n photons, a similar oscillation phenomenon was expected to occur between the states $|e,n\rangle$ and $|g,n+1\rangle$, at the larger frequency $\Omega_n = \Omega_0 \sqrt{(n+1)}$. If the cavity field was in a superposition of photon number states, we expected this Rabi oscillation to exhibit a beating between components evolving at the incommensurate $\Omega_0 \sqrt{(n+1)}$ frequencies, a phenomenon predicted theoretically by Eberly et al [30] that we were to observe only in 1996, as recalled above. In the early 1980, reaching this situation, now called the strong coupling regime of Cavity QED, became our Holy Grail.

We were, however, at a loss to improve the quality of our mirrors. The superconducting material was fine, but the mechanical properties of niobium made it difficult to get surfaces smooth enough to avoid photon scattering off small imperfections. While we were struggling with this technical problem, in 1984 we got news from Munich that a German colleague, Herbert Walther, working with his graduate student Dieter Meschede and G. Muller, a colleague expert

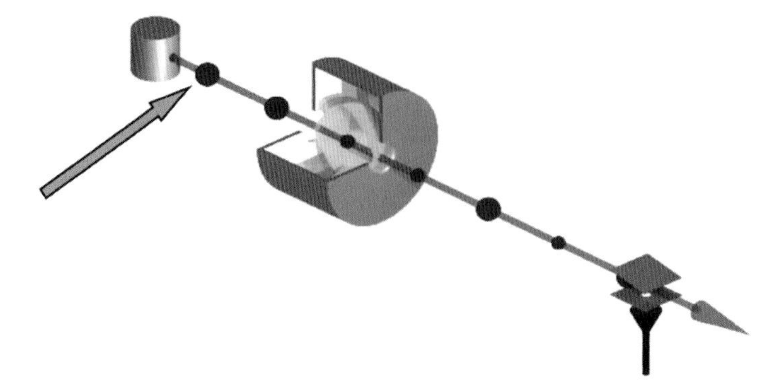

FIGURE 7. Sketch of the Munich micromaser: atoms excited by a laser into a Rydberg state (arrow at left) cross one by one a cylindrical superconducting cavity, before being detected by field ionization (at right).

in superconductivity, had reached our Holy Grail! They had had the bright idea to replace the open mirror structure by a closed cylinder, making it impervious to small irregularities in the inner walls and the life-time of the photons was increased by several orders of magnitudes.

Sending Rydberg atoms one at a time through this box, they realized a maser in which the field was building up into a steady state from the successive actions of single atoms crossing the apparatus one by one [31] (Figure 7). The system clearly operated in the strong coupling regime. Walther called it a "micromaser." Numerous studies of this device were made during the 1980s and the 1990s, both on the theoretical and on the experimental side [32,33,34]. Herbert Walther, who sadly died in 2006, was a great leader in quantum optics whose micromaser was an essential contribution to Cavity QED. Dieter Meschede became my postdoc when I was working for a while part time at Yale University during the 1980s. He is now a physics professor in Bonn, Germany, doing beautiful experiments in cold atom physics and quantum information. Our ENS and Bonn groups have been exchanging ideas and sharing many students and postdocs over the years.

During our common time at Yale, Dieter and I in collaboration with Edward Hinds, Luigi Moi and our students performed an experiment demonstrating the inhibition of spontaneous emission of an atom propagating between two mirrors separated by less than half the wavelength of the emitting transition [35]. This experiment, performed in 1986, was part of the thesis work of Won Jhe, a bright Korean student who has become since a professor in Seoul. It was a transposition to the optical domain of Kleppner's earlier experiment made in the microwave domain with Rydberg atoms [27]. During the same period, my

ENS group in Paris contributed to the micromaser studies by realizing, in 1987, a two-photon Rydberg atom maser operating with a closed cylindrical cavity, in which the light quanta were emitted by pairs [36] (related two-photon lasers had been realized a few years earlier [37]). The two-photon Rydberg maser was the PhD subject of Michel Brune, an outstanding student whom we have been lucky to keep in our group as a colleague since then.

At the time of these experiments, witnessing the successes of the Munich group, we were worried that our open cavity mirrors were leading us into a dead end. We were reluctant, though, to adopt definitively the closed cavity solution, because it had a serious problem for the experiments which we were planning: atoms had to enter and exit through small holes, passing near metallic surfaces producing stray electric fields. This proximity with metals was strongly perturbing the very sensitive Rydberg atoms, especially when they were prepared in superposition of states. The phase of the superposition was strongly affected by the stray fields, making it impossible to manipulate easily the atomic and field quantum states. The alternative solution was to stick to open cavity structures, in which atoms avoided passing close to metallic surfaces, but we had then to solve the surface smoothness issue. It took us about twenty years to solve this problem progressively and to realize the photon control and manipulation experiments which have been recognized by the Nobel Prize. In order to describe them, I will abandon the historical approach and adopt a more didactic point of view. A presentation of the early developments in Cavity QED can be found in a *Physics Today* article that I co-authored with Daniel Kleppner in 1989 [38].

COUNTING PHOTONS WITHOUT DESTROYING THEM: A NEW PHOTON BOX AND CIRCULAR RYDBERG ATOMS

Our ambition back at the end of the 1980s became to generate photons in a high Q cavity and to observe and manipulate these photons without destroying them. To achieve this feat, we had to overcome a difficulty: usually the detection of light belongs to the "postmortem" kind of procedure alluded to by Schrödinger in the quotation recalled in the first section of this Lecture. When you see light, the photons, which strike your eye, your camera, or your photodetector, are converted by the photoelectric effect into an electron, or some kind of electric current and the photon is destroyed in the process. The situation is not essentially different from the brute force collisions occurring in a cloud chamber or a particle accelerator. Seeing light amounts to collecting the debris of the photons colliding with a photosensitive surface. If you detect a click in your photon counter, the photon is no longer there and has been replaced by the vacuum, zero photons instead of one!

What we were looking for instead was a process by which a click in our non-destructive detector would leave the photon intact afterwards, allowing the measurement to be repeated again and again, on the same photon. The idea of such a quantum non-demolition (QND) measurement had been proposed for mechanical resonators by V. Braginsky and colleagues in the 1970s [39], in the context of gravitational wave detection. We merely needed to adapt the method for the detection of light quanta. For this purpose, a special kind of atomic detector was required, which was transparent to radiation but still sensitive enough to register an imprint of a single photon. And we needed also a very good photon box, an open structure keeping the photons for a very long time, enabling us to repeat the measurement again and again with successive atoms crossing the box one by one. This brought us back to the Einstein-Bohr photon box, but in a radically new context.

FIGURE 8. The ENS photon box (Photograph by Michel Brune). The mirrors have a diameter of 5 cm and are 2.7 cm apart (for a clear view, they are more separated in this picture than in the actual experimental set-up).

The box we came up with is not like the one drawn by Bohr eighty five years ago, but, interestingly enough, it is indeed coupled to a clock, a very special clock in fact, as described below. The solution to the cavity defects which had plagued us in the 1980s was found by successive improvements. We first succeeded in reaching photon lifetimes in the several hundred microsecond to a millisecond range by surrounding the niobium mirrors with an aluminum ring which reflected the field scattered on mirrors imperfections back into the cavity mode. But the real breakthrough came from collaboration with scientists at the French Atomic Energy Commission (CEA) center in Saclay. They worked on superconducting cavities for particle accelerator experiments and were experts in the art of designing high-Q microwave resonators, albeit at smaller frequencies than the ones (around 50 GHz) that we were using. With their help, we finally created mirrors out of copper, which can be machined with high precision, reducing the surface roughness to a few nanometers, and we sputtered a thin layer of superconducting niobium on top of them. By combining in this way a precise geometry with high electric conductivity, we obtained mirrors with quasi-spherical surfaces, whose quality steadily improved as we refined our procedures, without any longer requiring an aluminum ring around the mirrors. We finally reached a lifetime of 130 milliseconds in 2006 [40] with a fully open cavity structure (Figure 8). Photons travel between these mirrors over about 40.000 kms on a folded trajectory and survive long enough for thousand of atoms crossing the cavity one by one to interact with them.

In order to probe the photons, we send across the cavity a special kind of Rydberg atoms, called "circular," in which the outer electron orbits on a circle of large diameter, about a thousand times bigger than an ordinary ground state atom. These excited Rubidium atoms are prepared with lasers and radiofrequency excitation, using a modified version [41] of a procedure invented by Daniel Kleppner and Randy Hulet at MIT in 1983 [42]. According to quantum theory, the orbiting Rydberg electron is also a wave, which has a de Broglie wavelength, and the condition of a stable orbit is that there is an integer number of these wavelengths along the circumference. This number, called the principal quantum number of the Rydberg atom, is equal to 51 or 50 in our experiments (these Rydberg states are called e and g respectively in the following). The advantage of these circular Rydberg states of maximal angular momentum over the states of small angular momentum employed in our earlier experiments is their very long natural life-time, on the order of 30 milliseconds for the states with principal quantum number 50 or 51. This life-time, of the same order of magnitude as the photon life-time in our cavity, allows us to neglect in first approximation the atomic decay processes during the interaction time between the atoms and the cavity field.

In the e and g Rydberg states, the circulating de Broglie wave has a uniform amplitude, resulting in an electron charge density centered at the atomic nucleus, yielding a zero electric atomic dipole (see Figure 9a). In order to prepare an electric dipole, a pulse of resonant microwave can be applied to the atom, bringing it in a superposition of the two adjacent e and g states, with respectively 51 and 50 nodes in their wave function. This superposition of states can be referred to as a "Schrödinger cat" because it implies an atom at the same time in two levels, reminding us of the famous cat that Schrödinger imagined suspended between life and death. A better name should be a "Schrödinger kitten," because it is made of a single atom and thus very small. The two de Broglie waves making up this "kitten" interfere constructively at one end of the orbit and destructively at the other end, resulting in a net electric dipole, rotating in the orbital plane at 51 GHz (see Figure 9 b). This dipole behaves as a rotating antenna extremely sensitive to microwave radiation. It can also be described as the rotating hand of a clock, ticking at 51 GHz.

When microwave radiation, non-resonant with the transition between the two states e and g, impinges on the atom, it cannot absorb it and hence the photons remain intact, ensuring the non-demolition character essential to our experiments. However, the effect of this non-resonant light is to shift the atomic energy levels slightly and hence alter the rotating frequency of the atomic dipole, our clock hand. This light shift effect had been discovered in 1961 by Claude Cohen-Tannoudji in his seminal optical pumping studies [43]. Light shifts are proportional to the field energy, i.e. to the photon number. Being inversely

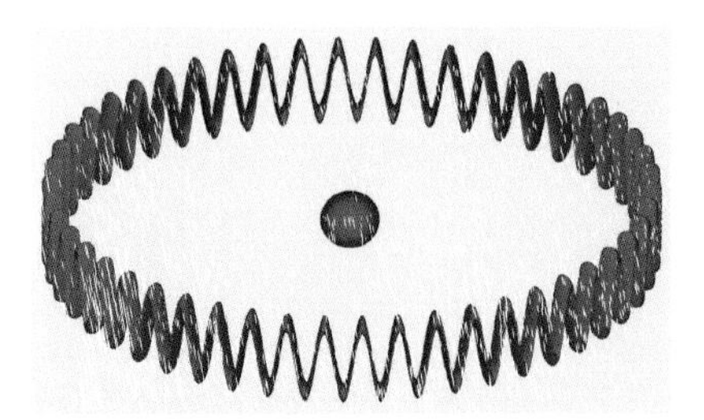

FIGURE 9. Sketch of circular Rydberg atom electron waves: a—left) In an energy eigenstate, the de Broglie electron wave has a uniform amplitude around the circular orbit, resulting in a zero electric dipole; b—right) in a superposition of adjacent circular Rydberg states, the de Broglie waves interfere positively on one side of the orbit and destructively on the other side, yielding a net electric dipole which rotates in the orbital plane.

proportional to the atom-cavity field detuning, they can be maximized by tuning the cavity close enough to resonance (typically a hundred kHz away from the atomic transition frequency in our experiments) but far enough to avoid any photon absorption or emission process. In the case of Rydberg atoms, the effect is then very large, resulting in a phase shift of the atomic dipole after the atom leaves the cavity which can reach the value of 180°, the dipole jumping in two opposite directions when the photon number changes by one unit. Measuring this phase shift amounts to counting the photon number without destroying the light quanta. Let us note that these light shifts play an essential role in other atomic physics and quantum optics experiments. They are at the heart of the methods used to trap and cool atoms in laser light, which were recognized by the Nobel Prize awarded to Claude Cohen-Tannoudji, William Phillips and Steven Chu in 1997 [44,45,46].

In order to measure these shifts, we followed a proposal that we made in 1990 [47]. We built an atomic interferometer around our photon storing cavity (Figure 10). The atoms, prepared in the circular state e in the box O, cross the cavity C one by one before being detected by field ionization in D. Essential to the experiment, two auxiliary microwave zones R_1 and R_2 are sandwiching the cavity C. In the first one, the atoms are prepared in the state superposition of e and g, a "Schrödinger Kitten" state. This procedure amounts to starting a stopwatch, giving to the atomic dipole, i.e. to the clock hand, its initial direction. The atomic

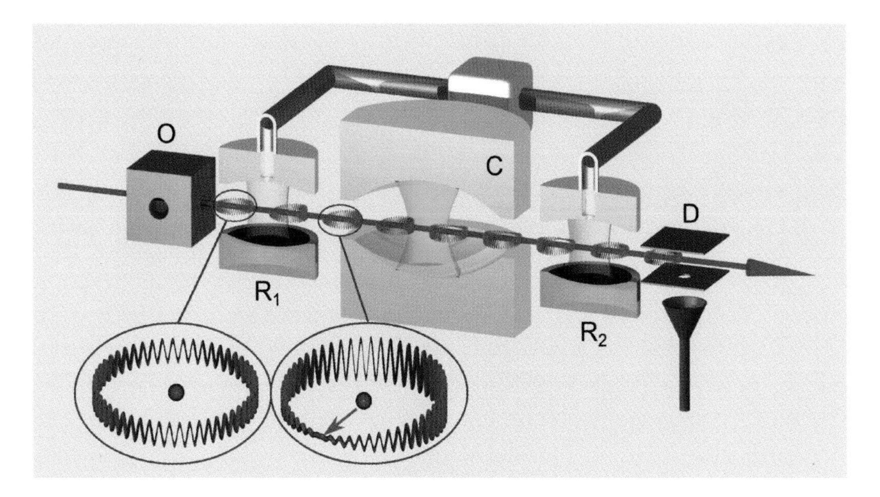

FIGURE 10. The Cavity QED Ramsey interferometer set-up. The insets show sketches of the circular atom in an energy eigenstate (left) and in a superposition state after interaction with the microwave pulse in R_1 (right). (Reprinted with permission from [49], © American Physical Society).

dipole then rotates as the atom crosses the cavity, until a second microwave flash, applied in R_2, is used to detect the direction of the atomic dipole at cavity exit, thus measuring the phase accumulation of the clock.

The combination of the two separated microwave resonators R_1 and R_2 is known as a Ramsey interferometer. The device had been invented in 1949 by Norman Ramsey [48] (who was later to become the PhD advisor of David Wineland). The method of separated field pulses is now used in all atomic clocks working on a hyperfine microwave transition between two atomic levels. The excitation by the two successive pulses induces a sinusoidal variation of the transition probability when the microwave frequency is scanned around resonance. This so-called "Ramsey fringe" signal is used to lock the microwave frequency to the atomic transition. In our experiment, the Ramsey interferometer is counting photons by detecting the perturbing effect they produce on the fringes of a special atomic clock, made of microwave sensitive Rydberg atoms [49]. If the phase shift per photon is set to 180°, the Ramsey fringes are offset by half a period when the number of photons changes by one. The interferometer is set at a fringe maximum for finding the atom in e if there is 1 photon in the cavity. The second pulse then transforms the state superposition of the atom exiting the cavity C either in state e (if there is 1 photon) or in state g (if there are 0 photons), this state being finally detected by the field ionization detector. The final atomic state g or e is thus correlated to the photon number, 0 or 1.

Figure 11 shows the detection events (g and e respectively represented by blue and red bars) as atoms cross a cavity cooled at 0.8K, which, according to Planck's law, contains either a vacuum (95% of the time) or one photon (5% of the time) [50]. One clearly sees when a photon pops inside the cavity, stays for a while and then disappears. Due to noise and imperfections, the correlation between the photon number and the atomic signal is not perfect, but a simple majority test allows us to reconstruct without ambiguity the evolution of the photon number (lower trace in the Figure). The sudden change in the photon number is a quantum jump, a phenomenon predicted long ago by quantum theory and observed in the 1980s in trapped ions, as described in David Wineland's lecture. It is observed here for the first time for light quanta. The figure also shows that hundreds of atoms see the same photon between two quantum jumps, which demonstrates that our detection method is quantum non-destructive (QND) for the field. Note that these field oscillator jumps bear a strong similarity to the quantum jumps between the cyclotron oscillator states of a single electron, which were also monitored by a QND procedure [51].

Our photon counting method can be extended to counting larger numbers of quanta [52]. We start by injecting inside the cavity a small coherent field, a

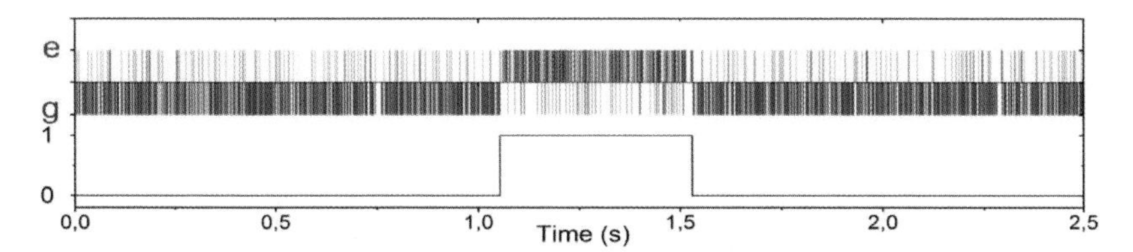

FIGURE 11. QND detection of a single photon: the sequence of single atom events (upper trace) detects the sudden change of the photon number (lower trace), revealing the birth, life and death of a single light quantum (reprinted with permission from [50], © Macmillan Publishers Ltd).

superposition of photon number states comprised between 0 and 7. This field is produced by scattering on the edges of the cavity mirrors a microwave pulse radiated by a classical source. This leads to the capture of a few photons which survive between the mirrors long after the source has been switched off. We then just need to send a sequence of atoms across the cavity, each carrying away a bit of information about the field. The phase shift per photon is optimally adjusted to a value such that different photon numbers correspond to well-separated atomic dipole directions at cavity exit. At the start of the experiment, we have no idea about the photon number and we assume a flat probability distribution, as shown by the histogram at the left of Figures 12a and 12b, giving equal weights to the probability of having from 0 to 7 photons in the cavity. As successive atoms provide information, our knowledge about the field evolves until finally a single photon number is pinned down (the single peaked histograms at the right of Fig. 12a and 12b correspond to n = 5 and 7 photons respectively).

The evolution of the inferred probability distribution is obtained by a Baysian argument developed in details in [4]: each atom's measurement provides information about the atomic dipole direction and allows us to update our knowledge of the photon number distribution. This experiment shows, so to speak, live, the measurement-induced "wave function collapse" which appears here as a progressive process transforming a flat histogram into a single peak. The field, initially in a superposition of different photon numbers, is projected by the mere acquisition of information into a photon number state, a so-called Fock state of well defined energy. The process is random, as shown by the two realizations of Fig12a and 12b, which have produced different results. The statistics of a large number of measurements reconstructs the photon probability distribution of the initial state (Figure 13). It obeys a Poisson law, as expected for a coherent state produced by a classical source of radiation [53].

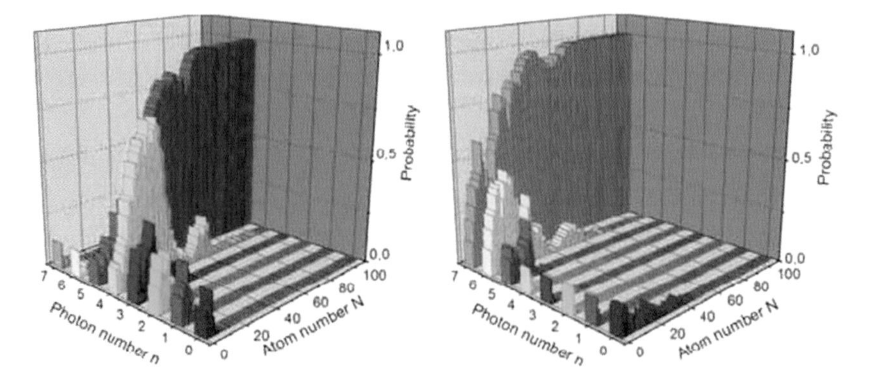

FIGURE 12. Progressive field state collapse induced by QND photon counting: (a) evolution of the inferred photon number probability distribution as the number of detected atom increases, in a realization of the experiment leading to the Fock state n = 5; (b) evolution of the same distribution in a realization leading to n = 7 (reprinted with permission from [52], © Macmillan Publishers Ltd).

Once a photon number has been pinned down, its ensuing evolution can be observed by continuing the measurement with subsequent atoms on the same realization of the experiment. We then observe the succession of quantum jumps leading the field inexorably back to vacuum, due to photon losses in the cavity walls [52]. Figure 14 shows the trajectories followed by fields containing initially different numbers of photons. A statistical analysis of these trajectories

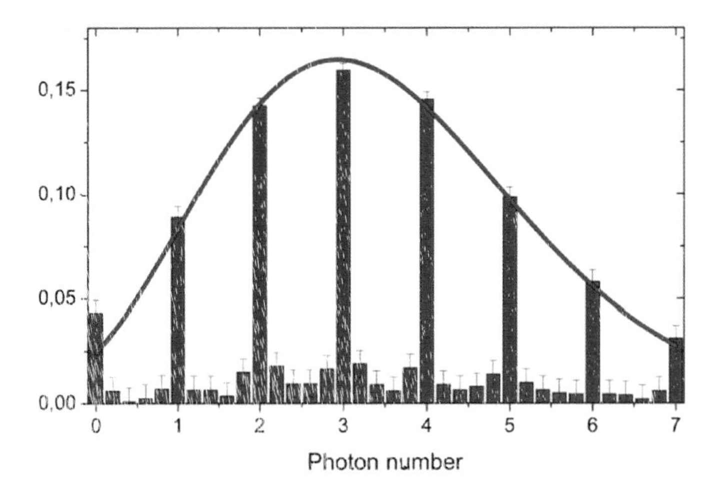

FIGURE 13. Photon number distribution of a coherent state with an average number of 3.4 photons reconstructed by statistical analysis of 3000 QND measurement sequences (reprinted with permission from [52], © Macmillan Publishers Ltd).

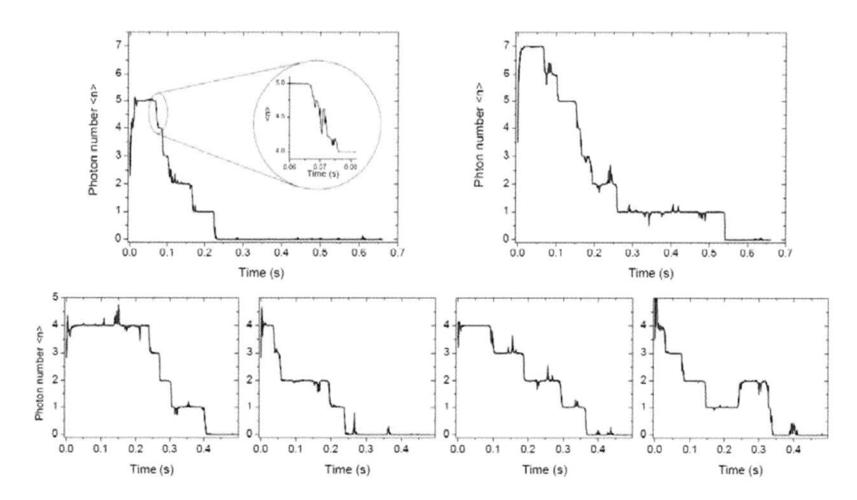

FIGURE 14. Field quantum jumps recorded by QND photon counting: (a) Evolution of the photon number following field collapse into n = 5 (left) and n = 7 (right). The zoom in the inset shows that, in order to record a quantum jump, the measurement requires the detection of several atoms, over a time interval of a few milliseconds. (b) Four random photon number "trajectories" following field collapse into n = 4. The different recordings illustrate the fluctuating nature of the field decay process. The trace at the right exhibits an upward jump due to the transient appearance of a thermal photon in the cavity (cooled at 0.8K) (reprinted with permission from [52], © Macmillan Publishers Ltd).

has allowed us to measure the Fock state life-times. Fluctuating randomly from one preparation to the next, the life-time of the n-photon number state is distributed according to an exponential probability law with the time constant Tc/n, where Tc is the cavity field energy damping time [54]. The 1/n variation of this life-time is a manifestation of the increasing fragility of these non-classical states of radiation when their energy increases. They share this feature with Schrödinger cat states of light (see below).

QUANTUM FEEDBACK

The ability to measure a quantum system without destroying it opens the possibility of reacting back on it and driving it deterministically towards a desired state. Soon after achieving our QND photon counting experiments, we have proposed to implement this quantum feedback strategy to "tame" a quantum field, preparing and maintaining it in a predetermined Fock state for an indefinite time [55]. We have then realized two versions of this experiment [56,57]. The Ramsey interferometer, acting as a QND photon counter, is used as a sensor providing information to a computer which estimates the field state in real time and decides how to react back on it in order to steer it toward the photon number "target" state. Once this state has been reached, the procedure monitors the quantum jumps and immediately corrects their effects by bringing back the system into the target state. The method is analogous to juggling. The juggler looks at the balls which he wants to keep on an ideal trajectory. His eyes are the sensors and the visual information, processed by his brain, is used to determine the correcting actions of his hands, the actuators. In the quantum version, the classical balls are replaced by photons bouncing between the mirrors. The quantum eye is the Ramsey interferometer, informing the computer "brain" which controls the experiment about the state of the field.

The quantum juggler faces a challenge, absent in the classical game. While looking at balls has no effect on them, the mere fact of observing the photons

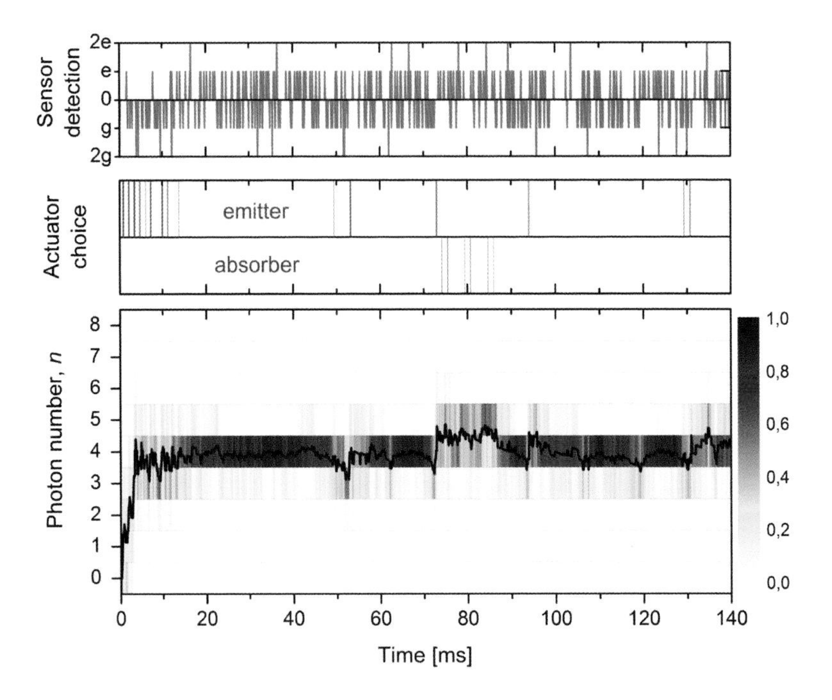

FIGURE 15. Stabilization by quantum feedback of the n = 4 Fock state. The upper and middle frames show the sequences of sensor atoms and actuator atoms detections (sensor, emitter and absorber atoms indicated by green, red and blue bars respectively). The lower frame shows the resulting evolution of the photon number distribution as reconstructed by the computer driving the experiment (probabilities represented by the color code indicated at right) (reprinted with permission from [57], © American Physical Society; online abstract at http://prl.aps.org/abstract/PRL/v108/i24/e243602).

has an unavoidable back-action, as required by the projection postulate of quantum physics. This back-action is random, since its effect depends upon the unpredictable outcome of measurements. Once an atom has been detected, though, the back-action on the field state is determined by projecting the field state according to the atomic measurement result. The computer can thus update the field state in real time, based on the results of successive measurements. At each step, it evaluates a distance between the actual state and the desired one, and it computes an adapted response to minimize this distance. As in classical feedback, the procedure is implemented in loops involving a succession of detection and corrections, until the target state is reached. It is then carried on to detect quantum jumps and correct for their effect.

What is the "hand" of this juggling game? In one version of the experiment [56], it is a microwave source used to inject periodically small microwave fields in C which, depending upon their phase, increase or decrease the field amplitude. In another version [57], the actuator is realized by atoms, tuned in exact resonance with C. They emit (if they enter C in e) or absorb one photon (if they enter in g). The actuator atoms belong to the same atomic beam as the sensors. To make them resonant during the correcting stages of the feedback loop, the detuning between the atoms and the cavity field is momentarily cancelled and the Ramsey pulses R_1 and R_2 are deactivated. A feedback loop involves a sample of a few sensor atoms followed by a few actuators. Quantum feedback signals corresponding to the stabilization with atomic actuators of the $n = 4$ Fock state are shown in Figure 15. The cavity is initially in vacuum and the computer orders the sending of several emitter atoms to build the field, which stops growing when the target is reached. Photon jumps occurring at later times are corrected by emitter atoms. When the correction overshoots, absorber atoms are used. On average, the photon number is stabilized, with small residual fluctuations around the target n-value. These experiments open interesting perspectives for the deterministic preparation and protection against decoherence of other non-classical states of radiation.

FIELD STATE RECONSTRUCTION

QND measurements, repeated on many realizations of the same field, have allowed us to reconstruct its photon number distribution P(n), which, for a coherent field, is a Poisson function centered around its mean photon number value $\langle n \rangle$ (see Figure 13). These P(n) histograms provide only partial information about the field. Describing the light intensity and its fluctuations, they are insensitive to field coherences. In general, a field state is described by a density matrix ρ, whose diagonal elements ρ_{nn} in the Fock state basis are the P(n) probabilities, and the off-diagonal ones, $\rho_{nn'}$, describe the field coherence. Expressed in terms of photon numbers, the P(n) probabilities are "1D" objects while the $\rho_{nn'}$ coherences are "2D" entities. Reconstructing coherences from the measurement of photon number probabilities, i.e. "going from 1D to 2D" in the representation of the field, is analogous to going from 2D to 3D in photography.

The photon number distribution of a field state is indeed like a flat photo, obtained by recording the light intensity that the object has scattered into the lens of a camera. To add an extra dimension and achieve a full reconstruction, one must realize a hologram by adding phase information to the one provided by the intensity recording [58]. In photography, this is achieved by interfering the scattered light with a reference beam—a small fraction reflected off the main laser beam illuminating the object. The interference pattern recorded on the hologram is a Fourier transform of the object. When illuminated by a laser beam similar to the one which has produced it, the hologram reproduces the appearance of the object by inverse Fourier transformation.

Similarly, the full "2D" $\rho_{nn'}$ information contained in the quantum state of a field can be reconstructed by mixing this field with reference fields of various phases and amplitudes and by reconstructing the photon number distributions of these interfering fields. This procedure is called quantum tomography [59]. In our cavity QED experiments, the Rydberg atom Ramsey interferometer is used to perform these state reconstructions [60]. Identical copies of the field are prepared, then admixed with reference coherent fields produced by a classical source. QND photon counting of the resulting "mixed fields" are then performed. From the data accumulated in many realizations with reference fields of different phases and amplitude, enough information is collected to reconstruct ρ.

To represent the field state, it is convenient to choose, instead of ρ, an alternative description. The field state is formally equivalent to the state of a mechanical oscillator evolving in a parabolic potential. Its state is represented by a real Wigner function [4,61] taking its values in the oscillator phase space (the coordinates in this space being the position x and momentum p of the fictitious oscillator, corresponding to "field quadratures"). This function, which generalizes for the quantum field the classical concept of probability distribution in phase space, contains the same information as ρ, to which it is related by Fourier transformation. To keep the holographic analogy, the Wigner function is to the density matrix what the hologram is to the direct image of an object. Its interfering patterns directly reveal the main features of the quantum field.

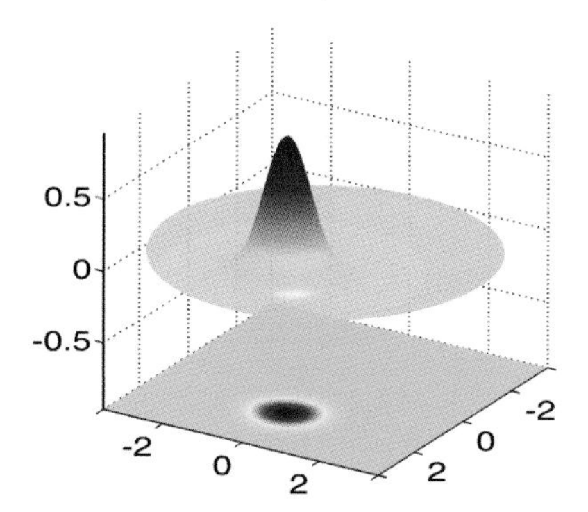

FIGURE 16. Experimental Wigner functions represented in 3D and 2D: (a) coherent state with <n> = 2.5; (b) n = 3 Fock state; (reprinted with permission from [60], © Macmillan Publishers Ltd)

Figure 16 shows the Wigner functions of two field states experimentally reconstructed by Ramsey interferometry [60]. Figure 16a shows the Wigner function of a coherent state with <n> = 2.5 photons on average. It is a Gaussian peak centered at the point in phase space whose polar coordinates correspond to the amplitude and to the phase of the field. This positive function, which contains more information than the photon probability distribution in Figure 13, can be interpreted as a classical probability distribution in phase space, with uncertainties in x and p reflecting the photon number and phase fluctuations of the coherent state. Figure 16b presents the experimental Wigner function of the n = 3 Fock state prepared by QND projective measurement, then reconstructed by our Cavity QED-adapted tomographic procedure. It consists in a set of circular ripples centered at phase space origin. Its interfering features, displaying negative values, are distinctive of this field state "quantumness." The central symmetry of this Wigner function reveals that a Fock state has a complete phase indeterminacy.

SCHRÖDINGER CAT STATES OF LIGHT AND DECOHERENCE STUDIES

By describing how we count and manipulate photons in a cavity, I have so far emphasized the 'particle aspect' of light. As was recalled above, however, light is also a wave. Which of the particle or the wave aspect manifests itself depends upon the kind of experiment, which is performed on the field. Let us describe now experiments in which the wave features of the field stored in the cavity is

FIGURE 17. The Schrödinger cat thought experiment (reprinted from "Science et Vie Junior").

essential. This will lead to the description of photonic Schrödinger cats and to decoherence experiments.

At this stage, it is appropriate to recall Schrödinger's thought experiment [62]. The Austrian physicist has imagined that a large system, a cat for instance, was coupled to a single atom, initially prepared in an excited state spontaneously decaying into a ground state by emitting a photon (or a radioactive particle). This emission triggered a lethal device, killing the cat. After half the life-time of the excited state, the atom has evolved into a superposition of two states, one of which would be associated with the dead cat and the other with the live cat (Figure 17). At this point, the atom and the cat would be entangled and the cat suspended between life and death. In our version of this experiment, we have a single atom in a superposition of two states and this atom controls the fate of a coherent field containing several photons (our Schrödinger cat), which takes two different phases at once, one that we can call "alive" and the other "dead." The way to perform this experiment was initially proposed in a paper written in 1991 together with our Brazilian colleagues Luiz Davidovich and Nicim Zagury [63]. A similar proposal for the preparation of Schrödinger cat states of light in the optical domain had been made earlier in another context [64].

Our method again employs the Ramsey interferometer. It starts with the preparation of a coherent field in the cavity, whose Wigner function is a Gaussian (Fig 16a). A single non-resonant atom is then prepared in a coherent superposition of two states, an atomic Schrödinger "kitten," as I have already called it. This atom crosses the cavity and its two components shift the phase of the field in different directions by a simple dispersive index effect. Here again, we take advantage of the huge coupling of Rydberg atoms to microwaves, which makes a single atom index large enough to have a macroscopic effect on the field phase. At the cavity exit, the atom and the field are entangled, each atomic state being correlated to a field state with a different phase (the phase difference being close to 135° in the experimental realization described below). We can consider that the field, with its small arrow in the complex plane, is a meter used to measure the atom's energy. After the atom has been exposed to the second Ramsey pulse and detected, there is no way to know in which state the atom crossed the cavity and the field collapses into a Schrödinger cat superposition. In other words, the atomic Schrödinger kitten has produced a photonic Schrödinger cat, which contains several photons on average.

By sending subsequent atoms across the cavity and achieving a tomographic field reconstruction with QND photon counting, we have been able in 2008 to reconstruct completely the Schrödinger cat state Wigner function [60]. Figure 18a exhibits clearly the two peaks corresponding to the two classical fields of the superposition. The fringes between these peaks account for the coherence of the superposition. The negative values of these fringes are a signature of the "quantumness" of the "cat" state. Theory shows that the coupling

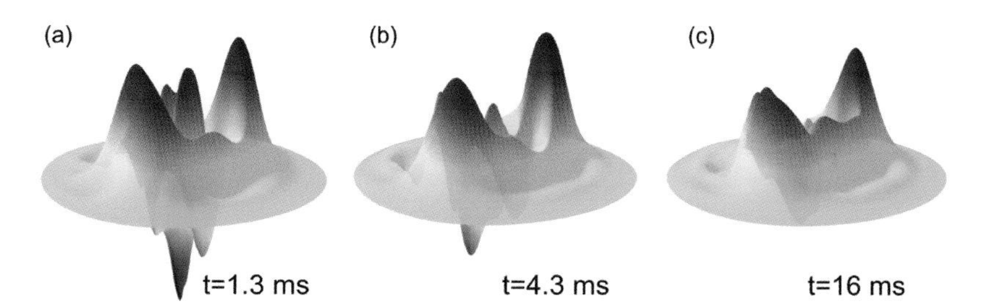

FIGURE 18. Schrödinger cat and decoherence: (a) Reconstruction of the Wigner function of a cat state with n = 3.5 photons on average, a short time (1.3 ms) after its preparation by a single atom crossing the cavity. (b) and (c): the same cat state after 4.3 and 16 ms: the vanishing of the fringe interfering features is a manifestation of decoherence (reprinted with permission from [49], © American Physical Society).

of the field to the environment very quickly washes out the quantum coherence of the cat, leading it into a mundane statistical mixture of states. W. Zurek has played an important role in elucidating the role of the environment in this process, which occurs faster and faster as the "size" of the cat, measured by the square of the distance of its components in phase space, is increased [65]. For a given phase difference between the Gaussian components, this size is proportional to the cat's mean photon number.

We have studied this decoherence phenomenon by reconstructing the field Wigner function at various times [60]. Figs 18b and 18c, which represent snapshots of the Wigner function taken at increasing times after the Schrödinger cat state preparation, show decoherence in action. Within a time much shorter than the energy damping time of 130 ms, the interfering features of the cat state are indeed suppressed, leaving the Wigner function as a sum of two quasi-Gaussian peaks. We have checked that decoherence occurs at a rate proportional to the size of the cat. It is important to stress that these Schrödinger cat state recordings, as all field state reconstruction, are obtained from analyzing many realizations of the experiment and performing complex statistical analysis of the data [60]. Acquiring knowledge about a quantum state always requires such a statistical procedure and these experiments rely on the fact that we can prepare an arbitrary number of copies of the state to be reconstructed and follow the subsequent evolution of all these copies.

An earlier version of this experiment had been performed in 1996, with a cavity having a much shorter damping time, in the hundred microsecond range [66]. Since it was not possible to send a sequence of measuring atoms across the cavity before its field had decayed, the experiment relied on the information provided by a single probe atom following the atom which had prepared the cat state. Instead of reconstructing the whole Wigner function, we used this single probe atom to get information about the Wigner function at phase space origin, where its value is very sensitive to the cat's coherence. Comparing the detection signal of the first atom which prepared the cat and that of the second, which probed its coherence, provided a two-atom correlation signal whose decay as a function of the delay between the two atoms measured the loss of quantum coherence of the cat versus time. Fig 19 shows this decay, for two cat states with different separations between their two components (as shown in the insets). The shortening of the decoherence time as the separation is increased was clearly demonstrated.

We have also prepared Schrödinger cat states of radiation by resonant atom-field interaction [67,68]. Letting a coherent field evolve under its coupling with

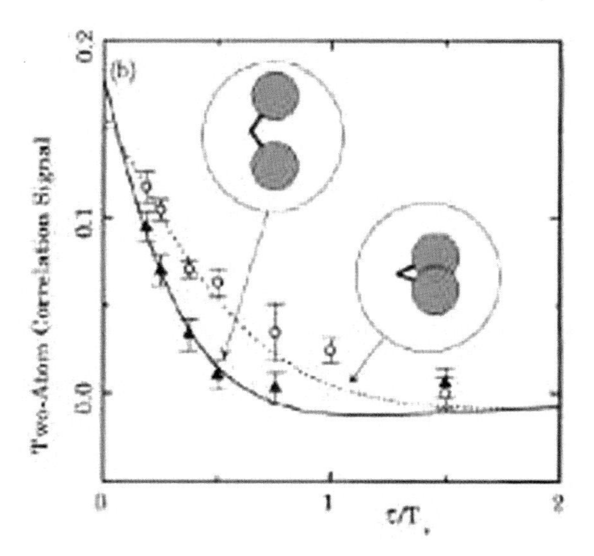

FIGURE 19. Observation of decoherence in a two-atom experiment. The correlation signal is plotted versus the delay between the atom preparing the "cat state" and the atom probing its coherence. This delay is measured in units of the cavity damping time Tc. (equal to 160 μs in this experiment). The points are experimental and the curves are theoretical. The two curves correspond to cat states with different separations between their components (as shown in the insets schematically representing their Gaussian components as circles) (Reprinted with permission from [66], © American Physical Society; online abstract at http://prl.aps.org/abstract/PRL/v77/i24/p4887_1).

a Rydberg atom at resonance turns the atom-field system, after some time, into an entangled atom-field state superposition involving two coherent fields with opposite phases. The two components of this cat merge together at a later time. This effect of field phase splitting and recombination is related to the collapse and revival of the Rabi oscillation phenomenon [30]. The Schrödinger cat experiments in Cavity QED illustrate the fragility of quantum coherences in systems made of increasing number of particles. They give us a glimpse at the boundary between the quantum world, where state superpositions are ubiquitous, and the classical one, where systems behave in a mundane classical way. A detailed study of the cavity QED Schrödinger cats along with a review of proposals for the generation of various Schrödinger cats in quantum optics can be found in the book "Exploring the quantum: atoms, cavities and photons" [4]. This book presents also a detailed analysis of the decoherence of these cat states.

QUANTUM INFORMATION IN CAVITY QED

The ENS photon trap and NIST ion trap experiments are representative of a very broad field of experimental physics. The non-destructive control and manipulation of single quantum particles now performed in many laboratories around the world is strongly motivated by the prospect of exploiting these systems in order to develop new ways to process quantum information. This domain of research became very active in the mid 1990s, at the time we were building our photon box experiment, with the non-destructive photon counting and manipulation of light as our main goal. Feynman's reflections about quantum simulators [69] and the propositions of cryptographic key distributions based on quantum laws [70–71] date back to the 1980s and early 1990s but they did not have any resonance at that time on our work at ENS. Things changed with the proposal of quantum computers working with qubits made of atomic or light particles [72,73] and with the discovery of quantum algorithms, which found an obvious use for such computers [74,75].

These ideas matured when our Cavity QED set-up became operational to test them, i.e. when the photon damping time in our open cavity reached a value (a couple hundred microseconds) exceeding the atom-cavity interaction time (about 20 microseconds). The convergence between these theoretical ideas and our experiments then became very fruitful. It is also around that time that we became influenced by theoretical studies on decoherence [65,76], and the possibility of checking them with simple experiments became a strong incentive for us. A review of the experimental methods we developed then to test basic steps of quantum information can be found in [77].

While the QND photon counting and the Schrödinger cat state experiments described above are based on the dispersive non-resonant interaction of atoms with the cavity field, our Cavity QED quantum information experiments have mostly exploited resonant atom-cavity field interactions. The basic phenomenon is then the Rabi oscillation, alluded to in the introduction of this Lecture. Figure 20 shows an experimental recording of the probability for finding the atom in level e as a function of the atom-cavity interaction time t, when the atom enters at time $t = 0$ in a cavity containing no photon. This signal, observed in 1996 [6], exhibits a damped oscillation at the vacuum Rabi frequency $\Omega_0/2\pi =$ 50 kHz (the damping being due to various experimental imperfections). This oscillation is the direct manifestation of the strong coupling regime of Cavity QED, whose observation had been our goal since the early 1980s. We first reached it in 1987 with the closed cylindrical cavity which we used for our two-photon

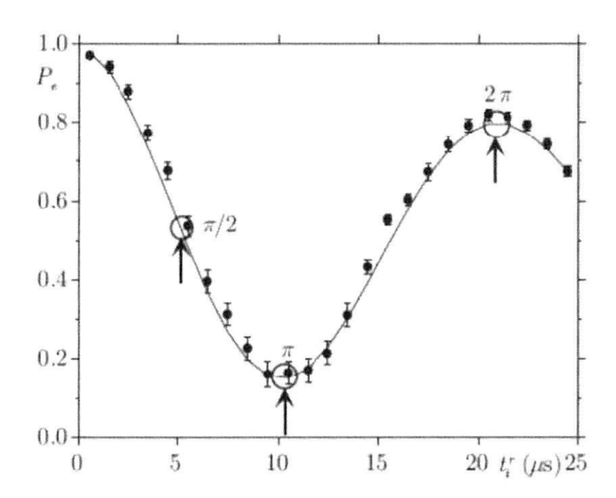

FIGURE 20. Vacuum Rabi oscillation representing the probability of finding the atom in its initial state e after interacting during time t with the cavity initially in vacuum and resonant on the e → g transition. Points are experimental and curve is a fit to a damped sinusoid. The three arrows mark the times corresponding to the $\pi/2$, π and 2π Rabi pulses.

Rydberg maser [35], then again in 1996 with our first open cavity able to keep a photon for a time longer than the atomic transit time across it.

It is only with this open cavity structure that we could easily vary the atom-cavity interaction time and thus directly record the Rabi oscillation. The method we use for this is to apply between the cavity mirrors an electric field which modifies by Stark effect the energies of the circular Rydberg states e and g. This allows us to tune the atomic transition in and out of resonance at a determined time. The atoms are prepared in the circular state by a pulsed process which also fixes their velocity, so that we know when they enter the cavity and can decide when the Stark pulse must be applied to fix the atom-cavity coupling time [77]. Our quantum information experiments, performed in a set-up similar to the one sketched in Figure 10, were based on the exploitation of the Rabi oscillation in the main cavity C, combined with classical microwave pulses produced in the auxiliary cavities R_1 and R_2. These experiments took thus advantage in various ways of the versatile properties of the Ramsey interferometer set-up.

Quantum information experiments manipulate two-level systems called qubits. In our case, these qubits are either atoms crossing the cavity while evolving between two Rydberg states, or the cavity field oscillating between the two states $|0>$ and $|1>$ containing 0 and 1 photon. The coupling between the qubits is based on the realization of Rabi oscillation pulses of various durations. The auxiliary cavities R_1 and R_2 sandwiching the photon storing cavity C are used to prepare and analyze the state of the atomic qubit, before or after their interaction with the cavity field.

When the atom-field interaction time is set so that $\Omega_0 t = \pi/2$, the Rabi oscillation maximally entangles the atom with the cavity field, realizing the transformation |e,0>→(|e0>+|g,1>)/√2. For $\Omega_0 t = \pi$, the atom and the field exchange a quantum of excitation, realizing the transformations |e,0> → |g,1> and |g,1> →–|e,0>. Another useful Rabi pulse corresponds to $\Omega_0 t = 2\pi$. It has the effect of inducing a π-phase shift of the atom+field state, realizing the transformations |e,0> →– |e,0>, and |g,1> →–|g,1>, while the atom in state |g> crossing under the same conditions the cavity in vacuum is not affected: |g,0> → |g,0>.

In order to maximally entangle two atoms, the first being prepared in state e and the second in state g, we have sent them successively in a cavity initially in vacuum, realizing a $\pi/2$ Rabi pulse on the first atom and a π Rabi pulse on the second [78]. The first pulse entangles the first atom with the cavity field, while the second exchanges the second atom and the cavity field excitations, finally leading to the two-atom entangled state (|e,g>– |g,e>)/√2. The cavity field, transiently entangled with the first atom, ends in its initial vacuum state, playing thus the role of a catalyst for the deterministic entanglement of the two atoms.

We have also realized a quantum memory, storing in the cavity the information initially carried by a first atom before copying it onto a second atom [79]. The first atom, prepared in a superposition state a|e> +b|g> in R_1 undergoes a π Rabi pulse in the initially empty cavity, preparing its field in the state a|1> + b|0>. A second atom, initially in |g> then undergoes a π-Rabi pulse in this field, exiting the cavity in the state –a|e>+b|g> which, within a phase shift, is identical to the initial state of the first atom. Applying to the second atom a classical microwave pulse in R_2 before detecting it allows us to analyze its final state, checking that the superposition state has been copied from the first atom to the second.

We have also demonstrated the operation of a quantum gate [80], an elementary device coupling a control and a target qubit in such a way that the target undergoes a unitary transformation if the control is in one state and remains unchanged if the control is in the other state (in all cases, the control state does not change). For this experiment, three circular atomic Rydberg states were relevant: the levels e and g with principal quantum numbers 51 and 50 (the transition between these states being resonant with the cavity mode) and a third circular Rydberg state with principal quantum number 49, called i in the following. The transition g to i being far off-resonant with the cavity, the atom in state i was totally insensitive to the presence of photons in the cavity. The control qubit was then the cavity field in state 0 or 1 while the target qubit was a Rydberg atom evolving between the two states g and i.

The interaction time between the atom and the cavity field was set to achieve a 2π Rabi pulse in vacuum for an atom evolving between states e and g. The coupling between the photon and atomic qubit then realized the mapping $|i,0> \rightarrow |i,0>$, $|g,0> \rightarrow |g,0>$, $|i,1> \rightarrow |i,1>$, $|g,1> \rightarrow -|g,1>$. This is the "truth table" of a "phase" gate inducing a conditional π-phase shift onto the target qubit, if and only if the control is in the state 1. By applying to the target atomic qubit two pulses in R_1 and in R_2, equivalent to so-called Hadamard one qubit gates, this "phase gate" was transformed into a "control-not" gate leaving the target state unchanged if the control was in state 0 and switching the two target states if the control was in state 1. Such a gate was operating as a 0–1 photon QND counter since the final state of the atom crossing the cavity (i or g) was determined by the photon number in it (0 or 1), which remained unchanged in the measurement process. This experiment, realized in 1999 [81], was a precursor of our later QND experiment [50]. It could not count photon numbers above 1 and could not repeat the counting procedure many times before the field decayed in the cavity.

The use of Rabi pulses of controlled duration acting on atoms crossing the cavity one by one can be described as a kind of deterministic quantum "knitting" procedure. By extending this procedure to three atoms, we have been able to prepare a triplet of entangled particles, a so-called GHZ state [82]. The preparation involves a $\pi/2$ Rabi pulse applied on a first atom, realizing an entangled first atom-cavity state, as described above. The state of the field is then read by applying a 2π Rabi pulse on a second atom, realizing a QND counting of the photon number of the field left by the first atom in the cavity. The state of the two atoms and the field then becomes $(|e,0,i> + |g,1,g>)\sqrt{2}$. A third atom, initially in g then crosses the cavity and undergoes a π pulse, thus copying the state of the field and bringing back the cavity in vacuum. The final three atom state is then $(|e,i,g> + |g,g,e>)/\sqrt{2}$ where the three symbols refer, in that order, to the three atoms having successively crossed the cavity.

In another entanglement experiment, we have tested the principle of complementarity by studying how the fringes of an atomic Ramsey interferometer vanish when information about the atomic path in the interferometer is stored in the field mixing the atomic levels [83]. Using again our Ramsey-Cavity QED set-up, we have also demonstrated how to exploit a cavity-assisted collision to entangle two atoms simultaneously crossing the field mode [84] and how to entangle two field modes via their resonant interaction with a single atom [85]. We have also shown how to exploit the dispersive interaction of atoms with the cavity field to detect atoms non-destructively by the phase shift they produce on a small coherent field [86]. All these experiments are described in detail in [4].

PERSPECTIVES FOR THE WORK OF ENS

In the course of our cavity QED studies we have learned how to measure quantum fields non-destructively, how to prepare and reconstruct various non-classical states of radiation and how to exploit the coupling of these fields to Rydberg atoms in order to demonstrate steps of quantum information procedures. We are presently developing these studies along several directions under a senior investigator grant obtained in 2009 from the European Research Council (ERC). The aim of this project is to push the quantum-classical boundary further by preparing non-classical Fock and Schrödinger cat states of light containing larger numbers of photons, by studying the non-local properties of entangled fields stored in two cavities and by investigating ways to protect efficiently non-classical fields against decoherence.

Another line of research will consist in tailoring the Hamiltonian ruling the field evolution by exploiting the Quantum Zeno effect. We have already demonstrated that it is possible to freeze the growth of a coherent field in the cavity by repetitive measurements of the field photon number [87]. We propose a generalization of this experiment in which we would use a repetitive QND procedure to find out whether the field contains a preset photon number n_0 in the cavity. By performing this measurement, we would prevent the field from reaching this number and force it to evolve in the subspaces spanned by Fock states with either $n<n_0$ or $n>n_0$. Restricting its evolution in this way would result in novel methods to generate photonic Schrödinger cats and more generally in new procedures to prepare and control arbitrary superpositions of coherent states in the cavity [88, 89]. For these and other experiments, we need a set-up in which atoms interact with the cavity field for times extending in the millisecond time range. An atomic-fountain set-up in which slow atoms will cross the cavity at velocities of the order of a few meters per second is in preparation in order to achieve this condition. The atoms, cooled in a magneto-optical trap [45], will be pushed upward by a laser kick and sent in a superconducting cavity located above the trap, at the turning point of the atomic parabolic trajectory. The slow atoms will be prepared into circular states and detected by field ionization inside the cavity.

RELATION TO OTHER DOMAINS IN QUANTUM INFORMATION SCIENCE

Our Rydberg atom-microwave superconducting cavity experiments are related to other experimental studies exploring, in different contexts, the atom-photon interaction in structures confining the electromagnetic field. Optical Cavity QED has developed in parallel with microwave Cavity QED and has reached the strong coupling regime in 1992 [90]. In this very active domain of research, many experiments have been realized over the last twenty years to test quantum properties of light or to demonstrate basic operations in quantum information processing. These experiments often involve the interaction of cold atoms with optical photons and are exploiting the techniques of laser cooling and trapping of atoms in conventional or chip-based devices. They use various cavity designs including high finesse Fabry-Perot, microsphere or micro-torroidal structures sustaining whispering gallery modes. The diversity of studies is so large that giving extensive reference to specific works is beyond the scope of this Lecture. A review of the early research in this domain can be found in [91] and a more recent description of the field is given in [92]. Other developments in solid state physics must be mentioned. There, atoms are replaced by quantum dots embedded in Bragg layers or in photonic band gap materials confining the photons [93]. Many interesting cavity QED effects have been observed with these systems. Strong coupling of light emitters with micro-cavity structures has been developed to achieve operations useful for quantum communication and quantum information processing purposes. Here again, I will not attempt to give more references to specific works.

A special mention should be made of the field of Circuit QED [94] because of its close connection with our microwave Cavity QED work. In this field, which has undergone very fast and spectacular development lately, atoms are replaced by small superconducting circuits with Josephson junctions which behave as two level artificial atoms coupled to strip line, coplanar structures or three dimension closed cavities sustaining radiofrequency fields. The coupling between the artificial atoms and the field obeys the same Jaynes-Cummings Hamiltonian as our Rydberg atom-superconducting cavity system [95]. Many groups in the world are working in this new field, and experiments very similar to the ones performed at ENS have been made with these devices, including the generation and study of Fock [96,97] and photonic Schrödinger cat states [98,99]. The very strong coupling of the giant artificial Josephson atoms to radiofrequencies leads to a faster dynamics than that of the Rydberg atom-superconducting cavity coupling. Quantum processing operations with these systems occur at the nanosecond time scale, instead of the microsecond time scale of the Rydberg atom experiments. These circuits can be produced by well established lithographic methods which makes them promising for scalable quantum information applications, provided their decoherence can be better controlled.

Ion trappers, such as photon ones, also form a wide community. David Wineland, in his Lecture, mentions experiments carried on by many teams around the world with ions in traps, which bear strong similarities with his own work. In a broader quantum information context, it is also important to mention, beyond Cavity QED and ion-trap physics, the wide variety of quantum optics experiments exploiting the detection of photon-correlations produced by various sources of twin-light beams. These experiments started with the pioneering demonstrations of Bell's inequality violations in the 1970s and early 1980s [100,101]. They have since then demonstrated various properties of non-classical light and explored their possible applications for quantum communication [102], quantum teleportation [103], and quantum computing [104]. The quantum non demolition studies of propagating optical beams [105] have not reached single quantum sensitivity, but they nevertheless share some of their features with our QND work. Schrödinger cat states of light presenting many similarities with our Cavity QED "cats" have also been generated in quantum optics experiments [106]. The domain is again too vast to mention more works specifically. Experiments with cold atoms, especially those which study the properties of atoms trapped in optical lattices in one, two or three dimensions, also aim at manipulating single particles in a controlled environment, with quantum information applications in mind [107,108]. Finally, the development of nano-mechanical oscillators coupled to atomic systems or to electromagnetic field resonators opens a new domain where the manifestations of state superposition and entanglement will soon become observable [109,110]. Many of the concepts and ideas exploited in these experiments have features in common with those used in cavity QED or ion-trap physics.

Without comparing the respective merits or limitations of the various lines of research aiming at controlling quantum systems, we can try to foresee the future of this physics at large. The all purpose quantum computer which would be able to implement arbitrary algorithms seems to be still very much a utopia [111]. The problem of decoherence is indeed extremely difficult to solve, in spite of the progress made in the implementation of error correction [112, 113]. Quantum simulators emulating condensed matter structures and dynamics with arrays of atoms in optical lattices [114] or ions in traps [115] seem more realistic. They will probably lead to important advances in the near future. Quantum communication with photons and the realization of quantum nodes in which information will be exchanged between atomic systems and photons in order to build efficient quantum repeaters are also likely to be developed [116]. The ideas of Cavity QED seem to be particularly well adapted in this context. Important progress in quantum metrology [117] using the tools of particle control can also be expected. But I am well aware that this exercise in prediction is largely futile. Most important applications of quantum physics have come in the past from the serendipitous convergence of disparate blue sky advances made by scientists who would never have predicted what their results were going to lead to. I guess the same will happen with the control of single quantum particles.

ACKNOWLEDGMENTS

Let me add a few more personal remarks. The work accomplished in my laboratory during the last three decades has been a collective endeavor in which my former students and now colleagues Jean-Michel Raimond and Michel Brune have played an essential role. I have been extremely lucky that we have stayed together over all these years and that we have been able to build such a strong and friendly collaboration (Figure 21 shows us in the lab, in the autumn of 2012). We have always discussed our projects openly and freely and decided together the way to follow. Many of the seminal ideas in our work have come

FIGURE 21. The author, Jean-Michel Raimond and Michel Brune (from left to right) in the lab in autumn 2012 (credit J.F. Dars, Photothèque de l'ENS).

FIGURE 22. The ENS team in October 2012: from left to right: M. Brune, J-M. Raimond, A. Facon, A. Signoles, E-K. Dietsche, S. Haroche, T. Rybarczyk, Nguyen Thanh Long, I. Dotsenko, C. Herman, S. Guerlich, B. Peaudecerf, S. Gleyzes, R. Texeira and J. Hare (a colleague, former group member).

from Jean-Michel or Michel and I am well aware that today's recognition is theirs as well as mine. I know they share this feeling!

This work has involved many people, almost a hundred students, postdocs and visitors altogether. The last generation is shown on the photograph of Figure 22 taken in October 2012, three days after the Nobel Prize was announced. Besides Jean-Michel and Michel, I would like to mention two young colleagues present in this photograph, Sebastien Gleyzes and Igor Dotsenko whose contribution to the photon detection, Schrödinger cat and quantum feedback experiments has been very important. Stefan Kuhr (now at the University of Strathclyde in Glasgow) was a postdoc in our group when we realized the first QND photon counting in 2006 and his contribution too has been essential. Gilles Nogues, after having been a graduate student, accompanied us until 2010 before taking up a position in Grenoble. Among the important participants to our work in earlier stages, I must recall the role of Philippe Goy whose technological skills and deep insight in physics have been essential. Michel Gross and Jean Hare played an important role in developing our circular Rydberg atom preparation procedure. Valerie Lefevre-Seguin was a member of our group when our QND ideas matured and Ferdinand Schmidt-Kaler, now in Mainz, helped us to realize the first experiments in the strong coupling regime with open cavities. On the theory side, the fruitful collaboration with our Brazilian colleagues Luiz Davidovich and Nicim Zagury has also been important. Other students, postdocs, visitors and collaborators are listed as coauthors in the references given at the end of this paper. I hope they will forgive me for not mentioning them here by names. All have by their successive contributions improved our expertise and cleverness in our photon juggling game step by step. Coming from many countries, they have been working hard and with an enthusiasm that the inevitable drawbacks of an experimenter's life could not diminish.

I will always remember some exhilarating moments when our efforts were rewarded and the difficulties forgotten: the 1987 Bastille day (or rather night) in the lab when we observed with our friend Luiz Davidovich the first evidence that our two-photon micromaser was working; the early morning in a Sydney hotel in July 1996 when I received by fax, for immediate presentation in the talk

FIGURE 23. Photograph taken on the day of Alfred Kastler's Nobel Prize announcement, in the same room as the photograph in Figure 22. From left to right: Franck Laloë, Claude Cohen-Tannoudji (Nobel 1997), Alfred Kastler (Nobel 1966), Serge Haroche, Jean Brossel and Alain Omont.

I was going to deliver at a conference, the long awaited Schrödinger cat signals that Michel, Jean-Michel and the students had just obtained while I was sleeping twenty thousand kilometers and ten time zones away; and the evening of September 11, 2006 (my birthday!) when Stefan Kuhr summoned me back by phone to the lab, just in time to observe the first recording of the birth and death of a single photon in our cavity.

I must also recall the unique context of the Kastler Brossel Laboratory, in which we have been working all these years in an atmosphere of trust, feeling always free to follow our ideas without having to justify our work by promising improbable applications. This spirit of freedom is the inheritance of great precursors, Alfred Kastler and Jean Brossel, who built this lab at the time they invented optical pumping, sixty years ago. Claude Cohen-Tannoudji followed in their footsteps and infused me with his passion for the quantum world. The photograph in Figure 23 was taken at ENS, in the same room as the previous one, 46 years almost to the day before, when the Nobel Prize of Alfred Kastler was announced. Claude and I have certainly changed physically, but our enthusiasm for physics has remained the same!

At the start of my career, I was funded directly by the laboratory, without ever being asked to write proposals or lengthy reports. Things changed progressively, when research became more project-oriented at the end of the 1980s. My group and its work were then recognized well enough to get grants from the CNRS and the Agence Nationale de la Recherche, as well as from European agencies, culminating with our ERC senior grant in 2009. I must also mention generous funding by the Japanese Science and Technology Corporation (JST) through a collaborative grant that I shared with Yoshisha Yamamoto from 1999 to 2009. It allowed us to freely pursue our research on quantum entanglement without any bureaucratic red tape.

I would not be complete without stressing that teaching has always been an important stimulation for me. At the University Paris VI and at ENS, I have found that preparing lectures and interacting with students has greatly helped me to understand physics better and has more than once given me new ideas for experiments. Since 2001 at College de France, I have had the challenging task of teaching a new course each year about topics in quantum physics directly related to my own research or to that of colleagues working in quantum information science. This experience has also been extraordinarily rewarding and fruitful.

Let me also acknowledge the influence on my work of so many friends and colleagues from all over the world. I have often been stimulated by attending their lectures, reading about their work or discussing with them in Paris or in their own laboratory. These friendships are the privilege of our lives as scientists. I cherish the opportunity we have to exchange with people of many different cultures sharing, in our diversity, the same curiosity and the same passion for discovering and understanding Nature. My friendship with David Wineland is especially worth mentioning here, since it has been a special pleasure for me to share this Nobel Prize with him.

Finally nothing would have been possible without Claudine, my wife, whom I met long before knowing what a photon is, and Julien and Judith, our children. They have illuminated my life with their love, their humor and their intellectual support. But this part of the story will be told elsewhere [118].

REFERENCES

1. N. Bohr, "Discussion with Einstein on epistemological problems in atomic physics," in *Albert Einstein, philosopher-scientist*, ed. Paul Arthur Shilpp, Harper (1949)
2. E. Schrödinger, "Are There Quantum Jumps?, Part II" in the *British Journal for the Philosophy of Science*, Vol **3**, No11, 233 (1952)
3. E.T. Jaynes and F.W. Cummings, *Proc. IEEE*, **51**, 89 (1963)
4. S. Haroche and J-M. Raimond, *Exploring the quantum: Atoms, Cavities and Photons*, Oxford University Press (2006)
5. D.M. Meekhof, C. Monroe, B.E. King, W.M. Itano and D.J. Wineland, *Phys.Rev.Lett.* **76**, 1796 (1996)
6. M. Brune, F. Schmidt-Kaler, A. Maali, J. Dreyer, E. Hagley, J-M. Raimond and S. Haroche, *Phys.Rev.Lett.* **76**, 1800 (1996)
7. S. Haroche and C. Cohen-Tannoudji, *J.Physique*, **30**, 125 (1969)
8. S. Haroche, Thèse d'Etat (PhD thesis), *Annales de Physique*, Paris, **6**, 189 and 327 (1971)
9. C. Cohen-Tannoudji and S. Haroche, *J.Physique*, **30**, 153 (1969)
10. C. Cohen-Tannoudji, S. Haroche, C. Audoin and J-P. Schermann, *Phys.Rev.Lett.* **24**, 861 (1970)
11. S. Haroche, J. Paisner and A.L. Schawlow, *Phys.Rev.Lett.* **30**, 948 (1973)
12. C. Cohen-Tannoudji and S. Haroche, C., *Rendus Acad.Sciences*, Paris, **261**, 5400 (1965)
13. S. Haroche, M. Gross and M. Silverman, *Phys.Rev.Lett.* **33**, 1063 (1974)
14. C. Fabre, P. Goy and S. Haroche, *J.Phys.B.Letters*, **10,** L-183 (1977)
15. C. Fabre, P. Goy and S. Haroche, *Phys.Rev.A* **18**, 229 (1978)
16. T.W. Dukas, M.G. Littman, R.R. Freeman and D. Kleppner, *Phys.Rev.Lett*, **35**, 366 (1975)
17. M. Gross, P. Goy, S. Haroche, C. Fabre and J-M. Raimond, *Phys.Rev.Lett.* **43**, 343 (1979)
18. R.H. Dicke, *Phys.Rev.* **93**, 99 (1954)
19. J-M. Raimond, P. Goy, M. Gross, C. Fabre and S. Haroche, *Phys.Rev.Lett.* **49**, 1924 (1982)
20. M. Gross and S. Haroche, *Physics Reports*, **93**, 302 (1982)
21. Y. Kaluzny, P. Goy, M. Gross, J-M. Raimond and S. Haroche, *Phys.Rev.Lett.* **51**, 1175 (1983)
22. R. Bonifacio, P. Schwendimann and F.Haake, *Phys.Rev.* A**4**, 302 (1971)
23. J-M. Raimond, P. Goy, M. Gross, C. Fabre and S. Haroche, *Phys.Rev.Lett.* **49**, 117 (1982)

24. E.M. Purcell, *Phys.Rev.* **69**, 681 (1946)
25. P. Goy, J-M. Raimond, M. Gross and S. Haroche, *Phys.Rev.Lett.* **50**, 1903 (1983)
26. D. Kleppner, *Phys.Rev.Lett.* **47**, 233 (1981)
27. R.G. Hulet, E.S. Hilfer and D. Kleppner, *Phys.Rev.Lett.* **55**, 2137 (1985)
28. G. Gabrielse and H. Dehmelt, *Phys.Rev.Lett.* 55, 67 (1985)
29. S. Haroche in *New Trends in Atomic Physics*, Les Houches Summer School Lecture Notes, G.Grynberg and R. Stora Editors, North Holland (1984).
30. J.H. Eberly, N.B. Narozhny and J.J. Sanchez-Mondragon, *Phys.Rev.Lett.* **44**, 1323 (1980)
31. D. Meschede, H. Walther and G. Muller, *Phys.Rev.Lett.* **54**, 551 (1985)
32. P. Meystre, G. Rempe and H. Walther, *Opt.Lett.* **13**, 1078 (1988)
33. G. Rempe, F. Schmidt-Kaler and H. Walther, *Phys.Rev.Lett.* **64**, 2783 (1990)
34. M.O. Scully, B-G. Englert and H. Walther, *Nature* (London), **351**, 111 (1991)
35. W. Jhe, A. Anderson, E. Hinds, D. Meschede, L. Moi and S. Haroche, *Phys.Rev.Lett.* **58**, 666 (1987)
36. M. Brune, J-M. Raimond, P. Goy, L. Davidovich and S. Haroche, *Phys.Rev.Lett.* 59, 1899 (1987)
37. B. Nikolaus, D.Z. Zangh and P. Toscheck, *Phys.Rev.Lett.* **47**, 171 (1981)
38. S. Haroche and D. Kleppner, *Physics Today* **42**, 24 (1989)
39. V.B. Braginsky, Y.I. Vorontosov and F.Y. Khalili, *Zh.Exp.Theor.Fiz.* 73, 1340 (1977) [SPJETP, **46**, 705 (1977)]; V.B. Braginsky and S.P. Vyatchanin, *Dokl. Akad. Nauk SSSR* **259**, 570 (1981) [Sov. phys. Dokl, 1981, 26, 686]; V.B. Braginsky and F.Y. Khalili, *Rev. Mod. Phys.* **68**, 1 (1996).
40. S. Kuhr, S. Gleyzes, C. Guerlin, J. Bernu, U-B. Hoff, S. Deléglise, S. Osnaghi, M. Brune, J-M. Raimond, S. Haroche, E.J acques, P. Bosland and B. Visentin, *Appl.Phys.Lett.* **90**, 164101 (2007)
41. P. Nussenzveig, F. Bernardot, M. Brune, J. Hare, J-M. Raimond, S. Haroche and W. Gawlik, *Phys.Rev.A* **48**, 3991 [1993]
42. R.G. Hulet and D. Kleppner, *Phys.Rev.Lett.* **51**, 1430 (1983)
43. C. Cohen-Tannoudji, *Annales de Physique (Paris)* **7**, 423 and 469 (1962)
44. S. Chu, *Reviews of Modern Physics*, **70**, 685 (1998)
45. C. Cohen-Tannoudji, *Reviews of Modern Physics*, **70**, 708 (1998)
46. W.D. Phillips, *Reviews of Modern Physics*, **70**, 721 (1998)
47. M. Brune, S. Haroche, V. Lefevre-Seguin, J-M. Raimond and N. Zagury, *Phys.Rev.Lett* **65**, 976 (1990)
48. N.F. Ramsey, *Phys.Rev.* **76**, 996 (1949)
49. S. Haroche, M. Brune and J-M. Raimond, *Physics Today*, **66**, 27 (2013)
50. S. Gleyzes, S. Kuhr, C. Guerlin, J. Bernu, S. Deleglise, U. Busk Hoff, M. Brune, J-M. Raimond and S. Haroche, *Nature* (London) **446**, 297 (2007)
51. S. Peil and G. Gabrielse, *Phys.Rev.Lett.* **83**, 1287 (1999)
52. C. Guerlin, J. Bernu, S. Deléglise, C. Sayrin, S. Gleyzes, S. Kuhr, M. Brune, J-M. Raimond and S. Haroche, *Nature* (London), **448**, 889 (2007)
53. R.J. Glauber, *Phys.Rev.* **131**, 2766 (1963)
54. M. Brune, J. Bernu, C. Guerlin, S. Deléglise, C. Sayrin, S. Gleyzes, S. Kuhr, I. Dotsenko, J-M. Raimond and S. Haroche, *Phys.Rev.Lett.* **101**, 240402 (2008)

55. I. Dotsenko, M. Mirrahimi, M. Brune, S. Haroche, J-M. Raimond and P. Rouchon, *Phys. Rev* A80, 013805 (2009)
56. C. Sayrin, I. Dotsenko, XX. Zhou, B. Peaudecerf, T. Rybarczyk, S. Gleyzes, P. Rouchon, M. Mirrahimi, H. Amini, M. Brune, J-M. Raimond and S. Haroche, *Nature* (London), **477**, 73 (2011)
57. XX. Zhou, I. Dotsenko, B. Peaudecerf, T. Rybarczyk, C. Sayrin, S. Gleyzes, J-M. Raimond, M. Brune and S. Haroche, *Phys.Rev.Lett.* **108**, 243602 (2012)
58. D. Gabor, Holography 1948–1971, Nobel Lecture on Nobelprize.org (1971)
59. D.T. Smithey, M. Beck and M.G. Raymer, *Phys.Rev.Lett.* **70**, 1244 (1993)
60. S. Deléglise, I. Dotsenko, C. Sayrin, J. Bernu, M. Brune, J-M. Raimond and S. Haroche, *Nature* (London) **455**, 510 (2008)
61. W. Schleich, *Quantum Optics in Phase Space*, Wiley (2005)
62. E.Schrödinger, *Naturwissenschaften* **23**, 844 (1935)
63. M. Brune, S. Haroche, J-M. Raimond, L. Davidovich and N. Zagury, *Phys.Rev.A* **45**, 5193 (1992)
64. C.M. Savage, S.L. Braunstein and D.F. Walls, *Optics Letters* **15**, 628 (1990)
65. W.H. Zurek, *Physics Today*, **44**, 36 (1991)
66. M. Brune, E. Hagley, J. Dreyer, X. Maïtre, A. Maali, C. Wunderlich, J-M. Raimond and S. Haroche, *Phys.Rev.Lett.* **77**, 4887 (1996)
67. A. Auffeves, P. Maioli, T. Meunier, S. Gleyzes, G. Nogues, M. Brune, J-M. Raimond and S. Haroche, *Phys.Rev.Lett.* **91**, 230405 (2003)
68. T. Meunier, S. Gleyzes, P. Maioli, A. Auffeves, G. Nogues, M. Brune, J-M. Raimond and S. Haroche, *Phys.Rev.Lett.* **94,** 010401 (2005)
69. R. Feynman, *Opt.News*, **11**, 11 (1985)
70. C.H. Bennett and G. Brassard, *Proceedings of IEEE International Conference on computers, systems and signal processing*, p 175, IEEE(New York) (1984)
71. A. Ekert, *Phys.Rev.Lett.* **67**, 661 (1991)
72. A. Ekert and R. Josza, *Rev.Mod.Phys.* **68**, 733 (1996)
73. M.A. Nielsen and I.L. Chuang, *Quantum Computation and quantum information*, Cambridge University Press, Cambridge (2000)
74. P.W. Shor, *Proceedings of the 35th Annual Symposium on the Foundations of Computer Science*, p.124, IEEE Computer Society Press, Los Alamitos CA (1994)
75. D. Deutsch and R. Josza, *Proc.Roy.Soc.London*, **A439**, 553 (1992)
76. D. Giulini, E. Joos, C. Kiefer, J. Kupsch, I.O. Stamatescu and H.D. Zeh, *Decoherence and the appearance of a classical world in quantum theory*, Springer, Berlin (1996)
77. J-M. Raimond, M. Brune and S. Haroche, *Reviews of Modern Physics*, **73**, 565 (2001)
78. E. Hagley, X. Maïtre, G. Nogues, C. Wunderlich, M. Brune, J-M. Raimond and S. Haroche, *Phys.Rev.Lett.* **79**, 1 (1997)
79. X. Maître, E. Hagley, G. Nogues, C. Wunderlich, P. Goy, M. Brune, J-M. Raimond and S. Haroche, *Phys.Rev.Lett.* **79**, 769 (1997)
80. A. Rauschenbeutel, G. Nogues, S. Osnaghi, P. Bertet, M. Brune, J-M. Raimond and S. Haroche, *Phys.Rev.Lett.* **83**, 5166 (1999)
81. G. Nogues, A. Rauschenbeutel, S. Osnaghi, M. Brune, J-M. Raimond and S. Haroche, *Nature* (London) **400**, 239 (1999)

82. A. Rauschenbeutel, G. Nogues, S. Osnaghi, P. Bertet, M. Brune, J-M. Raimond and S. Haroche, *Science*, **288**, 2024 (2000)
83. P. Bertet, S. Osnaghi, A. Rauschenbeutel, G. Nogues, A. Auffeves, M. Brune, J-M. Raimond and S.Haroche, *Nature* (London) **411**, 166 (2001)
84. S. Osnaghi, P. Bertet, A. Auffeves, P. Maioli, M. Brune, J-M. Raimond and S. Haroche, *Phys.Rev.Lett.* **87**, 037902-1 (2001)
85. A. Rauschenbeutel, P. Bertet, S. Osnaghi, G. Nogues, M. Brune, J-M. Raimond and S. Haroche, *Phys.Rev.A* **64**, 050301(R) (2001)
86. P. Maioli, T. Meunier, S. Gleyzes, A. Auffeves, G. Nogues, M. Brune, J-M. Raimond and S. Haroche, *Phys.Rev.Lett.* **94**, 113601 (2005)
87. J. Bernu, S. Deléglise, C. Sayrin, S. Kuhr, I. Dotsenko, M. Brune, J-M. Raimond and S. Haroche, *Phys.Rev.Lett.* **101**, 180402 (2008)
88. J-M. Raimond, C. Sayrin, S. Gleyzes, I. Dotsenko, M. Brune, S. Haroche, P. Facchi and S. Pascazio, *Phys.Rev.Lett.* **105**, 213601 (2010)
89. J-M. Raimond, P. Facchi, B. Peaudecerf, S. Pascazio, C. Sayrin, I. Dotsenko, S. Gleyzes, M. Brune and S. Haroche, *Phys.Rev.A* **86**, 032120 (2012)
90. R.J. Thompson, G. Rempe and H.J. Kimble, *Phys.Rev.Lett.* **68**, 1132 (1992)
91. H. Mabuchi and A.C. Doherty, Science **298**, 1372 (2002)
92. G. Rempe, "Optical Cavity Quantum Electrodynamics" in *Proceedings of Lasers and Electro-optics 2009 and the European Quantum Electronics Conference* CLEO Europe-EQEC (2009)
93. J.P. Reithmaier, "Strong exciton photon coupling in semiconductor quantum dot systems," *Semicond. Sci. Technol.* **23**, 12300& (2008)
94. R.J. Schoelkopf and S.M. Girvin, *Nature* (London), **451**, 664 (2008)
95. J.M. Fink, M. Göppl, M. Baur, R. Bianchetti, P.J. Leek, A. Blais and A. Wallraff, *Nature* (London) **454**, 315 (2008)
96. D.I. Schuster, A.A. Houck, J.A. Schreier, A. Wallraff, J.M. Gambetta, A. Blais, L. Frunzio, J. Majer, B. Johnson, M. Devoret, S.M. Girvin and R.J. Scheolkopf, *Nature* (London), **445**, 515 (2007)
97. M. Hofheinz, E.M. Weig, M. Ansman, R.C. Bialczak, E. Lucero, M. Neeley, A.D. O'Connell, H. Wang, J.M. Martinis and A.N. Cleland, *Nature* (London), **454**, 310 (2008)
98. M. Hofheinz, E.M. Weig, M. Ansman, R.C. Bialczak, E. Lucero, M. Neeley, A.D. O'Connell, D. Sank, J. Wenner, J.M. Martinis and A.N. Cleland, *Nature* (London), **459**, 546 (2009)
99. G. Kirchmair, B. Vlastakis, Z. Leghtas, S.E. Nigg, H. Paik, E. Ginossar, M. Mirrahimi, L. Frunzio, S.M. Girvin and R.J. Schoelkopf, *Nature* (London) to be published (2013)
100. S.J. Freedman and J.F. Clauser, *Phys.Rev.Lett.* **28**, 938 (1972)
101. A. Aspect, J. Dalibard and G. Roger, *Phys.Rev.Lett.* **49**, 1804 (1982)
102. N. Gisin, G.G. Ribordy, W. Tittel, and H. Zbinden, *Rev. Mod. Phys.* **74**, 145 (2002)
103. D. Bouwmeester, J.M. Pan, K. Mattle, M. Eibl, H. Weinfurter, and A. Zellinger, *Nature* **390**, 575 (1997)
104. P. Walther, K.J. Resch, T. Rudolph, E. Scheuk, H. Weinfurter, and A. Zellinger, *Nature* **434**, 169 (2005)
105. P. Grangier, J.A. Levenson and J.P. Poizat, *Nature* (London), **396**, 537 (1998)
106. A. Ourjoumtsev, H. Jeong, R. Tualle-Brouri and P. Grangier, *Nature* (London), **448**, 784 (2007)

107. J.J. Garcia-Ripoli and J.I. Cirac, *Phil. Trans. of the Royal Society* (London), **361**, 1537 (2003).
108. A. Negretti, P. Treutlein and T. Calarco, *Quantum Information Processing*, **10**, 721 (2011)
109. T. Rocheleau, T. Ndukum, C. Macklin, J.B. Hertzberg, A.A. Clerk and K.C. Schwab, *Nature* (London), **463**, 72 (2010)
110. J. Chan, T.P.M. Alegre, A.H. Safavi-Naeini, J.T. Hill, A. Krause, S. Groblacher, M. Aspelmeyer and O. Painter, *Nature* (London) **478**, 89 (2011).
111. S. Haroche and J-M. Raimond, *Physics Today*, **49**, 51 (1996)
112. P. Schindler, J.T. Barreiro, T. Monz, V. Nebendahl, D. Nigg, M. Chwalla, M. Hennrich and R. Blatt, *Science*, **332**, 1059 (2011)
113. M.D. Reed, L. DiCarlo, S.E. Nigg, L. Sun, L. Frunzio, S.M. Girvin and R.J. Schoelkopf, *Nature* (London) **482**, 382 (2012)
114. D. Jaksch and P. Zoller, *Annals of Physics*, **315**, 52 (2005)
115. C. Schneider, D. Porras and T. Schaetz, *Reports on Progress in Physics*, **75**, 024401 (2012)
116. H.J. Kimble, *Nature* (London), **453**, 1023 (2008)
117. V. Giovannetti, S. Lloyd and L. Maccone, "Advances in Quantum metrology," *Nature Photonics*, **5**, 222 (2011)
118. S. Haroche, *Autobiography*, to be posted on Nobelprize.org (2013)

David J. Wineland. © The Nobel Foundation. Photo: U. Montan

David J. Wineland

I was born in Wauwatosa, Wisconsin – just outside Milwaukee – in February 1944, two years after my sister Judy. My parents were born in the U.S. in 1903 and 1905; they both had independent careers until my sister was born; then my mother stayed at home to be with my sister and me. My father received an engineering degree from Cal Tech in 1931. He and my mother moved around quite often during the war; at that time my father was working for Allis Chalmers on jet engines. At the end of the war, we moved to Denver, where he returned to a job with the Bureau of Reclamation as a civil engineer. Shortly after that, in 1947 we moved to Sacramento, California where my sister and I grew up and my father spent the remainder of his career.

Growing up, I was always fascinated by mechanical things, particularly anything that had an engine in it. As a kid, this took the form of model airplanes, stimulated by the fact that we lived very near a major Air Force base that was filled with many airplanes from the war, which my friends and I could explore with little restriction. I also liked math from an early age; my father would play simple games like adding, multiplying, etc. in rapid succession – something I could beat my sister at! My parents were definitely products of the great depression; they emphasized the importance of frugality and getting a good education, which could enhance the prospects of finding a good job later on. So, from an early age I knew I was destined to go to college and I always kept up my grades to be able to do that. As a senior in high school, I took my first physics class. I immediately liked the idea that relatively simple mathematics could explain many of the things we see in the world. But at that time I didn't really jump in; my passion then was cars and motorcycles. With the permission of my father, I bought my first car at 14, a year and a half before I could have a driver's license, but it gave me time to take it apart and fix it up. Of course, this was partly social; my buddies and I all loved cars and motorcycles and having those close friends with a common interest was a great part of growing up.

After high school, in 1961, I attended the University of California, Davis, in part because of the prestige of the University of California and because Davis is fairly near to Sacramento. I started as a math major and was a bit unsure how I would measure up, but about half way through the first semester I realized that if I worked hard I could be near the top of my class. I now regret that I was too much motivated by achieving high grades, which meant I spent a disproportionate amount of time on subjects in the humanities that were difficult for me, rather than just enjoying those classes and getting B's. The math and physics came much more naturally.

At that time, Davis was still emerging from being primarily an agricultural school, so in my junior year I transferred to the Berkeley campus as a physics major – to be at the big time! Berkeley was a frightening, impersonal place to me at the time, but I loved it and the challenge it presented. Out of fear of not making good grades, I didn't become involved in research, so when I graduated, I still had no real idea of what experimental physics was like. My classical mechanics teacher at Berkeley, Fredrick Byron, had a great influence on me. His class was probably the most difficult one I took at Berkeley, but he had the knack of making us want to do the work. Because I respected him so much, I asked him about places to apply for graduate school. He recommended Harvard, so I applied there (and to Berkeley as insurance, because that would have been fine and I was pretty sure I would be accepted there).

FIGURE 1. Dave (the welder) and friends modifying a '36 Ford (1960).

In 1965, I started graduate school in physics at Harvard in a class of 26. Because of Harvard's reputation, I remember thinking uh-oh, 25 geniuses and me! In the end I think there were maybe a couple of geniuses, but the rest were like me, pretty ordinary, but motivated to do well. As much as I still liked how mathematics and physics combined so well together, I was beginning to see that theoretical physics was not for me, so I started to think about experiments. At that time, it appeared to me that the real excitement was in particle physics; it seemed that a new particle was being discovered every other week. The excitement was palpable and it was natural to look in that direction. What frightened me was the prospect of getting buried in a large group and not finding a way to distinguish myself – good or bad.

From looking at the low energy experimental groups, it was evident that I would probably be able to have my own experiment if I went that direction. There were several very good low energy experimental groups at Harvard, but in the end I asked to join Norman Ramsey's group. Ramsey, with his close colleague Dan Kleppner and student Mark Goldenberg had recently invented and demonstrated the first hydrogen masers. I was attracted to the idea of precision measurements and a not-insignificant factor was that Ramsey had the reputation of being a very nice person. After the initial results on hydrogen, Ramsey wanted to make precise measurements of the hyperfine frequencies of all three isotopes of hydrogen, so I chose to work on deuterium. The experiment was relatively straightforward, complicated a bit by the relatively long wavelength (~92 cm) of deuterium's hyperfine transition relative to that of hydrogen (~21 cm).

FIGURE 2. Norman Ramsey's group, Harvard (1966).

During that time, Ramsey served as president of the University Research Association, which managed the construction and operation of Fermilab in Batavia, Illinois. Therefore, he was away from Harvard much of the time. As a consequence, most of us graduate students would help each other along and I liked that experience very much. However, since we were left alone, it sometimes took a while for us to figure out when we made a wrong turn in our projects. In spite of Ramsey's commitments at Fermilab, he would nearly always make a point to return to Harvard for our weekly group meetings. Moreover, whether or not it was deserved, he always supported us and offered encouragement. Ramsey had a very intuitive approach to physics, and was able to make simple pictures that explained the essence of the physics without getting bogged down in mathematical complexity; a lot of the physics we explored could be explained by mapping the problem onto that of a spin-1/2 magnetic dipole in a magnetic field. By the time I finished graduate school, I was hooked on highresolution spectroscopy and precision measurements. However, I still wasn't completely immersed in physics and took many Sundays in graduate school and as a postdoc to realize a long-term dream of racing motorcycles. I had a lot of fun with it but it was clear it was not a way of the future for me.

During the time I was a graduate student, I read about the precision $^3He^+$ hyperfine spectroscopy and electron g-factor measurements of Hans Dehmelt and his colleagues at the University of Washington in Seattle. I was very attracted to ion spectroscopy, but I joined the group to work on Dehmelt's electron magnetic moment experiment. Fred Walls had started the electron experiment as a graduate student with Dehmelt and would later become a colleague at the National Bureau of Standards (NBS). After a while working on the experiment, it became clear that systematic shifts would be smallest for a single trapped electron, so obtaining that goal was my primary project while at Seattle. Dehmelt also had great intuition and was able to reduce all problems to simple semi-classical pictures. Similar to Ramsey, it seemed he could map nearly all problems onto a tuned circuit. I've tried my best to copy their intuitive approach and many times it has served me well. During my tenure at the University of Washington, I met and married my wife, Sedna. Her father, George Quimby, was an anthropologist and director of the Burke Museum, so she was used to and understood long hours spent at work. Her support of this has always been extremely important to me. We have two sons, Charles and Michael; neither are physicists – probably one in the family is enough! In Seattle, I also became good friends with fellow postdoc Bob Van Dyck, who later took a faculty position there and was the key person on the electron magnetic moment experiment. When I began looking for a permanent job, I had my heart set on finding an academic position. However, after about a year of looking, it became clear that the academic positions offered to me would not be able to provide sufficient funds to start a viable experimental program.

Fortunately, a position opened in the Time and Frequency Division of NBS in Boulder, Colorado, and with my background and interest in clocks, I was more than pleased to find a home there. When I arrived, NBS did not have an operating accurate Cesium beam clock, so there was some urgency to achieve this. David Glaze from the division had built the newest version of the Cesium beam clock, NBS-6, and he and I worked to calibrate the device and produce the unit of time, the second. This took about a year and a half. At that time, the Time and Frequency Division had very little research activity. Fortunately my boss, Helmut Hellwig, had a vision that the Division should be doing more basic research and prior to my being hired, he had hired Fred Walls, who had spent a postdoc at JILA, to start research on hydrogen masers. After the Cesium frequency standard was made operational, Helmut was able to get some internal support to start an experiment on laser cooling. Bob Drullinger, who had experience with dye lasers, joined in the project, and with my and Fred Walls' experience with ion traps, we were off and running. This was a very exciting time for me, because it was a project of our own choosing, and it would be great to realize an earlier idea of laser cooling that I had developed with Dehmelt. We started with an empty lab in the summer of 1977, but by the spring of 1978 we had our first results. I knew that we had competition because I was aware that Dehmelt had taken a sabbatical to work in Peter Toschek's lab in Heidelberg, with the same goal of demonstrating cooling. I didn't know at what stage they were at in their experiment and I don't believe they were aware of our experiment. Our paper was published a bit earlier than theirs, but in a near coincidence – even with no contact between the groups – our papers were received at Physical Review Letters within a day of each other (the Toschek group beat us by one day!).

Bob Drullinger and I were subsequently joined by Wayne Itano and Jim Bergquist on these projects and few years later by John Bollinger. We have spent nearly our entire careers working together or on closely related projects. This has been a great experience for me because they have been such great colleagues as well as friends. One of the primary and continuing goals our work has been to make better atomic clocks, and laser cooling has been important to suppress relativistic time dilation in these experiments. In 1985 we demonstrated the first clock that employed laser cooling, which was based on a hyperfine transition in $^{9}Be^{+}$ ions. In 2006, Jim Bergquist and colleagues demonstrated the first atomic clock whose systematic errors were smaller than those of Cesium. This clock was based on an optical transition in a single trapped $^{199}Hg^{+}$ ion, perhaps signaling the advent of high precision optical clocks.

Chris Monroe (now at the University of Maryland) was a very important part of our group from 1992 to 2000, during the time we were building up our experiments on quantum information. After Ignacio Cirac and Peter Zoller's 1995 proposal for a quantum processor using trapped ions, we were able to quickly make a demonstration of the two-qubit logic gate described in their proposal. Didi Leibfried joined our group shortly after that; first as a postdoc and later as a permanent staff member. Didi and I continue to work together to explore the use of atomic ions in quantum information processing. Till Rosenband, who joined the group as a permanent staff member in 2006, has masterfully developed an optical clock based on $^{27}Al^{+}$ ions; it is currently the clock with the lowest systematic error of around 1 part in 10^{17}. At this level it is possible to observe relativistic time dilation for ions moving at the speed of a fast runner.

FIGURE 3. NIST trapped-ion group (1979). Left to right Wayne Itano, Jim Bergquist, Dave Wineland, Bob Drullinger.

In 1988 NBS became NIST, the National Institute of Standards and Technology. In my 38 years at these institutions, our group has always had great support from our immediate supervisors Helmut Hellwig, Sam Stein, Don Sullivan and Tom O'Brian. We are also indebted to our laboratory director, Katharine Gebbie, for her support and encouragement. One measure of her success is that I am the fourth NIST person, after Bill Phillips, Eric Cornell, and Jan Hall, to receive a Nobel Prize during her tenure as lab director. Perhaps a good example of how this support has paid off is laser cooling, which started as a basic research project, but is now employed in all accurate atomic clocks.

It hasn't been all work with my lab-mates in Boulder. I've always liked outdoor activities and for example, Jim Bergquist and I played together on volleyball and softball teams and have ridden bicycles together since the mid 1980s. I still like things with motors on them; unfortunately I don't really have time for cars and motorcycles but I can still sometimes find an hour or so in the evening to build free-flight model airplanes.

Since receiving the Nobel Prize, I've often been asked for advice to give to young students. Of course there's no one answer that fits all, but for me, because of my upbringing, it's been pretty simple. I would suggest finding something interesting (even if you change your mind) and give it your best possible effort. That means hard work, and although not everybody above you will appreciate it, most of them will recognize it and support you. And, as nice as it is to be recognized for accomplishments, I think the biggest reward for me has been just to have the opportunity to explore new ideas. The physics has never been a job; it's more like a hobby – and just the process of doing research is extremely interesting and rewarding.

Superposition, Entanglement, and Raising Schrödinger's Cat

Nobel Lecture, December 8, 2012

by David J. Wineland
National Institute of Standards and Technology, Boulder, CO, USA; University of Colorado, Boulder, CO, USA.

I. INTRODUCTION

Experimental control of quantum systems has been pursued widely since the invention of quantum mechanics. In the first part of the 20th century, atomic physics helped provide a test-bed for quantum mechanics through studies of atoms' internal energy differences and their interaction with radiation. The advent of spectrally pure, tunable radiation sources such as microwave oscillators and lasers dramatically improved these studies by enabling the coherent control of atoms' internal states to deterministically prepare superposition states, as for example in the Ramsey method (Ramsey, 1990). More recently this control has been extended to the external (motional) states of atoms. Laser cooling and other refrigeration techniques have provided the initial states for a number of interesting studies, such as Bose-Einstein condensation. Similarly, control of the quantum states of artificial atoms in the context of condensed-matter systems is achieved in many laboratories throughout the world. To give proper recognition to all of these works would be a daunting task; therefore, I will restrict these notes to experiments on quantum control of internal and external states of trapped atomic ions.

The precise manipulation of any system requires low-noise controls and isolation of the system from its environment. Of course the controls can be regarded as part of the environment, so we mean that the system must be isolated from the uncontrolled or noisy parts of the environment. A simple example of quantum control comes from nuclear magnetic resonance, where the spins of a macroscopic ensemble of protons in the state $|\downarrow\rangle$ (spin antiparallel to an applied magnetic field) can be deterministically placed in a superposition state $\alpha|\downarrow\rangle + \beta|\uparrow\rangle$ ($|\alpha|^2 + |\beta|^2 = 1$) by application of a resonant RF field for a specified duration. Although the ensemble is macroscopic, in this example each spin is independent of the others and behaves as an individual quantum system.

But as early as 1935, Erwin Schrödinger (Schrödinger 1935) realized that in principle, quantum mechanics should apply to a macroscopic system in a more complex way, which could then lead to bizarre consequences. In his specific example, the system is composed of a single radioactive particle and a cat placed together with a mechanism such that if the particle decays, poison is released, which kills the cat. Quantum mechanically we represent the quantum states of the radioactive particle as undecayed = $|\uparrow\rangle$ or decayed = $|\downarrow\rangle$, and live and dead states of the cat as $|L\rangle$ and $|D\rangle$. If the system is initialized in the state represented by the wave function $|\uparrow\rangle\,|L\rangle$, then after a duration equal to the half life of the particle, quantum mechanics says the system evolves to a superposition state where the cat is alive and dead simultaneously, expressed by the superposition wave function

$$\psi = \frac{1}{\sqrt{2}}\left[|\uparrow\rangle|L\rangle + |\downarrow\rangle|D\rangle\right] \cdot \psi = \frac{1}{\sqrt{2}}\left[|\uparrow\rangle|L\rangle + |\downarrow\rangle|D\rangle\right]. \quad (1)$$

Schrödinger dubbed this an entangled state because the state of the particle is correlated with the state of the cat. That is, upon measurement, if the particle is observed to be undecayed, one can say with certainty that the cat is alive, and vice versa. But before measurement, the particle and cat exist in both states. This extrapolation of quantum mechanics from individual quantum systems to the macroscopic world bothered Schrödinger (and a lot of other people). As one way out of the dilemma, in 1952, Schrödinger (Schrödinger 1952b) wrote:

> "... we never experiment with just one electron or atom or (small) molecule. In thought experiments, we sometimes assume that we do; this invariably entails ridiculous consequences ..."

But of course these days, this argument doesn't hold and we can in fact experiment with individual or small numbers of quantum systems, deterministically preparing superpositions and entangled superpositions. Our control is best when we deal with very small numbers of particles, which enables us to realize many of the "gedanken experiments" that provided the basis for discussions between Schrödinger and the other founders of quantum mechanics. And, we can also make small versions of Schrödinger's cat, which are by no means macroscopic but have the same basic attributes. So far, it appears that our inability to make macroscopic "cats" is due just to technical, not fundamental, limitations. Admittedly, these technical limitations are formidable, but one can be optimistic about increasing the size of these states as technology continues to improve.

This contribution is based on the lecture I gave at the Nobel ceremonies in 2012. It is mostly a story about our group at the National Institute of Standards and Technology (NIST) in Boulder, Colorado, whose combined efforts were responsible for some of the contributions to the field of trapped-ion quantum control. It will be a somewhat personal tour, giving my perspective of the development of the field, while trying to acknowledge some of the important contributions of others. For me, the story started when I was a graduate student.

II. SOME EARLY STEPS TOWARDS QUANTUM CONTROL

From 1965 to 1970, I was a graduate student in Norman Ramsey's group at Harvard. Norman, with his close colleague Dan Kleppner and student Mark Goldenberg, had recently invented and demonstrated the first hydrogen masers (Goldenberg *et al.*, 1960; Kleppner *et al.*, 1962). As part of this program, Norman wanted to make precise measurements of the hyperfine frequencies of all three isotopes of hydrogen, so I chose to work on deuterium. The experiment was relatively straightforward, complicated a bit by the relatively long wavelength ($\lambda = 92$ cm) of deuterium's hyperfine transition relative to that of hydrogen ($\lambda = 21$ cm) (Wineland and Ramsey, 1972). Most importantly, this experiment taught me to pay close attention to, and control as best as possible, all environmental effects that would shift the measured transition frequency from that found for an isolated atom. In addition to enjoying the detective work involved in this, I also became hooked on the aesthetics of long coherence times of superposition states (~ 1 s in the masers), and their importance in atomic clocks. Norman received the 1989 Nobel Prize in physics for his invention of the separated oscillatory fields method in spectroscopy and development of the hydrogen maser (Ramsey, 1990).

During my time as a graduate student, I also read about and was intrigued by the experiments of Hans Dehmelt and his colleagues Norval Fortson, Fouad Major, and Hans Schuessler at the University of Washington. The trapping of ions at high vacuum presented some nice advantages for precision spectroscopy, including the elimination of the first-order Doppler shifts and relatively small collision shifts. The Washington group made high-resolution measurements of the $^3He^+$ hyperfine transition, which has internal structure analogous to hydrogen, by storing the ions in an RF (Paul) trap. One challenge was that detection by optical pumping was (and still is) not feasible because of the short wavelengths required. Therefore, in a heroic set of experiments, state preparation was accomplished through charge exchange with a polarized Cs beam that passed through the ions. Detection was accomplished through a charge-transfer process ($^3He^+ + Cs \rightarrow {}^3He + Cs^+$) that depended on the internal state of $^3He^+$, followed by detection of the depleted $^3He^+$ ion number by observing the ions' induced currents in the trap electrodes (Fortson *et al.* 1966; Schuessler *et al.* 1969).

Although these experiments were what first attracted me to ion trapping, my postdoctoral research with Dehmelt, starting in the fall of 1970, was focused on experiments where collections of electrons were confined in a Penning trap for a precise measurement of the electron's magnetic moment or g-factor. These experiments were started by Dehmelt's graduate student, Fred Walls, who later became my colleague at the National Bureau of Standards. After a while, it became clear that systematic effects in the experiment would be much better controlled if the experiment could be performed on single electrons. Therefore, a first task was to isolate a single trapped electron. This was accomplished by first loading a small number of electrons into the trap and driving their nearly harmonic motion (~ 60 MHz) along the magnetic field direction. This motion could be detected by observing the currents induced in the electrodes (proportional to the number of electrons). By adjusting the strength of the drive to a critical level, occasionally one of the electrons would gain enough energy to strike a trap electrode and be lost. Steps in the induced current level could then be used to determine when one electron was confined in the trap (Wineland *et al.* 1973). Subsequent experiments on single electrons by Robert Van Dyck, Paul Schwinberg and Dehmelt were used to make precision measurements of the electron's g-factor (Dehmelt 1990; Van Dyck *et al.* 1977). For this and the development of the ion-trapping technique, Dehmelt and Wolfgang Paul shared the Nobel Prize in 1989, along with Ramsey.

The modes of motion for a single charged particle in a Penning trap include one circular mode about the trap axis called the magnetron mode. For the electron g-factor experiments, it was desirable to locate the electron as close to the trap axis as possible by reducing the amplitude of this mode. This could be accomplished with a form of "sideband cooling" (Wineland and Dehmelt 1975a, 1976) as demonstrated in (Van Dyck *et al.* 1978). Around this time, I was also stimulated by the papers of Arthur Ashkin (Ashkin 1970a, b) on the possibilities of radiation pressure from lasers affecting the motion of atoms. In analogy with the electron sideband cooling, Dehmelt and I came up with a scheme for cooling trapped-ion motion with laser beams (Wineland and Dehmelt 1975b, see below). The cooling could also be explained in terms of velocity-dependent radiation pressure as in a concurrent proposal by Ted Hänsch and Art Schawlow (Hänsch and Schawlow 1975). We didn't anticipate all of the uses of laser cooling at the time, but it was clear that it would be important for high-resolution spectroscopy of trapped ions. For example, the largest systematic uncertainty in the $^3He^+$ experiment (Schuessler *et al.*, 1969) was the uncertainty in the time dilation shift, which would be reduced with cooling.

In the summer of 1975, I took a position in the Time and Frequency Division of NIST (then NBS, the National Bureau of Standards). My first task was to help make a measurement of the cesium hyperfine frequency, the frequency reference that defines the second. The apparatus, NBS-6, had been built by David Glaze of the Division. It was a traditional atomic beam apparatus but had a relatively long distance between Ramsey zones of 3.75 m. With it, we realized a fractional accuracy of 0.9×10^{-13} (Wineland *et al.*, 1976). At that time, the Division was more service oriented, with very little basic research. Fortunately my group leader, Helmut Hellwig, had a progressive view of the Division's future and was able to obtain NBS support to initiate laser cooling experiments. That support, along with some seed money from the Office of Naval Research (ONR), allowed us to start a project on laser cooling in the fall of 1977. With Robert Drullinger (a local laser expert) and Fred Walls, we chose to use $^{24}Mg^+$ because of its simple electronic structure and Penning traps and because of our prior experience with them. This was a very exciting time, being able to work on a project of our choosing. By the spring of 1978, we had obtained our first cooling results (Wineland *et al.* 1978). In our experiments we observed currents in the trap electrodes induced by the ions' thermal motion and hence had a direct measurement of the ions' temperature. Meanwhile, Peter Toschek's group in Heidelberg (joined by Dehmelt, who was on sabbatical), was working towards the same goal, using Ba+ ions confined in an RF Paul trap. They, with colleagues Werner Neuhauser and Martin Hohenstatt also observed the cooling at about the same time (Neuhauser *et al.* 1978), through the increased trapping lifetime of ions. In a near coincidence, although there was no contact between the groups, the manuscripts were received by Physical Review Letters within one day of each other (Peter Toschek's group "won" by one day!). The cooling observed in both experiments is typically called Doppler cooling, where the oscillation frequency of the ions' motion is less than the line width of cooling transition. Theoretical groups were becoming interested in the cooling, some of the earlier work is discussed in (Kazantsev 1978; Letokhov *et al.* 1977; Stenholm 1986).

To us, the cooling of course provided a start towards improving clocks and in 1985, working with John Bollinger, John Prestage, and Wayne Itano, we demonstrated the first clock that utilized laser cooling (Bollinger *et al.* 1985). But as physicists, we were excited by just the cooling process itself. So, in addition to clock applications, it would eventually lead to reaching and controlling the lowest quantized levels of motion for a trapped particle (below).

III. CONTROLLING THE QUANTUM LEVELS OF INDIVIDUAL TRAPPED IONS

One of the obvious next steps was to isolate single ions. In addition to the aesthetic appeal of this, as for single electrons, the systematic errors in spectroscopy would be smallest in this case (Dehmelt 1982). By observing steps in the ion laser fluorescence, the Heidelberg group was able to isolate Ba^+ single ions (Neuhauser *et al.* 1980). With Wayne Itano, we subsequently used this fluorescence "steps" method to observe single $^{24}Mg^+$ ions (Wineland and Itano 1981). The Heidelberg group also made photographs of a single ion, and because of its relatively long fluorescence wavelength (493 nm), with a magnifier, a single Ba^+ ion can be observed with the human eye!

In NIST single-ion experiments we chose to focus on Hg^+ because for frequency-standard applications, $^{199}Hg^+$ has a relatively high ground-state hyperfine clock transition frequency of 40.5 GHz (Cutler *et al.* 1982; Major and Werth 1973; Prestage *et al.* 1991) and also a narrow $^2S_{1/2} - {}^2D_{5/2}$ optical transition ($\tau(^2D_{5/2})$; 86 ms), which could potentially be used as an optical frequency standard (Bender *et al.*, 1976). Although optical pumping of $^{199}Hg^+$ could be achieved with radiation from isotopically selected Hg^+ fluorescence lamps (Cutler *et al.*1982; Major and Werth 1973; Prestage *et al.* 1991), laser excitation was made difficult because of the short (194 nm) wavelength involved. Jim Bergquist in our group, with colleagues Hamid Hemmati and Wayne Itano, first developed the required source by sum-frequency mixing a doubled Ar^+ laser at 515 nm with 792 nm from a dye laser in a potassium pentaborate crystal (Hemmati *et al.* 1983). We used an RF trap with a simple ring-and-endcap structure shown in Fig.1, similar to that used by the Heidelberg group.

By the mid-1980s ion trappers were able to directly address one of Schrödinger's questions, which formed the title for his publication "Are there quantum jumps?" (Schrödinger 1952a,b). Three similar demonstrations were made in 1986 (Bergquist *et al.* 1986; Blatt and Zoller 1988; Nagourney *et al.* 1986; Sauter *et al.* 1986); for brevity, we describe the experiment of Bergquist *et al.* Referring to Fig. 1, a binding potential called a pseudopotential (Paul 1990) is formed by applying an RF potential between the ring electrode and the endcap electrodes (held in common). The relevant optical energy levels of a Hg^+ ion are indicated in the upper left-hand part of the figure. The $^2S_{1/2} \rightarrow {}^2P_{1/2}$ electric

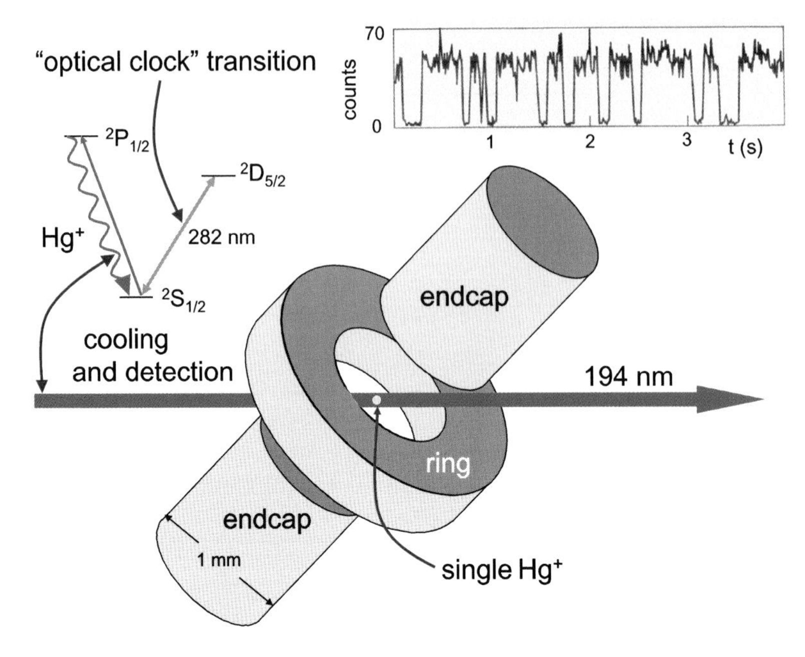

FIGURE 1. Schematic of the trap for single Hg^+ ion studies. An RF potential is applied between the ring electrode and endcap electrodes (which are in common), forming an RF "pseudopotential" for the ion. The relevant Hg^+ energy levels are indicated, including the narrow $^2S_{1/2} \rightarrow {}^2D_{5/2}$ "optical clock" transition. The data in the upper right-hand part of the figure show the number of 194 nm fluorescence photons detected in 10 ms detection bins vs. time when both transitions are excited simultaneously (Bergquist *et al.* 1986). Absence of detected counts indicates that the ion is in the $^2D_{5/2}$ state.

dipole transition ($\lambda = 194$ nm, $\tau(^2P_{1/2})$; 2.9 ns) was used for Doppler laser cooling. If continuously applied, a steady fluorescence from the ion would be observed and could be used to produce images of the ion. If $^2S_{1/2} \rightarrow {}^2D_{5/2}$ resonance radiation were applied simultaneously, one would expect the 194 nm fluorescence to decrease because of excitation to the $^2D_{5/2}$ state.

A density-matrix description, valid for an ensemble of atoms, would predict a reduced but steady fluorescence rate. But what would be observed for a single ion? (Cohen-Tannoudji and Dalibard 1986; Cook *et al.* 1985; Erber and Putterman 1985; Javanainen 1986; Kimble *et al.* 1986; Pegg *et al.* 1986; Schenzle *et al.* 1986). In fact the ion's fluorescence does not steadily decrease, but switches between the full value and no fluorescence, effectively indicating quantum jumps between the $^2S_{1/2}$ and $^2D_{5/2}$ states. For the data shown in the upper right-hand corner of Fig.1, the 194 nm fluorescence photon counts registered by a photomultiplier tube were accumulated in 10 ms time bins and plotted as a function of elapsed time to show the jumps. In a more general context, a measurement of the quantum system composed of the $^2S_{1/2}$ and $^2D_{5/2}$ states can be made by applying the 194 nm "measurement" beam for 10 ms and observing the presence or absence of fluorescence. The $^2S_{1/2} \rightarrow {}^2P_{1/2}$ transition is sometimes called a "cycling transition" because when the $^2S_{1/2}$ state is excited to the $^2P_{1/2}$ state, the ion decays back to the $^2S_{1/2}$ state, emitting a photon, and the excitation/decay process is then repeated. Neglecting the occasional decays of the $^2P_{1/2}$ state to the $^2D_{3/2}$ state (Itano *et al.* 1987), this procedure approximates an ideal measurement in quantum mechanics because the detection of the state is nearly 100% efficient and because the state of the Hg^+ ion, either the $^2S_{1/2}$ or $^2D_{5/2}$ state, remains in its original condition after the measurement. Dehmelt dubbed this "electron shelving" detection (Dehmelt 1982), where in this example, the ion is shelved to the $^2D_{5/2}$ state. Such measurements are also called quantum nondemolition (QND) measurements (Braginsky and Khalili 1996; Haroche and Raimond 2006). The method of detection by state-dependent fluorescence has now become rather ubiquitous in atomic physics.

To perform spectroscopy on the $^2S_{1/2} \rightarrow {}^2D_{5/2}$ transition (λ ; 282 nm), radiation was first applied near the transition frequency in the absence of the 194 nm beam; this avoids perturbations of the energy levels from the 194 nm beam. The 282 nm beam was then switched off, followed by measurement the ion's state with the 194 nm beam. This process was repeated many times, and by stepping the frequency of the 282 nm beam, spectra like that shown in Fig. 2 are obtained (Bergquist *et al.* 1987). To interpret this spectrum, we must consider the motion of the ion. Along any mode axis the motion is nearly harmonic, so in the frame of the ion, the laser beam appears to be sinusoidally frequency modulated due to the first-order Doppler shift. Thus the central feature or "carrier," which corresponds to the transition frequency, is surrounded by frequency-modulation sidebands spaced by the motional frequency of the ion (Dicke 1953). An equivalent picture is that the ion can absorb radiation while simultaneously gaining or losing one quantum of motion, which leads to absorption features spaced by the frequency of motion around the carrier.

As in many atomic physics experiments, by using highly coherent radiation, we can initialize an ion in an eigenstate and deterministically prepare superpositions; e.g., $|\downarrow\rangle \rightarrow \alpha\,|\downarrow\rangle + \beta\,|\uparrow\rangle$. To extract the values of $|\alpha|$ and $|\beta|$, we detect as described above. A single measurement indicates either the $|\downarrow\rangle$ or $|\uparrow\rangle$ state with respective probabilities $P = |\alpha|^2$ and $1 - |\alpha|^2$. Quantum fluctuations or "projection noise" in the measurements are characterized with a variance $\sqrt{P(1-P)/M}$, where M is the number of measurements on identically prepared atoms (Itano *et al.* 1993). Therefore, accurate measurements of P generally requires many repeated experiments. Similarly, Ramsey-type experiments where the two pulses are separated in time can measure the relative phase between α and β. From these types of measurements, many ion trap groups now routinely produce and verify superposition states of single ions that have coherence times exceeding 1 second. For ion ensembles, coherence times exceeding 10 minutes have been demonstrated (Bollinger *et al.* 1991; Fisk *et al.* 1995).

The Hg^+ clock project at NIST, led by Jim Bergquist, has been a long but very successful story. First, an accurate clock based on the 40.5 GHz hyperfine transition of a few $^{199}Hg^+$ ions confined in a linear Paul trap achieved systematic errors of about 4×10^{-14} (Berkeland *et al.* 1998). Although we felt these errors could be substantially reduced, we also realized that the future of high-performance clocks was in the optical domain, so we focused on the $^2S_{1/2} \rightarrow {}^2D_{5/2}$ optical clock transition. For many years it had been appreciated that higher frequency was advantageous in terms of measurement precision; basically the higher oscillation frequencies allows one to divide a time interval into finer units. But two things were needed: a laser with high enough spectral purity to take advantage of narrow optical transitions, and a practical means to count cycles of the "local oscillator," in this case the laser that would excite the clock transition. In our lab, Brent Young, Bergquist and colleagues were able to make a cavity-stabilized laser at 563 nm, which was doubled to produce the clock radiation. The 563 nm source had a line width of less than 0.2 Hz for an averaging time of 20 s (Young *et al.*, 1999). It is now understood that the line width was limited by thermal fluctuations in the mirror surface, currently still the limit for the most stable lasers. The solution to the second problem is by now well known. The relatively rapid development of optical combs by Jan Hall (Hall 2006), Ted Hänsch (Hänsch 2006), their colleagues, and other researchers meant that it was now possible to effectively count optical cycles. Including these developments, in 2006, Bergquist and colleagues demonstrated a $^{199}Hg^+$ optical clock with a systematic uncertainty of 7.2×10^{-17}, the first clock since the inception of atomic clocks that had smaller systematic errors than a cesium clock (Oskay *et al.* 2006).

IV. MANIPULATING ION MOTION AT THE QUANTUM LEVEL

An interesting next step would be to control an ion's motion at the quantum level. Since a cold trapped ion's motion along any mode axis is harmonic to a very good approximation, in a quantum description (Neuhauser *et al.* 1978; Stenholm 1986; Wineland and Itano 1979), we express its Hamiltonian in the usual way as $\hbar\omega_z a^\dagger a$ with ω_z the oscillation frequency (along the z axis here) and a and $a^\dagger$ the lowering and raising operators for the ion motion. The operator for

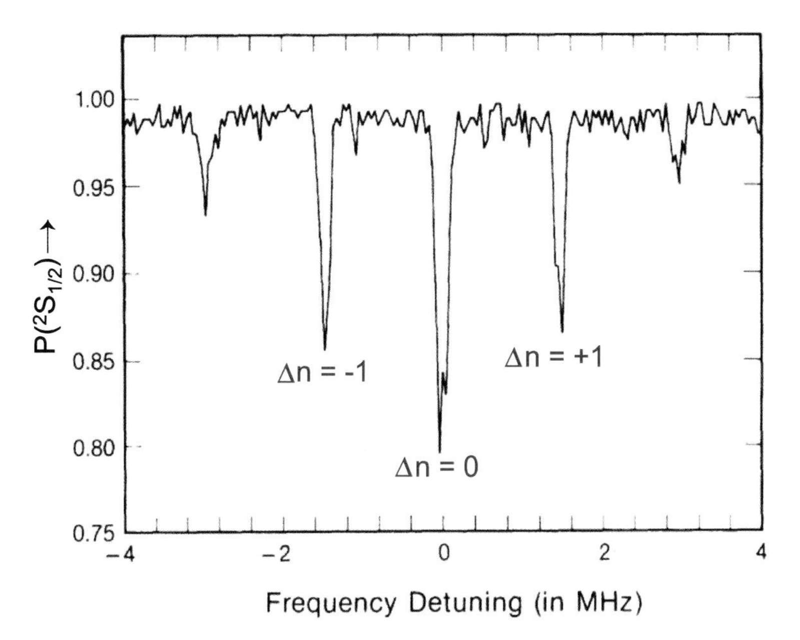

FIGURE 2. Spectroscopy of the $^2S_{1/2} \rightarrow {}^2D_{5/2}$ transition on a single $^{198}Hg^+$ ion. Referring to Fig.1, for each measurement cycle, the ion is prepared in $^2S_{1/2} \equiv |\downarrow\rangle$ state by allowing it to decay to that level. Then, application of a 282 nm "probe" laser beam is alternated with a 194 nm measurement beam. The $|\downarrow\rangle$ and $^2D_{5/2} \equiv |\uparrow\rangle$ states are detected with nearly 100% efficiency by observing presence or absence of 194 nm scattered light. By stepping the frequency of the probe beam and averaging over many measurements, we obtain the spectrum shown where we plot the probability of the ion remaining in the $^2S_{1/2}$ state $P(^2S_{1/2})$ vs. the 282 nm laser beam frequency. In a quantum picture of the motion, central feature or "carrier" denotes transitions of the form $|\downarrow\rangle|n\rangle \rightarrow |\uparrow\rangle|n\rangle$, where n denotes the motional Fock state quantum number. "Red" and "blue" sidebands correspond to $|\downarrow\rangle|n\rangle \rightarrow |\uparrow\rangle|n + \Delta n\rangle$ transitions with $\Delta n = -1$ or $+1$ respectively. The central feature or "carrier" is essentially unshifted by photon recoil, since the recoil is absorbed by the entire trap apparatus as in the Mössbauer effect; see e.g., (Dicke 1953; Lipkin 1973; Wineland *et al.* 1998).

the ion's position about its mean value is $z = z_0(a + a^\dagger)$, where $z_0 = \sqrt{\hbar/2m\omega_z}$ is the spread of the ground-state wave function, with m the ion's mass. In principle, we could detect the ion's motion through the current it induces in the trap electrodes, as was done for electrons. In practice however, a far more sensitive method is to map information about the motional states onto internal states of the ion and read those out as described above. For this, we need to efficiently couple an ion's internal states to its motion. To see how this works, consider a single trapped ion that has a single-electron electric-dipole transition with resonance frequency ω_0. If this transition is excited by a laser beam of frequency ω_L propagating along the z axis, the interaction is given by

$$H_I = -e\vec{r}\cdot\hat{\epsilon}\, E_0 \cos(kz - \omega_L t + \phi) = \hbar\Omega(\sigma_+ + \sigma_-)(e^{i(kz-\omega_L t+\phi)} + e^{-i(kz-\omega_L t+\phi)}), \qquad (2)$$

where $\vec{r}$ is the electron coordinate relative to the ion's core, e is the electron charge, $\hat{\epsilon}$, E_0 and k are respectively the laser beam's electric field polarization, amplitude, and wave vector, and φ is the electric-field phase at the mean position of the ion. The operators σ_+ ($= |\uparrow\rangle\langle\downarrow|$) and σ_- ($= |\downarrow\rangle\langle\uparrow|$) are the internal-state raising and lowering operators, and $\Omega \equiv -E_0\langle\uparrow|\vec{r}\cdot\hat{\epsilon}|\downarrow\rangle/2\hbar$ with $|\downarrow\rangle$ and $|\uparrow\rangle$ denoting the ion's ground and optically excited states as above. If we transform to an interaction picture for the ion's internal states ($\sigma_+ \to \sigma_+ e^{i\omega_0 t}$) and motion states ($a^\dagger \to a^\dagger e^{i\omega_z t}$) and assume $\omega_L \simeq \omega_0$, then neglecting terms that oscillate near 2 ω_0 (rotating wave approximation), Eq. (2) becomes

$$\begin{aligned} H_I &\simeq \hbar\Omega\sigma_+ e^{i(kz-(\omega_L-\omega_0)t+\phi)} + H.C. \\ &\simeq \hbar\Omega\sigma_+ e^{-i((\omega_L-\omega_0)t-\phi)} \\ &\times (1 + i\eta(ae^{-i\omega_z t} + a^\dagger e^{i\omega_z t})) + H.C. \end{aligned} \qquad (3)$$

Here, $H.C.$ stands for Hermitian conjugate and $\eta \equiv kz_0 = 2\pi\, z_0/\lambda$ is the Lamb-Dicke parameter, which we assume here to be much less than 1. For an ion of mass 40 u (e.g., $^{40}Ca^+$) in a well with $\omega_z/2\pi = 3$ MHz and $\lambda = 729$ nm, we have $z_0 =$ 6.5 nm and $\eta = 0.056$. For $\omega_L = \omega_0$ and $\eta\Omega \ll \omega_z$, to a good approximation we can neglect the non-resonant η term in Eq. (3) and obtain $H_I = \hbar\Omega e^{i\varphi} S_+ + H.C.$ This is the Hamiltonian for "carrier" transitions or equivalently, spin-vector rotations about an axis in the $x - y$ plane of the Bloch sphere. If we assume $\omega_L = \omega_0 - \omega_z$ (laser tuned to the "red sideband"), and absorb phase factors in the definition of Ω, the resonant term gives

$$H_I = \hbar\eta(\Omega\sigma_+ a + \Omega^*\sigma_- a^\dagger). \qquad (4)$$

This Hamiltonian describes the situation where a quantum of motion is exchanged with a quantum of excitation of the ion's internal state. It is most commonly known as the Jaynes-Cummings Hamiltonian from cavity QED, which expresses the exchange of energy between the internal states of an atom in a cavity and the photons confined by the cavity (Haroche and Raimond 2006; Jaynes and Cummings 1963). In the cavity-QED experiments of Serge Haroche, Jean-Michel Raimond, Michel Brune, and their colleagues in Paris, the atoms play much the same role as they do in the ion experiments; however, in the cavity-QED experiments, the relevant harmonic oscillator is that which describes a field mode of the cavity, whereas in the ion case, the relevant harmonic oscillator is that associated with the ion's motion (Blockley *et al.* 1992; Sauter *et al.* 1988). Over the years, this connection has led to some interesting and complementary experiments between the two types of experiments (Haroche and Raimond 2006).

In the trapped-ion world, this type of exchange at the quantum level was first used in the electron g-factor experiments of Dehmelt and colleagues, where a change of the electron's cyclotron quantum number was accompanied by spin flip of the electron, which could be detected indirectly (Dehmelt 1990). If we apply H_I to an atomic ion in the state $|\downarrow\rangle|n\rangle$, where n denotes the harmonic oscillator's quantum state (Fock state), we induce the transition $|\downarrow\rangle|n\rangle \rightarrow |\uparrow\rangle|n-1\rangle$. This corresponds to the absorption feature labeled $\Delta n = -1$ in Fig. 2, and reduces the energy of motion by $\hbar\omega_z$. When the ion decays, on average, the motion energy increases by the recoil energy $R = (\hbar k)^2/(2m)$, where $k = 2\pi/\lambda$. Typically, we can achieve the condition $R \ll \hbar\omega_z$, so that in the overall scattering process, the motional energy is reduced. In Fig. 2, the carrier absorption feature is labeled $\Delta n = 0$, indicating photon absorption without changing the motional state. This is a manifestation of the "recoilless" absorption of the Mössbauer effect (see, *e.g.*, (Dicke 1953; Lipkin 1973; Wineland *et al.* 1998)), but in the visible wavelength region.

Continuous application of the red sideband transition provides a relatively straightforward way to laser-cool the ion to near the ground state of motion. After many scattering events, the ion reaches the $|\downarrow\rangle|n=0\rangle$ state, a "dark state" in which scattering stops, since the $|\uparrow\rangle|n=-1\rangle$ state does not exist. The process is not perfect, since scattering in the wings of $\Delta n = 0, +1$ transitions leads to some residual recoil heating, but the condition $\langle n\rangle \ll 1$ can be achieved. This is easily verified because absorption on the $\Delta n = -1$ red sideband nearly disappears, but the $\Delta n = +1$ blue sideband absorption remains. In 1989, with Frank Diedrich, who was a postdoc in our lab, we achieved near-ground-state laser cooling in two dimensions, in essentially the way described here (Diedrich *et al.* 1989). Later in an experiment led by Chris Monroe, we achieved near-ground state cooling in 3-D using two-photon stimulated-Raman transitions (Monroe *et al.* 1995b).

In addition to suppressing Doppler shifts in spectroscopy to the highest degree possible (Wineland *et al.* 1987), one motivation for sideband cooling was the intrinsic appeal of (actively) placing a bound particle in its ground state of motion, the lowest energy possible within the limitations imposed by quantum mechanics. Here, the ground state is a Gaussian-shaped wave packet with spread $\sqrt{\langle z^2 \rangle} = \sqrt{\hbar/2m\omega_z} \equiv z_0$ and energy $\hbar\omega_z/2$. We were also interested in generating non-classical states of motion (Cirac *et al.* 1993a,b, 1996; Heinzen and Wineland 1990) or entangled states of spins (Bollinger *et al.* 1996; Wineland *et al.* 1992). For these experiments, cooling to the ground state of motion provides a clean starting point for motional state manipulation. (In the Paris experiments, the ground state of the cavity mode can be achieved either by thermally cooling to $\langle n \rangle \ll 1$ by operating at low temperature or by extracting photons with atoms sent through the cavity in a process analogous to ion sideband cooling (Haroche and Raimond 2006).

The red-sideband interaction of Eq. (4) and the "blue sideband" interaction ($H_I = \hbar\eta\Omega\sigma_+ a^\dagger + H.C.$, for $\omega_L = \omega_0 + \omega_z$) that induces $|\downarrow\rangle|n\rangle \rightarrow |\uparrow\rangle|n+1\rangle$ transitions, provide simple tools for the manipulation of an ion's motional states. For example, starting from $|\downarrow\rangle|n=0\rangle$, and applying a series of blue sideband, red sideband, and carrier π pulses, Fock states for a selected value of n can be deterministically prepared (Meekhof *et al.* 1996). From $|\downarrow\rangle|n=0\rangle$, we can also make coherent states ion motion by forcing the ion at its motion frequency with an oscillating classical uniform field (Carruthers and Nieto 1965) or by applying an oscillating optical dipole force (Meekhof *et al.* 1996), which results from spatial gradients of laser-beam-induced AC Stark shifts. A coherent state of a quantum particle is very much like an oscillating classical particle but, as opposed to a classical particle that can be point-like, the shape of the quantum particle's wave packet is the same as it is in the ground state. In a clever but straightforward scheme suggested by Chi Kwong Law and Joe Eberly (Law and Eberly 1996) arbitrary motional state superpositions can be prepared (Ben-Kish *et al.* 2003). As a final example, the red-sideband interaction applied for a "π-pulse" duration $t = \pi/(2\eta\Omega)$ provides internal-state to motion-state transfer

$$(\alpha|\downarrow\rangle + \beta|\uparrow\rangle)|0\rangle \rightarrow |\downarrow\rangle(\alpha|0\rangle + \beta|1\rangle). \tag{5}$$

V. SCHRÖDINGER'S CAT

The optical dipole force is interesting because the strength of the force can depend on the ion's internal state. In 1996 (Monroe *et al.* 1996), using state-dependent optical dipole forces, we were able to produce an analog to the Schrödinger's cat state in Eq. (1), which had the form

$$\Psi = \frac{1}{\sqrt{2}}\left[|\uparrow\rangle|\alpha\rangle + |\downarrow\rangle|-\alpha\rangle\right], \tag{6}$$

where $|\alpha\rangle$ denotes a coherent state. The amplitude of the particle's oscillatory motion is equal to $2\alpha z_0$. The spatial part of the state in Eq. (6), represents two wave packets that oscillate back and forth but are 180° out of phase with each other and therefore pass through each other at the center of the trap every half cycle of oscillation. Here, the analogy to Schrödinger's cat is that the spin states of the ion are like the states of the single radioactive particle and the coherent states of the ion, which follow more macroscopic classical trajectories, are like the state of the cat; e.g., the ion at its left extremum point = live cat, ion at its right extremum = dead cat. Figure 3 describes how this state was produced.

To analyze the experiment, in step (e) of Fig. 3, we can control the phase of the amplitude such that the coherent state is $e^{i\varphi}\alpha$ rather than $-\alpha$. Near the condition $\Phi = 0$, the probability $P(\downarrow)$ of the ion to be in state $|\downarrow\rangle$ oscillates as a function of Φ due to interference of the two wave packets. This verifies the coherence between the two components of the cat superposition state. These interference oscillations are very analogous to the fringe oscillations observed in Young's-slit-type experiments performed on individual photons, electrons, neutrons, or atoms, but in those experiments, the particle wave packets disperse in

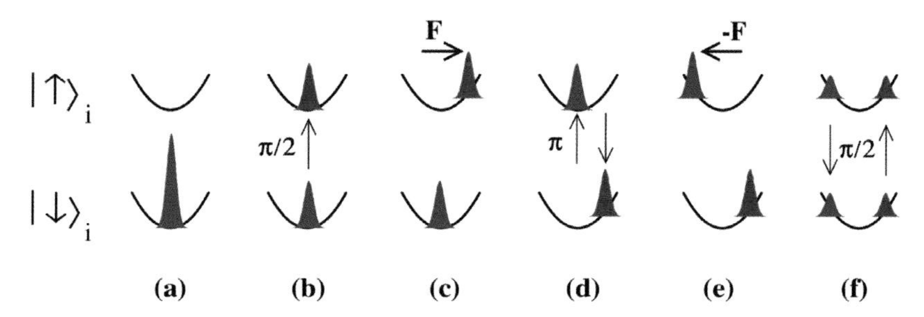

FIGURE 3. Depiction of the harmonic oscillator potential and the wave packets for each component of the ion's internal states, denoted $|\uparrow\rangle$ and $|\downarrow\rangle$. The images are snapshots in time; for images (c) through (f), the wave packets are shown at the extremes of their motion. The areas of the wave packets correspond to the probability of finding the atom in the given internal state. (a) The initial wave packet corresponds to the ground state of motion after laser cooling and preparation of the $|\downarrow\rangle$ internal state. (b) A $\pi/2$ carrier pulse creates the internal state superposition $\frac{1}{\sqrt{2}}(|\downarrow\rangle + |\uparrow\rangle)$. (c) An oscillating optical dipole force is applied that excites the $|\uparrow\rangle$ component of the superposition to a coherent state of amplitude α, creating the state $\frac{1}{\sqrt{2}}(|\downarrow\rangle|n=0\rangle + |\uparrow\rangle|\alpha\rangle)$. (d) The spin states are flipped by applying a carrier π-pulse. (e) The wave packet associated with the $|\uparrow\rangle$ state is excited by the optical dipole force to an amplitude of $-\alpha$, that is, out of phase with respect to the first excitation. This is the state of Eq. (6). (f) To analyze the state produced in step (e) and verify phase coherence between the components of the cat wave function, we apply a final $\pi/2$ carrier pulse and then measure the probability $P(\downarrow)$ of the ion to be in state $|\downarrow\rangle$ (see text).

time, whereas the wave packets in a harmonic oscillator do not, and in principle last arbitrarily long.

In (Monroe *et al.* 1996), for the condition described by Eq. (6), the maximum separation of the wave packets was $4\alpha z_0 = 83$ nm, while the size of the wave packets z_0, was 7.1 nm (see also (McDonnell *et al.* 2007; Poschinger *et al.* 2010)). Of course, one might object to dignifying the state produced by calling it a Schrödinger cat since it is so small. In fact as we tried to make $|\alpha|$ larger, the quality of the superposition became more susceptible to decoherence caused by noisy ambient electric fields (Myatt *et al.* 2000a,b; Turchette *et al.* 2000), limiting the size that was obtained. However, as far as we know, this is just a technical, not fundamental limitation and we should eventually be able to make a cat with $|\alpha|$ large enough that the wave packets are separated by macroscopic distances.

VI. ENTER QUANTUM INFORMATION

Following Peter Shor's development of a quantum mechanical algorithm for efficient number factoring (Shor 1994), there was a dramatic increase of activity in the field of quantum information science. The potential realization of general-purpose quantum information processing (QIP) is now explored in many settings, including atomic, condensed-matter, and optical systems.

At the 1994 International Conference on Atomic Physics held in Boulder, Colorado, Artur Ekert presented a lecture outlining the ideas of quantum computation (Ekert 1995), a subject new to most of the audience. This inspired Ignacio Cirac and Peter Zoller, who attended the conference and were very familiar with the capabilities (and limitations) of trapped-ion experiments, to propose a basic layout for a quantum computer utilizing trapped ions (Cirac and Zoller 1995). This seminal paper was the first comprehensive proposal for how a quantum information processor might be realized. In their scheme, quantum bits or "qubits" are realized with two internal states of the ion, e.g., the $|\downarrow\rangle$ and $|\uparrow\rangle$ states above. The ion qubits are held in a trap shown schematically in Fig. 4. The motion of the ions is strongly coupled by the Coulomb interaction and is best described by the normal modes of a kind of pseudo-molecule. Typically, the motion of each mode is shared among all the ions and can act as a data bus for transferring information between ions. A single-qubit gate or rotation (the relatively easy part) is implemented by applying a focused laser beam or beams onto that ion and coherently driving a carrier transition as described above. The harder part is to perform a logic gate between two selected ions. This can be accomplished by first laser-cooling all modes to the ground state. The internal qubit state of one ion is then transferred onto the qubit formed from the ground and first excited state of a particular mode of motion (laser beam 1 in Fig. 4), as indicated in Eq. (5). Laser beam 2 then performs a logic gate between the (shared) motion qubit state and a second selected ion. Since the second ion is generally in a superposition state, before the gate operation is performed, the wave function for the spin and motional state of the second qubit can be written as $\alpha|\downarrow\rangle|0\rangle + \beta|\downarrow\rangle|1\rangle + \xi|\uparrow\rangle|0\rangle + \zeta|\uparrow\rangle|1\rangle$. One type of logic gate imparts a minus sign to the $|\uparrow\rangle|1\rangle$ component of the wavefunction by coherently driving a 2π transition $|\uparrow\rangle|1\rangle \rightarrow |aux\rangle|0\rangle \rightarrow -|\uparrow\rangle|1\rangle$, where $|aux\rangle$ is a third "auxiliary" internal state of the ion (Cirac and Zoller, 1995). Flipping the sign of the $|\uparrow\rangle|1\rangle$ component of the wave function realizes an entangling two-qubit "π-phase" gate and is universal for computation. Finally, the initial transfer step on the first ion is reversed, restoring the motion to the ground state and effectively having performed the logic gate between the internal qubit states of the two laser-beam-selected ions. At NIST, since we had recently achieved ground-state cooling with stimulated Raman transitions on hyperfine qubit states, we were able to quickly demonstrate a universal gate between a hyperfine qubit and a motional mode qubit. (Monroe *et al.* 1995a). The complete Cirac/Zoller gate between two selected qubits was subsequently demonstrated by the Innsbruck group, led by Rainer Blatt (Schmidt-Kaler *et al.* 2003).

More streamlined gates were subsequently devised in which multiple ions are addressed simultaneously by the same laser beams (Milburn *et al.* 2000; Solano *et al.* 1999; Sørensen and Mølmer 1999, 2000; Wang *et al.* 2001). These gates also have the advantage that it is not necessary to prepare all modes in the ground state; it is only necessary that each ion is maintained well within the Lamb-Dicke regime ($\langle z^2\rangle \ll (\lambda/2\pi)^2$). These "geometric" gates can be viewed as arising from quantum phases that are acquired when a mode of the ions' motion is displaced in phase space around a closed path; the phases accumulated are proportional to the enclosed area in phase space. The different gates can be viewed in a common framework, the main difference being whether or not the forces act on the spin states in the z basis (eigenstates $|\downarrow\rangle$, $|\uparrow\rangle$) or in the x,y basis (eigenstates of the form $\frac{1}{\sqrt{2}}\left(|\downarrow\rangle+e^{i\xi}|\uparrow\rangle\right), \frac{1}{\sqrt{2}}\left(|\downarrow\rangle-e^{i\xi}|\uparrow\rangle\right)$ (Lee *et al.*, 2005). The forces required for the displacements are usually implemented with optical-dipole forces as in the Schrödinger cat example. Since the forces are state-dependent, the differential geometric phases generate entangling gates.

Two-qubit phase gates have been implemented in the z basis (Home *et al.* 2006; Leibfried *et al.* 2003b) and in the x–y basis (Benhelm *et al.* 2008; Haljan *et al.* 2005; Kim *et al.* 2009; Sackett *et al.* 2000). In the Innsbruck experiment of (Benhelm *et al.* 2008) a Bell state with fidelity 0.993(1) was produced, setting a

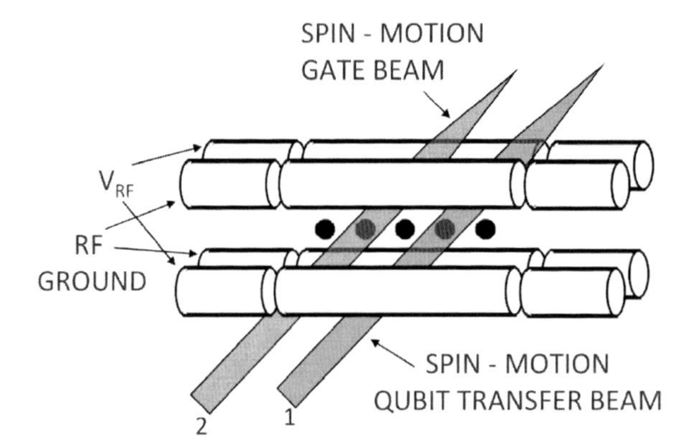

FIGURE 4. Scheme for quantum computation proposed by Cirac and Zoller (Cirac and Zoller, 1995). Quadrupolar electrodes are configured to produce a linear array of trapped ion qubits (filled black circles). Two diagonally opposite rods support an RF potential to realize a ponderomotive pseudopotential transverse to the trap's (horizontal) axis. Static potentials applied to the end segments of the electrodes confine ions along the axis. Ideally, all motional modes are laser-cooled to the ground state before logic operations. The quantized modes of motion can be used as a data bus to share information between the internal state qubits of ions that are selected by focused laser beams (see text).

standard for all QIP experiments. The use of single and multi-qubit gates has enabled the demonstration of several ion-based QIP algorithms; see for example (Blatt and Roos 2012; Blatt and Wineland 2008). At NIST most such demonstrations were led by Didi Leibfried. Chris Monroe's group at the University of Maryland is leading efforts on an entirely different scheme for ion entanglement generation based on performing joint measurements on photons that are first entangled with ion qubits (Moehring *et al.* 2007; Monroe *et al.* 2012; Olmschenk *et al.* 2010). This scheme has the advantage that the ions don't have to be in the Lamb-Dicke regime, and it also enables entanglement of widely separated qubits because of the relative ease of transferring photons over large distances.

The basic elements of the Cirac/Zoller proposal are carried forward in the different variations of trapped-ion QIP. This proposal rejuvenated the field of trapped ions and today there are over 30 groups in the world working on various aspects of quantum information processing. These include groups at the University of Aarhus; Amherst College; University of California, Berkeley; University of California, Los Angles; Duke University; ETH Zürich; University of Freiburg; Georgia Tech; Griffiths University; Imperial College; University of Innsbruck; Lincoln Laboratories; Mainz University; University of Hannover and PTB (Germany); MIT; NIST (USA); NPL (UK); Osaka University; Oxford University; Joint Quantum

Institute at the University of Maryland; Université de Paris; Saarland University (Saarbrücken); Sandia National Laboratory (USA); Siegen University; Simon Fraser University; National University of Singapore; Sussex University; University of Sydney; Tsinghua University; University of Ulm; University of Washington; Wabash College; and the Weizmann Institute.

A. Quantum simulation

In the early 1980s, Richard Feynman proposed that one quantum system might be used to efficiently simulate the dynamics of other quantum systems of interest (Feynman 1982; Lloyd 1996). This is now a highly anticipated application of QIP, and will likely occur well before useful factorization is performed. Of course, the universality of a large-scale quantum computer will allow it to simulate any quantum system of interest. However, it is also possible to use the built-in available interactions in a quantum processor to simulate certain classes of physical problems. For trapped ions, it has been possible to use the interactions employed in the various gates to simulate other systems of interest, for example nonlinear optical systems (Leibfried *et al.* 2002), motional quantum dynamics as in an electron's Zitterbewegung or trembling motion (Gerritsma *et al.* 2010) or the properties of a "quantum walk" (Schmitz *et al.* 2009b; Zähringer *et al.* 2010). Currently, efforts are underway in several laboratories to use QIP interactions to simulate various dynamics including those of condensed matter systems. Some of the basic ideas for how this might work with ions have been outlined in (Blatt and Roos 2012; Britton *et al.* 2012; Chiaverini and Lybarger 2008; Clark *et al.* 2009; Deng *et al.* 2005; Johanning *et al.* 2009; Korenblit *et al.* 2012; Pons *et al.* 2007; Porras and Cirac 2004, 2006; Schätz *et al.* 2007; Schmied *et al.* 2011; Schmitz *et al.* 2009a; Schneider *et al.* 2012; Taylor and Calarco 2008; Wunderlich and Balzer 2003). Here, logic gate interactions between ions *i* and *j* invoke a spin-spin like interaction of the form $\sigma_{\hat{u}i}\sigma_{\hat{u}j}$, where $\hat{u} \in \{\hat{x}, \hat{y}, \hat{z}\}$. Spin rotations about a direction û act like magnetic fields along $\hat{u}$. These basic interactions have been implemented on up to 16 ions in an RF trap (Edwards *et al.* 2010; Friedenauer *et al.* 2008; Islam *et al.* 2012; Kim *et al.* 2009, 2010; Korenblit *et al.* 2012; Schätz *et al.* 2007). One interesting aspect of this work is the study of quantum phase transitions by varying the relative strengths of the (simulated) spin-spin and magnetic field interactions. Under appropriate conditions, the effects of spin "frustration" are now becoming apparent. The basic interactions have also been implemented on over 100 spins in a Penning trap experiment led by John Bollinger at NIST (Britton *et al.* 2012), where the ions naturally form into a triangular array. In the Innsbruck group, simulations including engineered dissipation have also been implemented (Barreiro *et al.* 2011; Blatt and Roos 2012), and a striking demonstration of a digital quantum simulator has been made (Blatt and Roos 2012; Lanyon *et al.* 2011), in essence the first universal quantum computer.

B. Spectroscopy and quantum metrology

Some potential applications of quantum control and QIP are motivated by the idea of using entangled states to improve spectroscopic sensitivity (Bollinger *et al.* 1996; Goldstein *et al.* 2009; Leibfried *et al.* 2004; Roos *et al.* 2006; Wineland *et al.* 1994, 1992) and demonstrations of this increased sensitivity have been made (Leibfried *et al.* 2004, 2005; Leroux *et al.* 2010; Meyer *et al.* 2001; Monz *et al.* 2011; Roos *et al.* 2006). These demonstrations were made in the limit that noise was dominated by "projection noise," the fundamental noise arising from the fluctuations in which state the system is projected into upon measurement (Itano *et al.* 1993; Wineland *et al.* 1982). This might be the case in a spectroscopy experiment where the interrogation time is limited by a particular experimental constraint, like the duration of flight of atoms in a cesium fountain clock or by the desire to hold the temperature of ions below a certain value if they are heated during interrogation. However, if significant phase noise is present in either the atoms themselves (Huelga *et al.* 1997), or the interrogating radiation (André *et al.* 2004; Buzek *et al.* 1999; Rosenband 2012; Wineland *et al.* 1998), the gain from entanglement can be lost. This puts a premium on finding probe oscillators that are stable enough that the projection noise dominates for the desired probe duration.

Some ions of spectroscopic interest may be difficult to detect because they either don't have a cycling transition, or lack a cycling transition at a convenient wavelength. In some cases, this limitation can be overcome by simultaneously storing the ion(s) of spectroscopic interest with a "logic" ion or ions whose states can be more easily detected. Following the Cirac and Zoller scheme, we can use the internal-to-motion-state-transfer process described above. Here, the idea is to first transfer the two states of interest in the spectroscopy ion to the ground and first excited states of a mode of the ions' coupled motion. This is then followed by mapping the motional states onto the logic ion, which is subsequently measured (Wineland *et al.* 2002). In a project led by Till Rosenband at NIST, this technique has been used to detect optical transitions in $^{27}Al^+$ ions by transferring the relevant $^{27}Al^+$ states to a $^{9}Be^+$ or $^{25}Mg^+$ logic ion, which is then measured (Schmidt *et al.* 2005). It is now used routinely in an accurate optical clock based on $^{27}Al^+$ (Chou *et al.* 2010a; Rosenband *et al.* 2008) and might also be extended to molecular ions. Currently, the $^{27}Al^+$ single ion optical clock has the smallest systematic error of any clock at somewhat below 1 part in 10^{17} (Chou *et al.* 2010a). This level of precision has enabled observations of the predictions of Einstein's general theory of relativity on a human scale, such as time dilation for bicycling speeds and the gravitational red shift for height changes of around 30 cm (Chou *et al.* 2010b). Such clocks may become useful tools in geodesy.

The information transfer and readout process employed in the $^{27}Al^+/^9Be^+$ clock experiments typically had a fidelity of about 0.85, limited by errors caused by the ions' thermal motion in modes not used for information transfer (so-called "Debye-Waller" factors from Mössbauer spectroscopy (Lipkin 1973; Wineland *et al.* 1998)). However, the quantum logic detection process is a QND-type of measurement in that it doesn't disturb the detected populations of the $^{27}Al^+$ ion. It can therefore be repeated to gain better information on the $^{27}Al^+$ ion's (projected) state. By use of real-time Bayesian analysis on successive detection cycles, the readout fidelity was improved from 0.85 to 0.9994 (Hume *et al.* 2007). This experiment shares similarities with those of the Paris cavity-QED group, where successive probe atoms are used to perform QND measurements of the photon number in a cavity (Deléglise *et al.* 2008). In Hume *et al.* 2007, the same atom ($^9Be^+$) is reset after each detection cycle and used again. Also, because the detection was accomplished in real time, the procedure was adaptive, requiring on each run a minimum number of detection cycles to reach a certain measurement fidelity.

VII. SUMMARY

I have tried to give a brief account of some of the developments that have taken place in the area of quantum state manipulation of small numbers of trapped atomic ions. With apologies, I have omitted several aspects of this subject and for the topics discussed here, I primarily used examples from the NIST, Boulder group. Much of the other work has been discussed in various comprehensive articles and reviews; see for example (Blatt and Roos 2012; Blatt and Wineland 2008; Cirac *et al.* 1996; Duan and Monroe, 2008, 2010; Häffner *et al.*, 2008; Kielpinski, 2008; Korenblit *et al.*, 2012; Lee *et al.*, 2005; Leibfried *et al.* 2003a; Monroe and Lukin 2008; Monroe *et al.* 2012; Schneider *et al.* 2012; Sasura and Buzek 2002; Wineland *et al.* 1998). Reviews on advanced clocks including those based on ions are contained in Gill 2005, 2011; Maleki 2008; and Margolis 2009. See also Madej *et al.* 2012 and references therein.

VIII. ACKNOWLEDGEMENTS

Certainly my role in this work is very small when compared to that of my colleagues both at NIST and around the world, who have made so many important contributions. Having been recognized by the Nobel Foundation is really more recognition of our field rather than individual accomplishment; many others are at least as deserving. Just the work of the NIST group was due to the combined efforts of a very large number of people. I have been lucky to work with NIST permanent staff members Jim Bergquist, John Bollinger, Bob Drullinger, and Wayne Itano for my entire career, and we have been fortunate to be joined by Didi Leibfried and Till Rosenband in the last decade. Chris Monroe was a very important part of our group from 1992 to 2000 and now has his own group at the University of Maryland. Of course our successes would not have happened if not for the dedication of many students, postdocs, and visiting scientists to our group, numbering over 100 people. Having a group working directly together or on related problems has been a source of strength for us, and the congenial atmosphere over the years has made our efforts so enjoyable.

Throughout my career, our group has enjoyed the support and encouragement of NBS/NIST management. My direct supervisors over the years, Helmut Hellwig, Sam Stein, Don Sullivan, and Tom O'Brian have always supported our goals and desires as much as possible. More recently, we have also enjoyed the support of Carl Williams, who heads NIST's quantum information program. We are all indebted to our laboratory director, Katharine Gebbie, for her support and encouragement. Perhaps one measure of her success is that I am the fourth person, after Bill Phillips, Eric Cornell, and Jan Hall, to receive a Nobel Prize during her tenure as lab director. We are also grateful for the support of agencies outside of NIST, such as AFOSR, ARO, DARPA, ONR, and various intelligence agencies who have supported our work on quantum information. I have great respect for the leaders of some of our group's strongest competition such as Rainer Blatt (Innsbruck) and Chris Monroe (University of Maryland) and have enjoyed their friendship for many years. It was also a great pleasure to share this recognition with Serge Haroche. I have known Serge for about 25 years and have enjoyed both his group's elegant science and also the mutual friendship that my wife and I have shared with him and his wife, Claudine. Most importantly, I have been very fortunate to have the support, understanding, and patience of my wife Sedna and sons Charles and Michael.

I thank John Bollinger, Wayne Itano, Didi Leibfried, and Till Rosenband for helpful suggestions on the manuscript. This is a contribution of NIST and is not subject to U.S. copyright.

REFERENCES

1. André, A., A. Sørensen, and M. D. Lukin (2004), *Phys. Rev. Lett.* **92** (23), 230801.
2. Ashkin, A. (1970a), *Phys. Rev. Lett.* **24** (4), 156.
3. Ashkin, A. (1970b), *Phys. Rev. Lett.* **25** (19), 1321.
4. Barreiro, J. T., M. Müller, P. Schindler, D. Nigg, T. Monz, M. Chwalla, M. Hennrich, C. F. Roos, E. Solano, and R. Blatt (2011), *Nature* **470**, 486.
5. Ben-Kish, A., B. DeMarco, V. Meyer, M. Rowe, J. Britton, W. M. Itano, B. Jelenkovic, C. Langer, D. Leibfried, T. Rosenband, and D. J. Wineland (2003), *Phys. Rev. Lett.* **90**, 037902.
6. Bender, P. L., J. L. Hall, R. H. Garstang, F. M. J. Pichanick, W. W. Smith, R. L. Barger, and J. B. West (1976), *Bull. Am. Phys. Soc.* **21**, 599.
7. Benhelm, J., G. Kirchmair, C. F. Roos, and R. Blatt (2008), *Nature Physics* **4**, 463.
8. Bergquist, J. C., R. G. Hulet, W. M. Itano, and D. J. Wineland (1986), *Phys. Rev. Lett.* **57** (14), 1699.
9. Bergquist, J. C., W. M. Itano, and D. J. Wineland (1987), *Phys. Rev. Lett.* **36**, 428.
10. Berkeland, D. J., J. D. Miller, J. C. Bergquist, W. M. Itano, and D. J. Wineland (1998), *Phys. Rev. Lett.* **80** (10), 2089.
11. Blatt, R. and C. F. Roos (2012), *Nature Phys.* **8**, 277.
12. Blatt, R. and D. J. Wineland (2008), *Nature* **453**, 1008.
13. Blatt, R. and P. Zoller (1988), *Eur. J. Phys.* **9** (14), 250.
14. Blockley, C. A., D. F. Walls, and H. Risken (1992), *Eur. Phys. Lett.* **17** (6), 509.
15. Bollinger, J. J., D. J. Heinzen, W. M. Itano, S. L. Gilbert, and D. J. Wineland (1991), *IEEE Trans. Instr. Meas.* **40** (2), 126.
16. Bollinger, J. J., W. M. Itano, D. J. Wineland, and D. J. Heinzen (1996), *Phys. Rev.* A **54** (6), R4649.
17. Bollinger, J. J., J. D. Prestage, W. M. Itano, and D. J. Wineland (1985), *Phys. Rev. Lett.* **54**, 1000.
18. Braginsky, V. B. and F. Y. Khalili (1996), *Rev. Mod. Phys.* **68**, 1.
19. Britton, J. W., B. C. Sawyer, A. C. Keith, C. C. J. Wang, J. K. Freericks, H. Uys, M. J. Biercuk, and J. J. Bollinger (2012), *Nature* **484**, 489.
20. Buzek, V., R. Derka, and S. Massar (1999), *Phys. Rev. Lett.* **82**, 2207.
21. Carruthers, P. and M. M. Nieto (1965), *Am. J. Phys.* **7**, 537.
22. Chiaverini, J. and W. E. Lybarger (2008), *Phys. Rev.* A **77**, 022324.
23. Chou, C. W., D. B. Hume, J. C. J. Koelemeij, D. J. Wineland, and T. Rosenband (2010a), *Phys. Rev. Lett.* **104**, 070802.
24. Chou, C. W., D. B. Hume, T. Rosenband, and D. J. Wineland (2010b), *Science* **329**, 1630.
25. Cirac, J. I., R. Blatt, A. S. Parkins, and P. Zoller (1993a), *Phys. Rev. Lett.* **70** (6), 762.
26. Cirac, J. I., A. S. Parkins, R. Blatt, and P. Zoller (1993b), *Phys. Rev. Lett.* **70**, 556.
27. Cirac, J. I., A. S. Parkins, R. Blatt, and P. Zoller (1996), *Adv. Atom. Mol. Phys.* **37**, 238.
28. Cirac, J. I. and P. Zoller (1995), *Phys. Rev. Lett.* **74** (20), 4091.
29. Clark, R. J., T. Lin, K. R. Brown, and I. L. Chuang (2009), *J. Appl. Phys.* **105**, 013114.
30. Cohen-Tannoudji, C. and J. Dalibard (1986), *Europhys. Lett.* **1**, 441.
31. Cook, R. J., D. G. Shankland, and A. L. Wells (1985), *Phys. Rev.* A **31**, 564.

32. Cutler, L. S., R. P. Giffard, and M. D. McGuire (1982), in *Proceedings of the 13th Annual PTTI Applications and Planning Meeting*, Vol. No. 2220 (NASA, Washington, DC) p. 563.
33. Dehmelt, H. G. (1982), *IEEE Trans. Instrum. Meas.* **IM-31** (2), 83.
34. Dehmelt, H. G. (1990), *Rev. Mod. Phys.* **62** (3), 525.
35. Deléglise, S., I. Dotsenko, C. Sayrin, J. Bernu, M. Brune, J. M. Raimond, and S. Haroche (2008), *Nature* **455**, 510.
36. Deng, S. L., C. Porras, and J. I. Cirac (2005), *Phys. Rev.* A **72**, 063407.
37. Dicke, R. H. (1953), *Phys. Rev.* **89**, 472.
38. Diedrich, F., J. C. Bergquist, W. M. Itano, and D. J. Wineland (1989), *Phys. Rev. Lett.* 62, 403.
39. Duan, L. M. and C. Monroe (2008), *Adv. Atom. Mol. Phys.* **55**, 419.
40. Duan, L. M. and C. Monroe (2010), *Rev. Mod. Phys.* **82**, 1209.
41. Edwards, E. E., S. Korenblit, K. Kim, R. Islam, M. S. Chang, J. K. Freericks, G. D. Lin, L. M. Duan, and C. Monroe (2010), *arXiv:1005.4160.*
42. Ekert, A. (1995), in *Atomic Physics* 14, edited by D. J. Wineland, C. E. Wieman, and S. J. Smith (American Institute of Physics, New York) pp. 450–466.
43. Erber, T., and S. Putterman (1985), *Nature* **318**, 41.
44. Feynman, R. P. (1982), *Int. J. Th. Phys.* **21** (6/7), 467.
45. Fisk, P. T. H., M. J. Sellars, M. A. Lawn, C. Coles, A. G. Mann, and D. G. Blair (1995), *IEEE Trans. Instrum. Meas.* **44**, 113.
46. Fortson, N., F. Major, and H. Dehmelt (1966), *Phys. Rev. Lett.* **16**, 221.
47. Friedenauer, A., H. Schmitz, J. T. Glueckert, D. Porras, and T. Schätz (2008), *Nature Physics* **4**, 757.
48. Gerritsma, R., G. Kirchmair, F. Zähringer, E. Solano, R. Blatt, and C. F. Roos (2010), *Nature* **463**, 68.
49. Gill, P. (2005), *Metrologia* **42**, S125.
50. Gill, P. (2011), *Phil.Trans. R. Soc.* A **369**, 4109.
51. Goldenberg, H. M., D. Kleppner, and N. F. Ramsey (1960), *Phys. Rev. Lett.* **5**, 361.
52. Goldstein, G., P. Cappellaro, J. R. Maze, J. S. Hodges, L. Jiang, A. S. Sørensen, and M. D. Lukin (2009), *arXiv:1001.0089.*
53. Häffner, H., C. F. Roos, and R. Blatt (2008), *Physics Reports* **469**, 155.
54. Haljan, P. C., P. J. Lee, K. A. Brickman, M. Acton, L. Deslauriers, and C. Monroe (2005), *Phys. Rev.* A **72**, 062316.
55. Hall, J. L. (2006), *Rev. Mod. Phys.* **78**, 1279.
56. Hänsch, T. W. (2006), *Rev. Mod. Phys.* **78**, 1297.
57. Hänsch, T. W. and A. L. Schawlow (1975), *Opt. Commun.***13**, 68.
58. Haroche, S. and J. M. Raimond (2006), *Exploring the quantum* (Oxford University Press, Oxford, U.K.).
59. Heinzen, D. J. and D. J. Wineland (1990), *Phys. Rev.* A **42** (5), 2977.
60. Hemmati, H., J. C. Bergquist, and W. M. Itano (1983), *Opt. Lett.* **8** (2), 73.
61. Horne, J. P., M. J. McDonnell, D. M. Lucas, G. Imreh, B. C. Keitch, D. J. Szwer, N. R. Thomas, S. C. Webster, D. N. Stacey, and A. M. Steane (2006), *New J. Phys.* **8**, 188.
62. Huelga, S. F., C. Macchiavello, T. Pellizzari, A. K. Ekert, M. B. Plenio, and J. I. Cirac (1997), *Phys. Rev. Lett.* **79**, 3865.

63. Hume, D. B., T. Rosenband, and D. J. Wineland (2007), *Phys. Rev. Lett.* **99**, 120502.
64. Islam, R., C. Senko, W. C. Campbell, S. Kornblit, J. Smith, A. Lee, E. E. Edwards, C. D. Wang, J. k Freericks, and C. Monroe (2012), *arXiv:1210.0142.*
65. Itano, W. M., J. C. Bergquist, J. J. Bollinger, J. M. Gilligan, D. J. Heinzen, F. L. Moore, M. G. Raizen, and D. J. Wineland (1993), *Phys. Rev.* A **47** (5), 3554.
66. Itano, W. M., J. C. Bergquist, R. G. Hulet, and D. J. Wineland (1987), *Phys. Rev. Lett.* **59** (24), 2732.
67. Javanainen, J. (1986), *Phys. Rev.* A **33**, 2121.
68. Jaynes, E. T. and F. W. Cummings (1963), *Proceedings of the IEEE* **51**, 89.
69. Johanning, M., A. F. Varon, and C. Wunderlich (2009), *J. Phys. B: At. Mol. Opt. Phys.* 42, 154009.
70. Kazantsev, A. P. (1978), *Sov. Phys. Usp.* **21**, 58.
71. Kielpinski, D. (2008), *Front. Phys. China* **3** (4), 365.
72. Kim, K., M. S. Chang, R. Islam, S. Korenblit, L. M. Duan, and C. Monroe (2009), *Phys. Rev. Lett.* **103**, 120502.
73. Kim, K., M. S. Chang, S. Korenblit, R. Islam, E. E. Edwards, J. K. Freericks, G. D. Lin, L. M. Duan, and C. Monroe (2010), *Nature* **465**, 590.
74. Kimble, H. J., R. J. Cook, and A. L. Wells (1986), *Phys. Rev.* A **34**, 3190.
75. Kleppner, D., H. M. Goldenberg, and N. F. Ramsey (1962), *Phys. Rev.* **126**, 603.
76. Korenblit, S., D. Kafri, W. C. Campbell, R. Islam, E. E. Edwards, Z. X. Gong, G. D. Lin, L. M. Duan, J. Kim, and C. Monroe (2012), *New. J. Phys.* **14**, 095024.
77. Lanyon, B. P., C. Hempel, D. Nigg, M. Müller, R. Gerritsman, F. Aähringer, P. Schindler, J. T. Barreiro, M. Rambach, G. Krichmair, M. Hennrich, P. Zoller, R. Blatt, and C. F. Roos (2011), *Science* **334**, 57.
78. Law, C. K., and J. H. Eberly (1996), *Phys. Rev. Lett.* **76** (7), 1055.
79. Lee, P. J., K. A. Brickman, L. Deslauriers, P. C. Haljan, L. M. Duan, and C. Monroe (2005), *J. Opt. B: Quantum Semiclass Opt.* **7**, S371.
80. Leibfried, D., M. D. Barrett, T. Schätz, J. Britton, J. Chiaverini, W. M. Itano, J. D. Jost, C. Langer, and D. J. Wineland (2004), *Science* **304**, 1476.
81. Leibfried, D., R. Blatt, C. Monroe, and D. Wineland (2003a), *Rev. Mod. Phys.* **75**, 281.
82. Leibfried, D., B. DeMarco, V. Meyer, D. Lucas, M. Barrett, J. Britton, W. M. Itano, B. Jelenkovi´c, C. Langer, T. Rosenband, and D. J. Wineland (2003b), *Nature* **422**, 412.
83. Leibfried, D., B. DeMarco, V. Meyer, M. Rowe, A. BenKish, J. Britton, W. M. Itano, B. Jelenkovi´c, C. Langer, T. Rosenband, and D. J. Wineland (2002), *Phys. Rev. Lett.* **89**, 247901.
84. Leibfried, D., E. Knill, S. Seidelin, J. Britton, R. B. Blakestad, J. Chiaverini, D. Hume, W. M. Itano, J. D. Jost, C. Langer, R. Ozeri, R. Reichle, and D. J. Wineland (2005), *Nature* **438**, 639.
85. Leroux, I. D., M. H. Schleier-Smith, and V. Vuletic (2010), *Phys. Rev. Lett.* **104**, 250801.
86. Letokhov, V., V. G. Minogin, and B. D. Pavlik (1977), *Sov. Phys. JETP* **45**, 698.
87. Lipkin, H. J. (1973), *Quantum Mechanics* (North-Holland, New York).
88. Lloyd, S. (1996), *Science* **273** (4), 1073.
89. Madej, A. A., P. Dubé, Z. Zhou, J. E. Bernard, and M. Gertsvolf (2012), *Phys. Rev. Lett.* **109**, 203002.
90. Major, F. G. and G. Werth (1973), *Phys. Rev. Lett.* **30** (23), 1155.

91. Maleki, L., Ed. (2008), *Frequency Standards and Metrology, Proceedings of the 7th Symposium* (World Scientific, Singapore).
92. Margolis, H. S. (2009), *J. Phys. B: At. Mol. Opt. Phys.* **42**, 154017.
93. McDonnell, M. J., J. P. Home, D. M. Lucas, G. Imreh, B. C. Keitch, D. J. Szwer, N. R. Thomas, S. C. Webster, D. N. Stacey, and A. M. Steane (2007), *Phys. Rev. Lett.* **98**, 063603.
94. Meekhof, D. M., C. Monroe, W. M. Itano, B. E. King, and D. J. Wineland (1996), *Phys. Rev. Lett.* **76** (11), 1796.
95. Meyer, V., M. A. Rowe, D. Kielpinski, C. A. Sackett, W. M. Itano, C. Monroe, and D. J. Wineland (2001), *Phys. Rev. Lett.* **86**, 5870.
96. Milburn, G. J., S. Schneider, and D. F. V. James (2000), *Fortschr. Physik* **48**, 801.
97. Moehring, D. L., P. Maunz, S. Olmschenk, K. C. Younge, D. N. Matsukevich, L. M. Duan, and C. Monroe (2007), *Nature* **449**, 68.
98. Monroe, C., and M. Lukin (2008), *Physics World* (August), 32.
99. Monroe, C., D. M. Meekhof, B. E. King, W. M. Itano, and D. J. Wineland (1995a), *Phys. Rev. Lett.* **75** (25), 4714.
100. Monroe, C., D. M. Meekhof, B. E. King, S. R. Jefferts, W. M. Itano, D. J. Wineland, and P. Gould (1995b), *Phys. Rev. Lett.* **75** (22).
101. Monroe, C., D. M. Meekhof, B. E. King, and D. J. Wineland (1996), *Science* **272**, 1131.
102. Monroe, C., R. Raussendorf, A. Ruthven, K. R. Brown, P. Maunz, L. M. Duan, and J. Kim (2012), *arXiv:1208.0391*.
103. Monz, T., P. Schindler, J. T. Barreiro, M. Chwalla, D. Nigg, W. A. Coish, M. Harlander, W. Hänsel, M. Hennrich, and R. Blatt (2011), *Phys. Rev. Lett.* **106**, 130506.
104. Myatt, C. J., B. E. King, Q. A. Turchette, C. A. Sackett, D. Kielpinski, W. M. Itano, C. Monroe, and D. J. Wineland (2000a), *J. Mod. Opt.* **47**, 2182.
105. Myatt, C. J., B. E. King, Q. A. Turchette, C. A. Sackett, D. Kielpinski, W. M. Itano, C. Monroe, and D. J. Wineland (2000b), *Nature* **403**.
106. Nagourney, W., J. Sandberg, and H. G. Dehmelt (1986), *Phys. Rev. Lett.* **56** (26), 2797.
107. Neuhauser, W., M. Hohenstatt, P. Toschek, and H. Dehmelt (1978), *Phys. Rev. Lett.* **41** (4), 233.
108. Neuhauser, W., M. Hohenstatt, P. Toschek, and H. Dehmelt (1980), *Phys. Rev.* A **22** (3), 1137.
109. Olmschenk, S., D. Hayes, D. N. Matsukevich, P. Maunz, D. L. Moehring, and C. Monroe (2010), *Int. J. Quant. Information* **8**, 337.
110. Oskay, W. H., S. A. Diddams, E. A. Donley, T. M. Fortier, T. P. Heavner, L. Hollberg, W. M. Itano, S. R. Jefferts, M. J. Delaney, K. Kim, F. Levi, T. E. Parker, and J. C. Berqquist (2006), *Phys. Rev. Lett.* **97**, 020801.
111. Paul, W. (1990), *Rev. Mod. Phys.* **62**, 531.
112. Pegg, D. T., R. Loudon, and P. L. Knight (1986), *Phys. Rev.* A **33**, 4085.
113. Pons, M., V. Ahufinger, C. Wunderlich, A. Sanpera, S. Braungardt, A. Sen(De), U. Sen, and M. Lewenstein (2007), *Phys. Rev. Lett.* **98**, 023003.
114. Porras, D., and J. I. Cirac (2004), *Phys. Rev. Lett.* **92**, 207901.
115. Porras, D., and J. I. Cirac (2006), *Phys. Rev. Lett.* **96**, 250501.
116. Poschinger, U., A. Walther, K. Singer, and F. Schmidt-Kaler (2010), *Phys. Rev. Lett.* **105**, 263602.
117. Prestage, J. D., G. J. Dick, and L. Maleki (1991), *IEEE Trans. Instrum. Meas.* **40**, 132.

118. Ramsey, N. F. (1990), *Rev. Mod. Phys.* **62**, 541.
119. Roos, C. F., M. Chwalla, K. Kim, M. Riebe, and R. Blatt (2006), *Nature* **443**, 316.
120. Rosenband, T. (2012), *arXiv:1203.0288.*
121. Rosenband, T., D. B. Hume, P. O. Schmidt, C. W. Chou, A. Brusch, L. Lorini, W. H. Oskay, R. E. Drullinger, T. M. Fortier, J. E. Stalnaker, S. A. Diddams, W. C. Swann, N. R. Newbury, W. M. Itano, D. J. Wineland, and J. C. Bergquist (2008), *Science* **319**, 1808.
122. Sackett, C. A., D. Kielpinski, B. E. King, C. Langer, V. Meyer, C. J. Myatt, M. Rowe, Q. A. Turchette, W. M. Itano, D. J. Wineland, and C. Monroe (2000), *Nature* 404.
123. Sauter, T., H. Gilhaus, I. Siemers, R. Blatt, W. Neuhauser, and P. E. Toschek (1988), *Z. Phys.* D **10**, 153.
124. Sauter, T., W. Neuhauser, R. Blatt, and P. E. Toschek (1986), *Phys. Rev. Lett.* **57** (14), 1696.
125. Schätz, T., A. Friedenauer, H. Schmitz, L. Petersen, and S. Kahra (2007), *J. Mod. Opt.* **54**, 2317.
126. Schenzle, A., R. G. DeVoe, and R. G. Brewer (1986), *Phys. Rev.* A **33**, 2127.
127. Schmidt, P. O., T. Rosenband, C. Langer, W. M. Itano, J. C. Bergquist, and D. J. Wineland (2005), *Science* **309**, 749.
128. Schmidt-Kaler, F., H. Häffner, M. Riebe, S. Gulde, G. P. T. Lancaster, T. Deuschle, C. Becher, C. Roos, J. Eschner, and R. Blatt (2003), *Nature* **422**, 408.
129. Schmied, R., J. Wesenberg, and D. Leibfried (2011), *New J. Phys.* **13**, 115011.
130. Schmitz, H., A. Friedenauer, C. Schneider, R. Matjeschk, M. Enderlein, T. Huber, and T. Schätz (2009a), *Appl. Phys.* B **95**, 195.
131. Schmitz, H., R. Matjeschk, C. Schneider, J. Glueckert, M. Enderlein, T. Huber, and T. Schätz (2009b), *Phys. Rev. Lett.* **103**, 090504.
132. Schneider, C., D. Porras, and T. Schaetz (2012), *Rep. Prog. Phys.* **75**, 024401.
133. Schrödinger, E. (1935), *Naturwissenschaften* **23**, 807.
134. Schrödinger, E. (1952a), *British Journal for the Philosophy of Science* **3** (10), 109.
135. Schrödinger, E. (1952b), *British Journal for the Philosophy of Science* **3** (11), 233.
136. Schuessler, H. A., E. N. Fortson, and H. G. Dehmelt (1969), *Phys. Rev.* **187** (1), 5.
137. Shor, P. W. (1994), in *Proceedings of the 35th Annual Symposium on the Foundations of Computer Science*, Vol. 35, edited by S. Goldwasser (IEEE Computer Society Press, Los Alamitos, CA) p. 124.
138. Solano, E., R. L. de Matos Filho, and N. Zagury (1999), *Phys. Rev.* A **59**, 2539.
139. Sørensen, A., and K. Mølmer (1999), *Phys. Rev. Lett.* **82**, 1971.
140. Sørensen, A., and K. Mølmer (2000), *Phys. Rev.* A **62**, 02231.
141. Stenholm, S. (1986), *Rev. Mod. Phys.* **58**, 699.
142. Taylor, J. M., and T. Calarco (2008), *Phys. Rev.* A **78**, 062331.
143. Turchette, Q. A., C. J. Myatt, B. E. King, C. A. Sackett, D. Kielpinski, W. M. Itano, C. Monroe, and D. J. Wineland (2000), *Phys. Rev.* A **62**, 053807.
144. Van Dyck, R. S., P. Schwinberg, and H. Dehmelt (1977), *Phys. Rev. Lett.* **38** (7), 310.
145. Van Dyck, R. S., P. B. Schwinberg, and H. G. Dehmelt (1978), in *New Frontiers in High-Energy Physics*, edited by B. M. Kursunoglu, A. Perlmutter, and L. F. Scott (Plenum, New York) p. 159.
146. Sasura, M. and V. Buzek (2002), *J. Mod. Opt.* **49**, 1593.
147. Wang, X., A. Sørensen, and K. Mølmer (2001), *Phys. Rev. Lett.* **86**, 3907.

148. Wineland, D. and H. Dehmelt (1975a), *Int. J. Mass Spect. and Ion Phys.* **16**, 338.
149. Wineland, D. and H. Dehmelt (1975b), *Bull. Am. Phys. Soc.* **20**, 637.
150. Wineland, D. and H. Dehmelt (1976), *Int. J. Mass Spect. and Ion Phys.* **19**, 251.
151. Wineland, D., P. Ekstrom, and H. G. Dehmelt (1973), *Phys. Rev. Lett.* **31** (21), 1279.
152. Wineland, D. J., D. W. Allan, D. J. Glaze, H. W. Hellwig, and S. Jarvis (1976), *IEEE Trans. Instrum. Meas.* **IM-25** (4), 1976.
153. Wineland, D. J., J. C. Bergquist, J. J. Bollinger, R. E. Drullinger, and W. M. Itano (2002), in *Proc. 6th Symposium Frequency Standards and Metrology*, edited by P. Gill (World Scientific, Singapore) pp. 361–368.
154. Wineland, D. J., J. J. Bollinger, W. M. Itano, and D. J. Heinzen (1994), *Phys. Rev.* A **50** (1), 67.
155. Wineland, D. J., J. J. Bollinger, W. M. Itano, F. L. Moore, and D. J. Heinzen (1992), *Phys. Rev.* A **46** (11), R6797.
156. Wineland, D. J., R. E. Drullinger, and F. L. Walls (1978), *Phys. Rev. Lett.* **40** (25), 1639.
157. Wineland, D. J. and W. M. Itano (1979), *Phys. Rev.* A **20** (4), 1521.
158. Wineland, D. J. and W. M. Itano (1981), *Phys. Lett.* **82A** (2), 75.
159. Wineland, D. J., W. M. Itano, J. C. Bergquist, and R. G. Hulet (1987), *Phys. Rev.* A **36** (5), 2220.
160. Wineland, D. J., W. M. Itano, J. C. Bergquist, and F. L. Walls (1982), in *Proc. 35th Ann. Frequency Control Symposium*, Vol. NTIS No. AD-A110870 (Electronic Industries Assn., Washington, DC) pp. 602–611.
161. Wineland, D. J., C. Monroe, W. M. Itano, D. Leibfried, B. E. King, and D. M. Meekhof (1998), *J. Res. Nat. Inst. Stand. Tech.* **103**, 259.
162. Wineland, D. J. and N. F. Ramsey (1972), *Phys. Rev.* A 5 (2), 821.
163. Wunderlich, C. and C. Balzer (2003), *Adv. Atom. Mol. Phys.*
164. Young, B. C., F. C. Cruz, W. M. Itano, and J. C. Bergquist (1999), *Phys. Rev. Lett.* **82**, 3799.
165. Zähringer, F., G. Kirchmair, R. Gerritsma, E. Solano, R. Blatt, and C. F. Roos (2010), *Phys. Rev. Lett.* **104**, 100503.

Physics 2013

François Englert and Peter W. Higgs

"for the theoretical discovery of a mechanism that contributes to our understanding of the origin of mass of subatomic particles, and which recently was confirmed through the discovery of the predicted fundamental particle, by the ATLAS and CMS experiments at CERN's Large Hadron Collider"

The Nobel Prize in Physics

Speech by Professor Lars Brink of the Royal Swedish Academy of Sciences.

Translation of the Swedish text.

Your Majesties, Your Royal Highnesses, Ladies and Gentlemen,

"The world is so big so big . . .
Bigger than you can ever imagine . . ."

Since Zacharias Topelius wrote those words almost 150 years ago, we have understood that the world is even bigger and more complex than people could imagine at that time. We can now explore space billions of light-years away, and with the biggest microscopes, particle accelerators, we can study physics at incredibly small distances. Is it then not presumptuous of us to try to understand all the physics that the world holds, to try to find laws that cover a multitude, perhaps all of those phenomena. No! Albert Einstein once wrote that the most fantastic about Nature is that it is possible to comprehend.

Edith Södergran once wrote: "Can you catch in your hands a star that is rising towards the zenith, can you measure its flight?"

Yes we can measure its flight. Isaac Newton explained to us that the law we should use for the movement of a star is the same as the one that governs an apple falling on Earth. A law of physics describes many phenomena that seem to have completely different origin. What do the stars and the apple have in common? Well, they have a mass and the force of gravity is determined by the masses of the objects, nothing else. Nature is rational.

When the new quantum physics was developed a hundred years ago, we also understood that Nature is quantised. Matter can be divided into its smallest constituents. Exactly one hundred years ago Niels Bohr explained that the hydrogen atom consists of a small nucleus, a single proton, with an electron circling around it. We could then understand that the unified theory for the world must be sought for in the Microcosm.

The large particle accelerators developed since the 1950s are the tools we use to survey the Microcosm. The complexity of these modern constructions corresponds to that of the large cathedrals built one thousand years ago. We have found not only a multitude of new particles but also two forces that are only apparent in the Microcosm; the Strong Nuclear Force that binds the nuclei and the Weak Force responsible for radioactive decay. How can we describe such forces that only act over very small distances? The rules of the game are very restrictive.

The prototype of a successful theory was quantum electrodynamics, which describes a force with long range. Just think of two charged balls. If we simply change the force from long range to short range the result is inconsistent because important symmetries of the theory are destroyed. However, it turns out that there is a way to make the force short range and still keep such symmetries in the equations of the theory by choosing a ground state that breaks the symmetry.

Could that be used? It required the existence of a new massless particle together with a massive one. The massless particle would lead to a new type of radiation that has never been seen. Could one get rid of this by modifying the theory or were scientists on the wrong path?

The solution came almost fifty years ago in two papers, one by François Englert and Robert Brout and the other by Peter Higgs—two short papers roughly a page long, that changed the world. They simply combined a theory like electromagnetism with the two fields which lead to the two particles just mentioned, and showed that the unwanted massless particle, which would lead to unobserved radiation, combined with the electromagnetic field to produce a short range force. However the new massive particle still appeared in the theory. We now had a theory analogous to electromagnetism, but with short range forces and massive particles. We had unified the Electromagnetic and the Weak forces. What did particle physicists do now? They ignored the new theory! Could it really be consistent? Brout, Englert and Higgs believed so, but it was not until seven years later that a young Dutch student, Gerhard 't Hooft indeed could show that the theory works, in a work of immense complexity leading to the Nobel Prize in 1999.

Now the physics world understood, and very quickly developed a unified theory of the forces that act in Microcosm (the strong, electromagnetic and weak forces). This is called "the Standard Model of Particle Physics," a theory that in practice governs the world as long as the force of gravity can be ignored. It demanded new particles, quarks and leptons and force particles, which were speedily discovered. But the new particle that Brout, Englert and Higgs had predicted, which was now understood to give masses to all massive fundamental particles was not found. Thirty years ago we had essentially found everything except this particle. Was it really there?

It was understood that the discovery of such a particle would require a very large accelerator, one that could produce collisions with an energy comparable to that arising in the very early universe. It took some thirty years of preparation and then construction at CERN in Geneva before scientists could search for the wanted particle. In 2010 6,000 scientists began two large experiments, and on July 4, 2012 they spread the news that they had indeed found the particle. The Standard Model was complete and it had been found that Nature follows precisely that law that Brout, Englert and Higgs had created. A fantastic triumph for Science.

Professeur Englert,

Vous avez, avec votre collègue le Professeur Brout tristement décédé et avec le Professeur Higgs, trouvé la clé pour comprendre la masse des particules élémentaires, découverte pour laquelle le Prix Nobel vous est décerné. Au nom de l'Académie Royale des Sciences de Suède, j'ai le privilège de vous transmettre mes plus chaleureuses félicitations pour votre travail exceptionnel. Je vous demande maintenant de bien vouloir vous avancer pour recevoir votre prix Nobel des mains de Sa Majesté le Roi.

Professor Higgs,

You have together with Professors Englert and Brout found the key to understanding the masses of elementary particles for which you have been awarded the Nobel Prize. On behalf of the Royal Swedish Academy of Sciences it is my privilege to convey to you my warmest congratulations for your outstanding work. I now ask you to step forward to receive your Nobel Prize from the hands of His Majesty the King.

François Englert. © Nobel Media AB. Photo: A. Mahmoud

François Englert

I was born in Belgium on 6 November 1932. I am married to Mira Nikomarow and have five children: Michèle, Anne, Georges from a first marriage with Esther Dujardin and Sarah, Hélène from a second one with Danielle Vindal.

My parents emigrated from Poland in 1924 with my brother, who was a few months old. They were from a simple family of Polish Jews. They were looking, I suppose, for a better economic life and were escaping from an anti-Semitic environment. They worked hard, set up a textile shop and managed to reach a rather decent life when in May 1940, Nazi Germany invaded Belgium, less than one year after the invasion of Poland. I was seven years old and quite aware of the situation. Persecution came gradually. After two years of relatively normal life, I was compelled to wear, as all Jews who could not hide their identity, the distinctive Star of David. A few months later, the Nazis started the deportation of Jews to concentration camps where they were murdered.

My parents, my brother and I survived the war in Belgium. We were helped and hidden by people who did not even know us, people who in those times of darkness took the great risk of displaying generosity, humanity and courage.

My parents were hiding in a place unknown to me. Separated from them to increase my chances of survival, I was taken care of by Camille and Louise Jourdan, the owners of a cafe-restaurant in Lustin, a village in the Ardennes. I want to pay tribute to their memory and of that of their daughter Yvonne, whose tenderness in initiating me to music and piano was like a glimmer of hope in a world of desperation. And to the memory of priest Warnon of Annevoie, another small village in the Ardennes, where our family, fleeing from Lustin after a denunciation, reunited and stayed to the end of the war; he presented us to the village inhabitants as Christians; he went as far as baptising me so that I could attend the Catholic College "Notre-Dame de Bellevue" in Dinant as an ordinary student; he enrolled my brother, dressed in a fake soutane, at a seminarist school as if preparing him for priesthood. And also to the memory of many others who helped us. Without these wonderful people we could not have escaped the persecution and I would not be here to tell about it.

I have always present in my mind the courage of my parents, their perspicacity in finding the right move for escape at many critical moments, and most of all their love: in confronting the barbarity of the German Nazis and their many complicities, they always put the life of their children before the preservation of their own. Their survival was tragically marred by the complete disappearance of their Polish family, murdered in Poland by the German Nazis.

After the war, we attempted and largely succeeded in resuming a normal life. I went to secondary school and, while memories of past years still haunted the nightmares of my sleep, I functioned well at school. I developed an interest in literature, music and mathematics. My teacher of mathematics, whom I deeply appreciated, recommended studies in polytechnics for its extended program in mathematics at the University. My parents also pushed me in the same direction, out of concern about my future well-being. So in 1955 I got my degree in electrical-mechanical engineering. I realised however that my interest was less in practical applications than in the understanding of the underlying theoretical structure and I decided to learn physics. As an assistant in the polytechnic department, I was able to finance new studies and got my Physics Masters Degree in 1958 and my PhD in 1959.

I had discovered a passion for research and I was thrilled when the same year, based on recommendations and a few previous publications in *Condensed Matter Physics*, I was offered a two-year position in the United States at Cornell University, Ithaca (NY), as Research Associate for the young Professor Robert Brout. I immediately accepted and left for Ithaca.

Our first contact was unexpectedly warm. During my stay the convergence of our vision of science and life laid the groundwork for lasting collaboration and a lifelong friendship. In Ithaca, we worked together in condensed matter physics and in the statistical theory of phase transitions, mainly on ferromagnetism and superconductivity. We realised the importance of spontaneous symmetry breaking in phase transitions and we were extremely impressed when Yoichiro Nambu showed how this notion could be transferred to elementary particle physics to explain the small pion mass on the hadron scale. This work and his beautiful analysis of superconductivity in field theoretic terms drove us later to study the fate of spontaneous symmetry breaking in the context of gauge theory.

In fall 1961, I was scheduled to return to Belgium. By that time our collaboration and our friendship had become deeply rooted. I was offered a University Professorship at Cornell but I was missing Europe very much. I decided not to accept it and to return to Belgium. Robert and his wife Martine had a similar attraction for the Old Continent; Robert got a Guggenheim fellowship and they joined me with their children in Belgium. After a few months, the social life there and our personal relationship persuaded Robert to resign from his professorship at Cornell University and to settle permanently at the Université Libre de Bruxelles in Brussels. He eventually acquired Belgian nationality and together we directed the theoretical physics group at the ULB.

In Brussels, we resumed our analysis on spontaneous symmetry breaking, both in the statistical theory of phase transition and, inspired mainly by Nambu's work, in field theory. This is how we discovered the mass generating mechanism that may now indeed be viewed as a phase transition from a high temperature phase in the early Universe, where elementary particle were massless, to the present low temperature phase where their mass arises from a generalisation of spontaneous symmetry breaking to Yang-Mills fields, namely the BEH mechanism.

At the ULB, Brout and I initiated a research group in fundamental interactions, that is, in the search for the general laws of nature. Joined by brilliant students, many of them becoming world renowned physicists, our group contributed to the many fields at the frontier of the challenges facing contemporary physics. While the mechanism discovered in 1964 was developed all over the world to encode the nature of weak interactions in a "Standard Model," our group contributed to the understanding of strong interactions and quark confinement, general relativity and cosmology. There we introduced the idea of a primordial exponential expansion of the universe, later called inflation, which we related to the origin of the universe itself, a scenario, which I still think may possibly be conceptually the correct one. During these developments, our group extended our contacts with other Belgian universities and got involved in many international collaborations.

With our group and many other collaborators I analysed fractal structures, supergravity, string theory, infinite Kac-Moody algebras and more generally all tentative approaches to what I consider as the most important problem in fundamental interactions: the solution to the conflict between the classical Einsteinian theory of gravitation, namely general relativity, and the framework of our present understanding of the world, quantum theory. Although this conflict appears experimentally to affect known results only at very tiny scales of the order of 10^{-33} cm, transcending it would amount to overcoming a conceptual mistake. As such, a solution of this conflict might affect our understanding of the laws of nature at all scales and is crucial for attempting to reach a rational understanding of the origin of the Universe.

Robert was less interested in these new developments and concentrated more on cosmology. Our collaboration became less frequent but our friendship was unaffected. He passed away on May 3, 2011 after a prolonged illness and missed the remarkable discovery of the Standard Model scalar boson at CERN and the awarding of the Nobel Prize.

The BEH Mechanism and its Scalar Boson

Nobel Lecture, 8 December 2013

by François Englert
Université Libre de Bruxelles, Brussels, Belgium

1 INTRODUCTION: SHORT AND LONG RANGE INTERACTIONS

Physics, as it is conceived today, attempts to interpret diverse phenomena as particular manifestations of testable general laws. Since its inception in the Renaissance, mainly through Galileo's revolutionary concepts, this has been an extraordinarily successful adventure—to the point where after impressive developments in the first half of the twentieth century, one might have even conceived that all phenomena, from the atomic scale to the edge of the visible universe, are governed solely by two fundamental laws, and two known laws. Namely classical general relativity, Einstein's generalisation of Newtonian gravity, and quantum electrodynamics, the quantum version of Maxwell's electromagnetic theory.

Gravitational and electromagnetic interactions are long range interactions, meaning they act on objects no matter how far they are separated from each other. The progress in the understanding of such physics applicable to large scales is certainly attributable to the fact they can be perceived without the mediation of highly sophisticated technical devices. But the discovery of subatomic structures had revealed the existence of other fundamental interactions that are short range, that is, negligible at larger distance scales. In the early 1960s, there was no consistent theoretical interpretation of short range fundamental interactions, nor of the "weak interactions" responsible for radioactive decay, nor of the "strong interactions" responsible for the formation of nuclear structures.

Robert Brout and I [1], and independently Peter Higgs [2], constructed a mechanism to describe short range fundamental interactions. Robert Brout passed away in 2011 and left me alone to tell our story. I will explain how we were led to propose the mechanism, and how it allows for consistent fundamental theories of short range interactions and for building elementary particle masses. It became a cornerstone of the Standard Model and was recently confirmed by the magnificent discovery at CERN of its predicted scalar boson.

We became convinced that a consistent formulation of short range interactions would require a common origin for both short and long range interactions.

While both classical general relativity and quantum electrodynamics describe long range interactions and are both built upon very large symmetries, labelled "local symmetries," they have very different structures: in contradistinction to general relativity, long range quantum electrodynamics is fully consistent at the quantum level and was experimentally verified at that level, in particular by the successful inclusion of chemistry in the realm of known physics. As a valid theory of short range interactions clearly required quantum consistency, we were naturally driven to take, as a model of the corresponding long range interactions, the generalisation of quantum electrodynamics, known as Yang-Mills theory.

The quantum constituents of electromagnetic waves are "photons," massless neutral particles travelling with the velocity of light. Their massless character implies that the corresponding waves are polarised only in directions perpendicular to their propagation. These features are apparently protected by local symmetry, as the latter does not survive the explicit inclusion of a mass term in the theory. Yang-Mills theory is built upon similar local symmetries, enlarged to include several massless interacting quantum constituents, neutral and charged ones. These massless objects are labelled gauge vector bosons (or often simply gauge bosons).

To transmute long range interactions into short range ones in the context of Yang-Mills theory it would suffice to give these generalised photons a mass, a feature that, as we just indicated, is apparently forbidden by local symmetries. Momentarily leaving aside this feature, let us first recall why massive particles transmit in general short-range interactions.

Figure 1 is a Feynman diagram whose intuitive appearance hides a precise mathematical content. Viewing time as running from bottom to top, it describes the scattering of two electrons resulting from the exchange of a massive particle labeled Z of mass m_Z. Classically such process could not occur, as the presence of the Z particle would violate energy conservation. Quantum mechanically it is allowed if the violation takes place within a time span of the order $\hbar/mc^2$. This process then describes in lowest order perturbation theory a short-range interaction cut-off at a range $\sim \hbar/mc$.

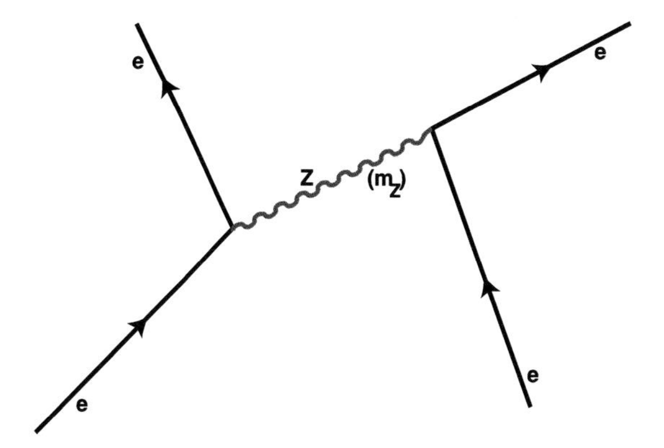

FIGURE 1. Massive particle mediating short-range interactions.

As local symmetries apparently prevent the introduction of massive gauge bosons in the theory, we turn our attention to a class of theories where the state of a system is asymmetric with respect to the symmetry principles that govern its dynamics. This is often the case in the statistical physics of phase transitions [3]. This is not surprising, since more often than not energetic considerations dictate that the ground state or low lying excited states of a many body system become ordered. A collective variable such as magnetisation picks up expectation value, which defines an order parameter that otherwise would vanish by virtue of the symmetry encoded in the formulation of the theory (isotropy in the aforementioned example). This is an example of Spontaneous Symmetry Breaking (SSB) which frequently occurs in the statistical theory of second order phase transitions. Could mass of gauge bosons arise through a similar SSB? This question arises naturally from the seminal work of Yoichiro Nambu, who showed that SSB could be transferred from the statistical theory of phase transitions to the realm of relativistic quantum field theory [4–6], the mathematical framework designed to analyse the world of elementary particles.

This raises a deeper question: could SSB be the agent of the transmutation of long range interactions mediated by massless gauge fields to short range interactions mediated by massive ones, without impairing the validity of the quantum behaviour that characterise the simplest Yang-Mills theory, namely quantum electrodynamics?

As we shall see, the answer is yes to both questions *provided that the notion of SSB is traded for a more subtle one: the BEH mechanism* [1, 2]. To prepare for the discussion of the mechanism, I will first review how SSB can be transferred from the theory of phase transitions to relativistic quantum field theory.

2 SPONTANEOUS SYMMETRY BREAKING

2.1 Spontaneous symmetry breaking in phase transitions

Consider a condensed matter system, whose dynamics is invariant under a continuous symmetry. As the temperature is lowered below a critical one, the symmetry may be reduced by the appearance of an ordered phase. The breakdown of the original symmetry is always a discontinuous event at the phase transition point, but the order parameters may set in continuously as a function of temperature. In the latter case the phase transition is second order. Symmetry breaking by a second order phase transition occurs in particular in ferromagnetism, superfluidity and superconductivity.

I will first discuss the ferromagnetic phase transition, which illustrates three general features of the SSB which set in at the transition point in the low temperature phase: ground state degeneracy, the appearance of a "massless mode" when the dynamics is invariant under a *continuous* symmetry, and the occurrence of a "massive mode" characterising the rigidity of the order parameter.

In the absence of external magnetic fields and of surface effects, a ferromagnetic substance below the Curie point displays a global orientation of the magnetisation, while the dynamics of the system is clearly rotation invariant; namely, the Hamiltonian of the system is invariant under the full rotation group. This is SSB.

A ferromagnetic system is composed of microscopic atomic magnets (in simplified models such as the Heisenberg Model these are spin 1/2 objects) whose interactions tend to orient neighbouring ones parallel to each other. No global orientation appears at high temperature where the disordering thermal motion dominates. Below a critical "Curie temperature" energy considerations dominate and the system picks up a global magnetisation. The parallel

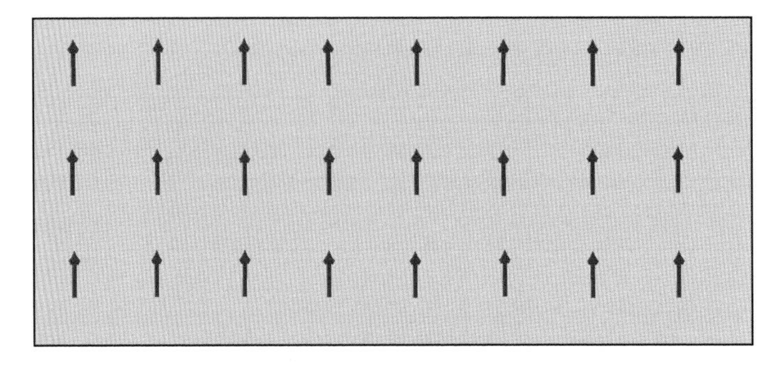

FIGURE 2. Classical representation of a ferromagnet ground state.

orientation of neighbouring magnets propagates, ending up in a macroscopic magnetisation. This selects a direction, which for an infinite isolated ferromagnet is arbitrary. It is easily proven that for an infinite system any pair of possible orientations defines orthogonal ground states and any local excitations on top of these ground states are also orthogonal to each other. Thus the full Hilbert space of the system becomes split into an infinity of disjointed Hilbert spaces. This is ground state degeneracy (Figure 2).

The effective thermodynamical potential V, whose minimum yields magnetisation in the absence of an external magnetic field, is depicted in Figure 3. Above the Curie point T_C the magnetisation $\vec{M}$ vanishes. Below the Curie point the potential develops in a plane $V\,M_z$ a double minimum which generate a valley in the M_x, M_y directions. Each point of the valley defines one of the degenerate ground states with the same $|\vec{M}|$.

At a given minimum, say, $\vec{M} = M^z \vec{1}_z$, the curvature of the effective potential measures the inverse susceptibility which determines the energy for infinite wavelength fluctuations. This is the analogue of mass in relativistic particle physics. The inverse susceptibility is zero in directions transverse to the order parameter and positive in the longitudinal direction. One thus obtains from the transverse susceptibility a "massless" transverse mode characteristic of broken continuous symmetry: these are the "spin-waves" whose quantum constituents are interacting bosons called "magnons." The longitudinal susceptibility yields a (possibly unstable) "massive" longitudinal mode which corresponds to fluctuations of the order parameter. In contradistinction to the massless mode which exists only in continuous SSB for which there is a valley, the massive mode is present in any SSB, continuous or discrete, and measures the rigidity of the ordered structure.

The structure of Figure 3 is common to many second order phase transitions and leads to similar consequences. However, in superconductivity a new

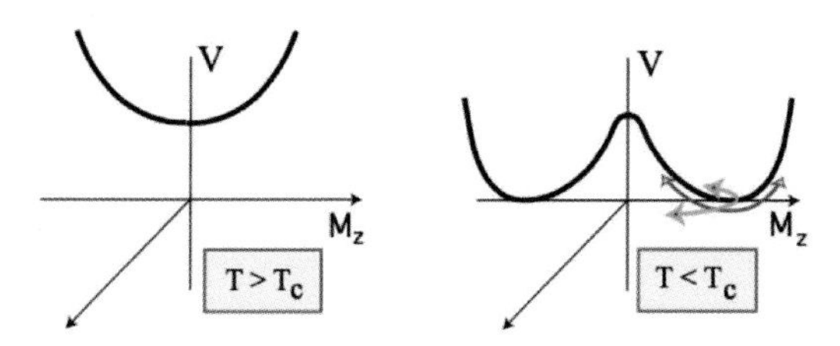

FIGURE 3. Effective thermodynamical potential of a ferromagnet above and below the Curie point.

phenomenon occurs. The quantum phase symmetry is broken by a condensation of electron pairs bounded by an attractive force due to phonon exchange in the vicinity of the Fermi surface. The condensation leads to an energy gap at the Fermi surface. For neutral superconductors, this gap would host a massless mode and one would recover the general features of SSB. But the presence of the long-range Coulomb interactions modifies the picture. The *massless mode disappears*: it is absorbed in electron density oscillations, namely in the "massive" plasma mode. As will be apparent later, this is a precursor of the BEH mechanism [7, 4, 8].

2.2 Spontaneous symmetry breaking in field theory

Spontaneous symmetry breaking was introduced in relativistic quantum field theory by Nambu in analogy with the BCS theory of superconductivity [4]. The problem studied by Nambu [5] and Nambu and Jona-Lasinio [6] is the spontaneous breaking of the $U(1)$ symmetry of massless fermions resulting from the arbitrary relative (chiral) phase between their decoupled right and left constituent neutrinos. Chiral invariant interactions cannot generate a fermion mass in perturbation theory but may do so from a (non-perturbative) fermion condensate: the condensate breaks the chiral symmetry spontaneously. Nambu [5] showed that such spontaneous symmetry breaking is accompanied by a massless pseudoscalar. This is interpreted as the chiral limit of the (tiny on the hadron scale) pion mass. Such an interpretation of the pion constituted a breakthrough in our understanding of strong interaction physics. The massless pseudoscalar is the field-theoretic counterpart of the "massless" spin-wave mode in ferromagnetism. In the model of reference [6], it is shown that SSB also generates a massive scalar boson which is the counterpart of the "massive mode" measuring in phase transitions the rigidity of the order parameter in the spontaneously broken phase.

The significance of the massless boson and of the massive scalar boson occurring in SSB is well illustrated in a simple model devised by Jeffrey Goldstone [9]. The potential $V(\phi_1, \phi_2)$ depicted in Figure 4 has a rotational symmetry in the plane of the real fields (ϕ_1,ϕ_2), or equivalently is invariant under the $U(1)$ phase of the complex field $\phi = (\phi_1 + i\phi_2)/\sqrt{2}$. This symmetry is spontaneously broken by the expectation value $\langle\phi\rangle$ of the ϕ-field acquired at a minimum of the potential in some direction of the (ϕ_1, ϕ_2) plane, say $\langle\phi_1\rangle$. Writing $\phi = \langle\phi\rangle + \varphi$

$$\phi_1 = \langle\phi_1\rangle + \varphi_1, \tag{2.1}$$

$$\phi_2 = \varphi_2. \tag{2.2}$$

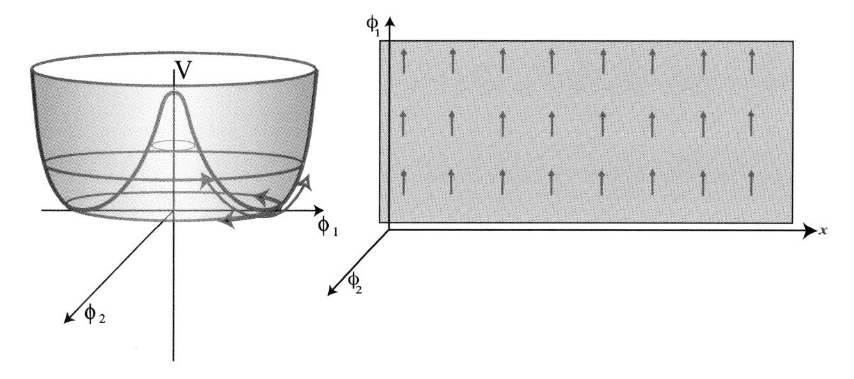

FIGURE 4. Spontaneous symmetry breaking in the Goldstone model

For small φ_1 and φ_2 we may identify the quantum fluctuation φ_1 climbing the potential as the massive mode measuring the rigidity of the SSB ground state selected by $\langle\phi_1\rangle$, and the quantum fluctuation φ_2 in the orthogonal valley direction as the massless mode characteristic of a continuous SSB.

Their significance is illustrated in Figure 5 and Figure 6 depicting respectively classical φ_2 and φ_1 wave modes, on the classical background $\langle\phi_1\rangle$. The corresponding massless and massive bosons are the quantum constituents of these waves.

Figure 5 (a) represents schematically a lowest energy state (a "vacuum") of the system: a constant non-zero value of the field $\phi_1 = \langle\phi_1\rangle$ pervades space-time. Figure 5 (b) depicts the excitation resulting from the rotation of half the fields in the (ϕ_1,ϕ_2) plane. This costs only an energy localised near the surface separating

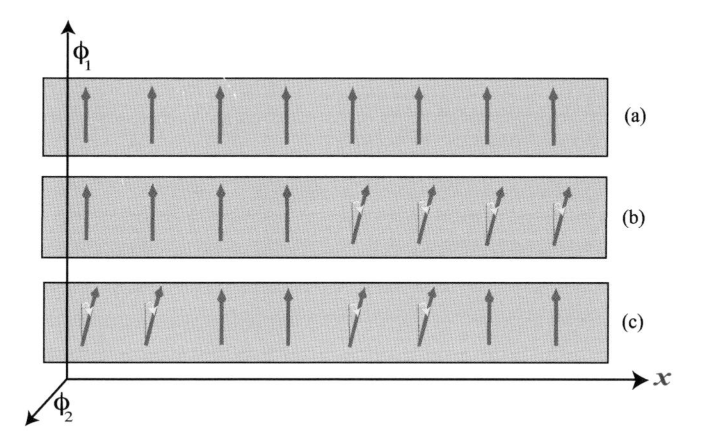

FIGURE 5. Massless Nambu-Goldstone mode φ_2.

the rotated fields from the chosen vacuum. SSB indeed implies that rotating all the fields would cost no energy at all: one would merely trade the initial chosen vacuum for an equivalent one with the same energy. This is the characteristic *vacuum degeneracy* of SSB. Figure 5 (c) mimics a wave of φ_2. Comparing 5 (c) with 5 (b), we see that as the wavelength of the wave increases indefinitely, its energy tends to zero, and may be viewed as generating in that limit a motion along the valley of Figure 4. Quantum excitations carried by the wave reach thus zero energy at zero momentum and the mass m_{φ_2} is zero. Figure 5 can easily be generalised to more complex spontaneous symmetry breaking of continuous symmetries. Massless bosons are thus a general feature of such SSB already revealed by Nambu's discovery of the massless pion resulting from spontaneous chiral symmetry breaking [5]. They will be labelled massless Nambu-Goldstone (NG) bosons. Formal proofs corroborating the above simple analysis can be found in the literature [10].

Figure 6 depicts similarly a classical wave corresponding to a stretching of the vacuum fields. These excitations in the φ_1 direction describe fluctuations of the order parameter $\langle\varphi_1\rangle$. They are volume effects and their energy does not vanish when the wavelength becomes increasingly large. They correspond in Figure 4 to a climbing of the potential. The quantum excitations φ_1 are thus now massive. These considerations can be again extended to more general SSB (even to discrete ones) to account for order parameter fluctuations. Lorentz invariance imposes that such massive excitations are necessarily scalar particles. They were also already present in reference [6] and will be denoted in general as *massive scalar bosons*.

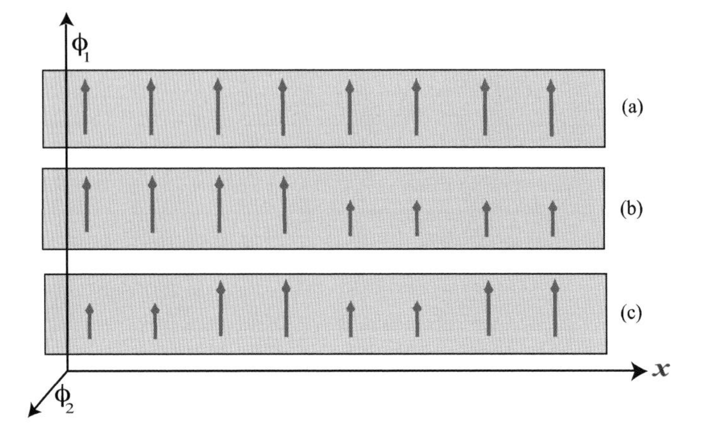

FIGURE 6. Massive scalar mode φ_1.

To summarise, φ_2 describes massless bosons, φ_1 massive ones, and the "order parameter" $\langle\phi_1\rangle$ may be viewed as a condensate of φ_1 bosons.

3 THE BEH MECHANISM

The above considerations are restricted to spontaneous symmetry breaking of global continuous symmetries. Global means that the symmetry operations are independent of the space-time point x. For instance in the Goldstone model, the global rotations of the fields in Figure 5 (a) in the (ϕ_1,ϕ_2) plane by angles independent of the space-time point x are symmetries of the theory (they describes motion in the valley of Figure 4): these rotations cost no energy and simply span the degenerate vacua. We will now discuss the fate of SSB when the global symmetry is extended to a local one.

3.1 The fate of the Nambu-Goldstone boson and vector boson masses

We extend the $U(1)$ symmetry of the Goldstone model from global to local. Thus the rotation angle in the (ϕ_1,ϕ_2) plane in Figure 5, or equivalently the rotation in the valley of Figure 4, can now be chosen independently at each space-time point (x) with no cost of energy and no physical effect. To allow such feature, one has to invent a new field whose transformation would cancel the energy that such motion would generate in its absence. This is a "gauge vector field" A_μ. It has to be a vector field to compensate energy in all space directions and it has to transform in a definite way under a rotation in the (ϕ_1,ϕ_2) plane: this is called a gauge transformation and results in a large arbitrariness in the choice of the A_μ field corresponding to arbitrary "internal" rotations at different points of space. The consequence of this gauge symmetry is that the waves are polarised in directions perpendicular to their direction of propagation and that their quantum constituents have to be introduced as massless objects.

Local $U(1)$ symmetry is the simplest gauge field theory and is the symmetry group of quantum electrodynamics. In the local generalisation (the gauging) of the Goldstone model, the introduction of the potential of Figure 4 will deeply affect the "electromagnetic potential" A_μ.

As in the Goldstone model of Section 1.2, the SSB Yang-Mills phase is realised by a non vanishing expectation value for $\phi = (\phi_1 + i\phi_2)/\sqrt{2}$, which we choose to be in the ϕ_1-direction. Thus

$$\phi = \langle\phi\rangle + \varphi, \tag{3.1}$$

with $\phi_1 = \langle\phi_1\rangle + \varphi_1$ and $\phi_2 = \varphi_2$. As previously φ_2 and φ_1 appear to describe a NG massless boson and a massive scalar boson.

However a glance on Figure 5 depicting the NG mode immediately shows that Figure 5 (b) and Figure 5 (c) differ from Figure 5 (a) only by local rotations and hence in the local Goldstone model they are just symmetry (or equivalently gauge) transformations. They cost no energy and therefore the NG boson has disappeared: the corresponding fluctuations in the valley are redundant (gauge transformed) descriptions of the same gauge invariant vacuum. It is easy to see that this argument remains valid for any local symmetry and hence *Nambu-Goldstone bosons do not survive the gauging of a global SSB to a local symmetry*. The vacuum is no longer degenerate and strictly speaking there is no spontaneous symmetry breaking of a local symmetry. The reason why the phase with non vanishing scalar expectation value is often labelled SSB is that one uses perturbation theory to select at zero gauge field coupling a scalar field configuration from global SSB; but this *preferred* choice is only a convenient one.

The disappearance of the NG boson is thus an immediate consequence of local symmetry. The above argument [11] was formalised much later [12] but formal proofs not directly based on the gauge invariance of the vacuum were already presented in 1964 [13, 14].

One may now understand in qualitative terms the consequence of the disappearance of the NG boson. Clearly, one does not expect that the degrees of freedom carried by the NG ϕ_2 field could vanish. As the NG boson disappears because of its coupling to the gauge field, one expects that these degrees of freedom should be transferred to it. This can only occur by adding to the transverse polarisation of the gauge field a longitudinal one. But such polarisation is forbidden as mentioned earlier, for a massless field. Therefore the coupling of the would-be NG boson to the gauge field must render the latter massive! This is the essence of the BEH mechanism.

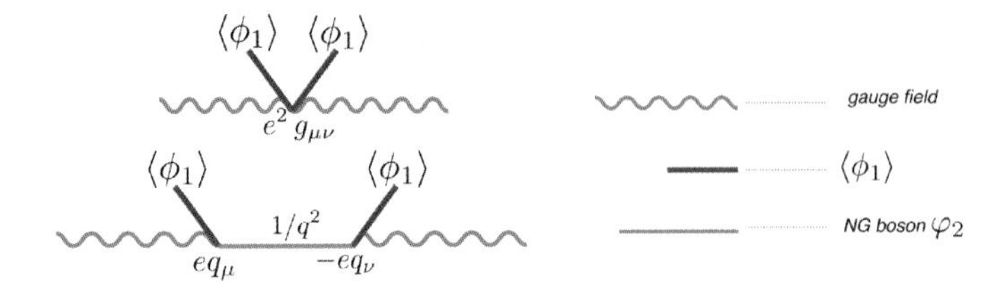

FIGURE 7. Interaction of the gauge field with the condensate.

These qualitative considerations can be made quantitative [1] by considering the Feynman graphs (time runs horizontally) describing the propagation of the A_μ gauge field in the vacuum with non vanishing scalar field expectation value, say $\langle \phi_1 \rangle \neq 0$. This propagation is depicted in lowest order in Figure 7 (time runs horizontally) and the interaction of A_μ with the condensate $\langle \phi_1 \rangle$ amounts to a "polarisation" of the vacuum. The first graph shows the local interaction of the gauge field with the condensate while the second one gives a non-local interaction due to the propagation of a NG boson. Here e is the coupling of the gauge vector to matter, q_μ is a four-momentum (q_0 = energy; $\vec{q}$ = momentum), $q^2 = q_0^2 - \vec{q}^{\,2}$ and $g_{\mu\nu}$ has only non-zero values if $\mu \neq \nu$: 1,–1,–1,–1. The two graphs add up to

$$\Pi_{\mu\nu} = (g_{\mu\nu} - q_\mu q_\nu / q^2)\Pi(q^2), \tag{3.2}$$

where

$$\Pi(q^2) = e^2 \langle \phi_1 \rangle^2. \tag{3.3}$$

The second factor of Equation (3.2) does not vanish when $q^2 = 0$. In field theory this means that the gauge field has acquired a mass

$$(M_V^2) = e^2 \langle \phi_1 \rangle^2. \tag{3.4}$$

The first factor describes the projection at $q^2 = m_V^2$ of $g_{\mu\nu}$ on a three-dimensional space of polarisations, which, as explained in qualitative terms above, is required for a massive vector. Its transversality (i.e. its vanishing under multiplication by q^μ) is characteristic of a "Ward Identity" which expresses the fact that the local gauge symmetry has not been broken and is identical to the analogous factor in quantum electrodynamics, an important fact that will be commented on in the following section.

The generalisation of these results to more complicated symmetries yields (for real fields) a mass matrix

$$(M_V^2)^{ab} = -e^2 \langle \phi^B \rangle T^{aBC} T^{bCA} \langle \phi^A \rangle, \tag{3.5}$$

where T^{aBC} is a real anti-symmetric generator coupled to a gauge field A_μ^a and $\langle \phi^A \rangle$ designates a non vanishing expectation value.

In these cases, some gauge fields may remain massless. Consider for instance instead of the invariance of the Goldstone model on a circle in the plane (ϕ_1, ϕ_2), an invariance on a sphere in a 3-dimensional space (ϕ_1, ϕ_2, ϕ_3) broken by $\langle \phi_1 \rangle \neq$

0. There are now three gauge fields associated to the rotations on the sphere, and while A_μ^2 and A_μ^3 acquire mass A_μ^1 remains massless. This can be understood in the following way: rotation generators around the directions 2 and 3 would move $\langle\phi_1\rangle$ if the symmetry were global and would thus give rise, as in Figure 5, to NG bosons; their degrees of freedom are transferred in local symmetries to the massive gauge vector fields A_μ^2 and A_μ^3, providing their third degree of polarisation. The expectation value $\langle\phi_1\rangle$ is not affected by rotation generators around the direction 1 and does not generate NG bosons in the global symmetry case and hence the corresponding A_μ^1 remains massless.

Thus the BEH mechanism can unify long and short range interactions in the same theory by leaving unbroken a subgroup of symmetry transformation (e.g. rotation around the direction 1) whose corresponding gauge fields remain massless.

3.2 The fate of the massive scalar boson

A glance at Figure 5 shows that the stretching of (classical) scalar fields is independent of local rotations of the ϕ-field in the (ϕ_1, ϕ_2) plane. This translates the fact that the modulus of the φ-field is gauge invariant. Hence the scalar bosons survive the gauging and their classical analysis is identical to the one given for the Goldstone model in Section 1.2. The coupling of the scalar boson φ_1 to the massive gauge bosons follows from the Figure 7, by viewing the Feynman diagrams with time going from top to bottom and using Equation (3.1). One gets the two vertices of Figure 8 where the heavy wiggly lines on the right hand side represent the massive gauge propagators. The vertex couplings follow from Equation (3.4).

3.3 Fermion masses

Let us couple the Yang-Mills fields to massless fermions in a way that respects Yang-Mills symmetry. This coupling preserves the chiral symmetry of the

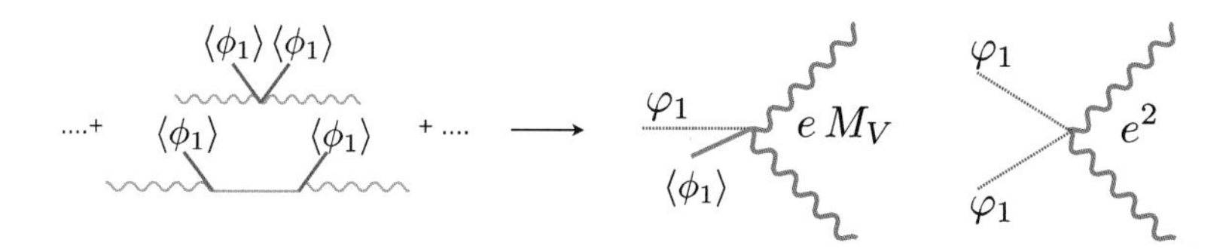

FIGURE 8. Coupling of the scalar boson φ_1 to massive gauge bosons.

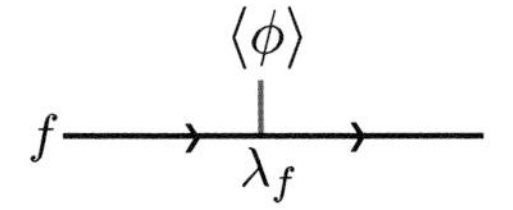

FIGURE 9. Mass generation $m_f = \lambda_f\langle\phi\rangle$ from a coupling λ_f of fermions to the scalar field ϕ.

massless fermions and fermion mass requires SSB. In the Nambu theory of spontaneous breaking of chiral symmetry, this gives rise to NG bosons which are eaten up here by massive gauge fields. This can be done by suitable couplings of the scalar fields whose expectation value breaks the symmetry. Mass generation for fermions is depicted in Figure 9.

3.4 Why is the mechanism needed?

Equation (3.2) expresses the fact that the mass generation from the BEH mechanism does not destroy local symmetry, in contradistinction to a mass term introduced by hand *ab initio*. This equation remains valid at higher orders in perturbation theory and has the same form as the polarisation in quantum electrodynamics. As in the latter case, it implies that in covariant gauges, the gauge vector boson propagator tames the quantum fluctuations, and therefore suggests that the theory is renormalisable [15]. However it is a highly non trivial matter to prove that it does not introduce contributions from unphysical particles, and it is therefore a very difficult problem to prove quantum consistency to all orders. That this is indeed the case has been proven by 't Hooft and Veltman [16] (see also Ref. [17])

The quantum consistency of the BEH mechanism is the basic reason for its success. Precision experiments can be predicted and were indeed verified. Quantum consistency played a critical role in the analysis of the production of the scalar boson at the LHC and of its decay products, leading to the confirmation of the detailed validity of the mechanism.

3.5 Dynamical symmetry breaking

The symmetry breaking giving mass to gauge vector bosons may also arise from a fermion condensate. This is labelled *dynamical symmetry breaking*. If a spontaneously broken global symmetry is extended to a local one by introducing gauge fields, the massless NG bosons disappear as previously from the physical spectrum and their absorption by gauge fields renders these massive. In contradistinction with breaking by scalar field condensate, it is very difficult in this way to give mass in a renormalisable theory simultaneously to both gauge vector fields and fermions.

3.6 The electroweak theory and the Standard Model

The most impressive success of the BEH mechanism is the electroweak theory for weak and electromagnetic interactions [18] applied to all particles of the Standard Model. These encompass all known particles. These are a) the fermions which are listed in Figure 10, b) γ and W^+, W^-, Z, the gauge vector bosons transmitting the electromagnetic and the weak interactions, c) eight "gluons," the gauge vectors bosons of the "colour group" $SU(3)$ mediating the strong interactions, and d) last but not least, one massive scalar boson which was recently discovered and identified as the scalar predicted by the BEH mechanism.

The first row in Figure 10 contains the basic constituents of the atom, namely the electron, the three up and down coloured quarks building the proton and neutron bonded by the gluons, to which is added the electron neutrino. The second [19] and third row [20] were completed as predictions in the 1970s and verified afterwards. Colour was also introduced in the 60s. The particles in the first and the second row are called leptons. To all fermions of the table, one must of course also add their antiparticles.

All the fermions are chiral and their chiral components have different group quantum numbers. Hence they are, as the gauge vector bosons, massless in absence of the BEH mechanism, i.e. in absence of the scalar condensate. The condensate $\langle\phi\rangle \neq 0$ gives mass to the W^+, W^-, Z bosons and to all fermions except

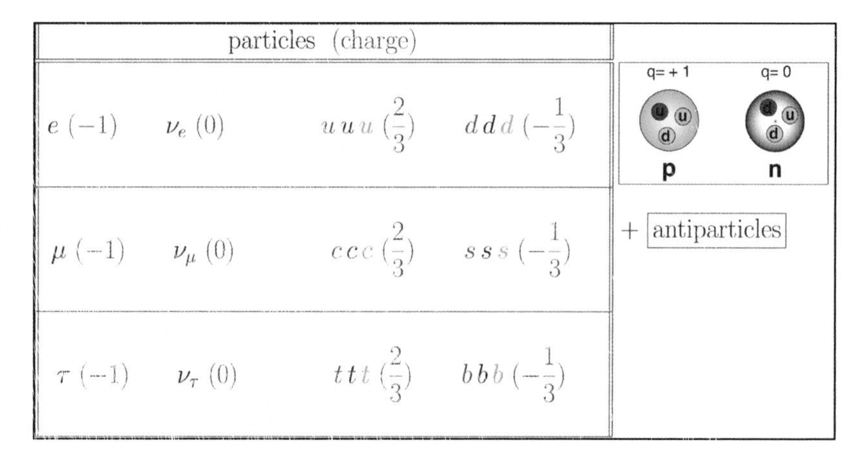

FIGURE 10. Fermion constituents of the Standard Model.

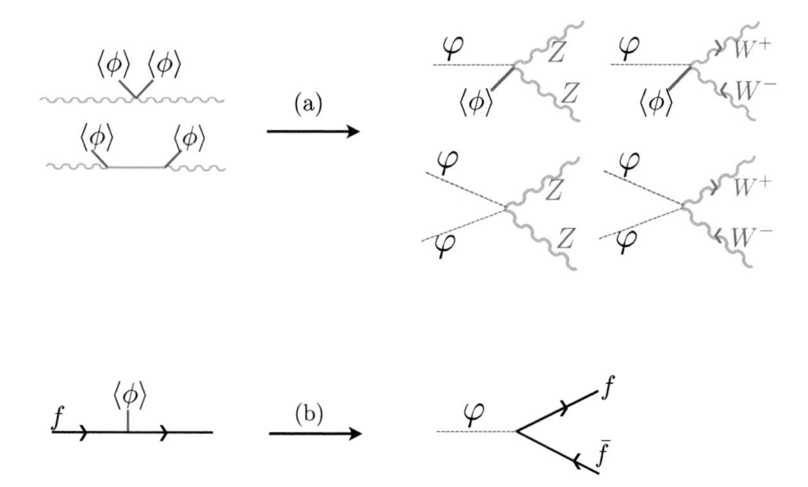

FIGURE 11. Coupling of the scalar boson φ to massive gauge bosons and to elementary fermions.

the three chiral neutrinos which have no opposite chirality counterpart in the conventional Standard Model. The photons and the gluons remain massless but the latter become short range due in the conventional description to a highly non-perturbative vacuum (resulting from a mechanism somehow dual to the BEH mechanism).

The discovery of the Z and W bosons in 1983 and the precision experiments testing the quantum consistency of the Standard Model established the validity of the mechanism, but it was still unclear whether this was the result of a dynamical symmetry breaking or of a particle identifiable as an elementary boson at the energy scale considered.

4 THE DISCOVERY

In the Standard Model, there is one real massive scalar boson φ (also labelled H). It couples to the massive W and Z bosons. This follows from Figure 8 and the couplings are depicted in Figure 11 (a). Its coupling to elementary fermions similarly follows from the couplings in Figure 9 as shown in Figure 11 (b). The coupling to the massless photons is a genuine quantum effect involving loops, even in the lowest order, as indicated in Figure 12.

The LHC site circling under the French-Swiss border is schematically indicated in the picture of Figure 13. The 27 km circular tunnel containing two

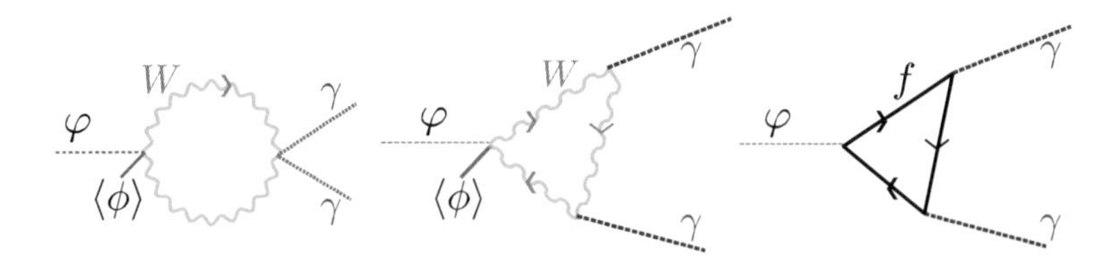

FIGURE 12. Coupling of the scalar boson φ to photons.

opposite beams of protons surrounded by guiding superconducting electromagnets cooled by superfluid helium is pictured in Figure 14. Figures 15 and 16 are pictures of the ATLAS and CMS detectors at diametrical opposite sites of the tunnel. There collisions occur and were used primarily to detect and identify the scalar boson of the Standard Model (and possibly other ones). At the end of 2012 proton-proton collisions occurred at the rate of nearly $10^9 s^{-1}$ and the proton energy reached 8 TeV. At these energies, all quarks of Figure 10 and the gluons connecting them may contribute to the production of the scalar boson. The leading production processes are represented in Figure 17.

As an example of the data gathered by CMS and ATLAS, Figure 18 presents the data obtained by the CMS group of observed decays into 4 leptons at the end

FIGURE 13. Schematic location of the LHC.

FIGURE 14. The LHC dipole magnets.

FIGURE 15. The Atlas detector.

FIGURE 16. The CMS detector.

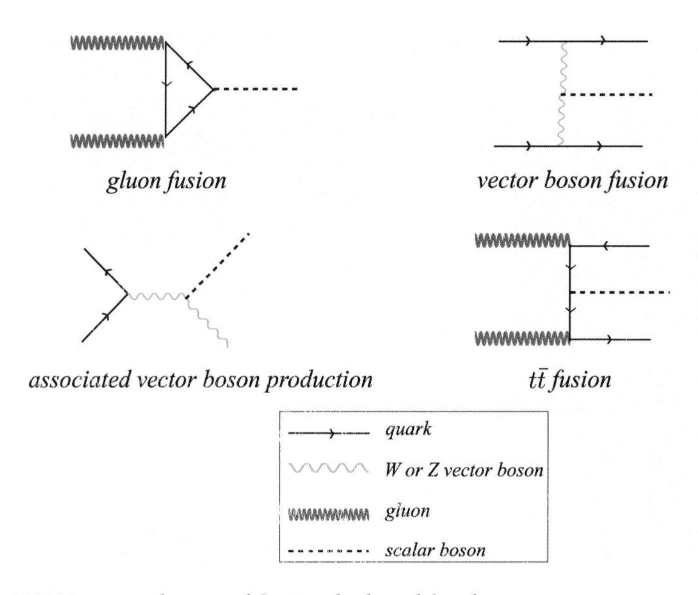

FIGURE 17. Production of the Standard Model scalar.

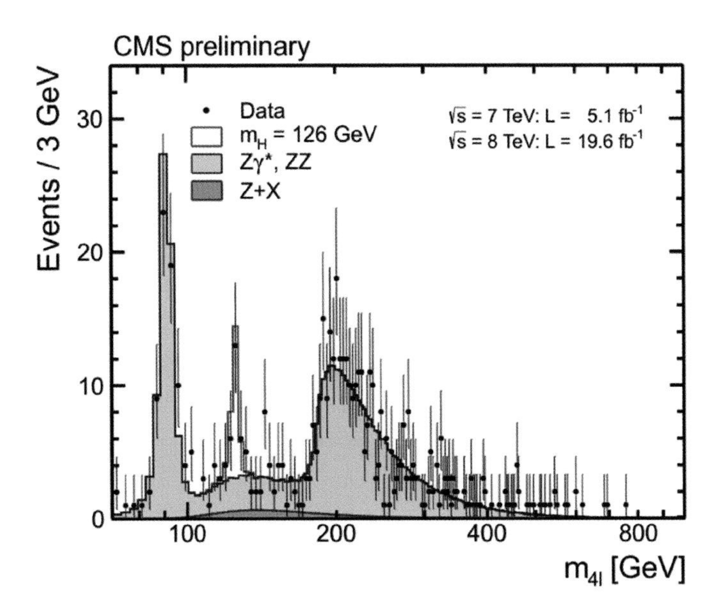

FIGURE 18. Decay of the scalar boson into 4 leptons from two *Z,s*.

of the 2012 run. The blue area is the expected background, namely those decays which would follow from the Standard Model if, at given total mass, there would be no contribution from the scalar boson. The red curve measures the contribution that could be due to the scalar decaying into two *Z* vector bosons which further decay into leptons, as was confirmed by further analysis. Note that one of the *Z* is real but the other is "virtual," meaning that this decay is forbidden by energy conservation but may contribute in the quantum theory. Consideration of other decay channels and spin analysis show that the particle detected is consistent with the Standard Model scalar with a mass $m_H \simeq 125$ GeV. The absence of new particles at comparable energies, as well as the success of Feynman graph analysis including loops, points towards an elementary particle, at least up to the energy range considered. This is the first elementary spin zero particle ever detected. It raises the interesting possibility of supersymmetry broken at attainable energies, although there is no indication of it so far.

The elementary character of the scalar already eliminates many dynamical models of symmetry breaking and raises interesting possibilities for extrapolation beyond presently known energies, up to those close to the Planck scale where quantum gravity effects might play a dominant role. The analysis of these speculations is beyond the scope of this talk.

REFERENCES

1. F. Englert and R. Brout, "Broken symmetry and the mass of gauge vector mesons," *Phys. Rev. Letters* **13** (1964) 321.
2. P.W. Higgs, "Broken symmetries and the masses of gauge bosons," *Phys. Rev. Lett.* **13** (1964) 508.
3. L.D. Landau, "On the theory of phase transitions I," *Phys. Z. Sowjet.* **11** (1937) 26 [JETP **7** (1937) 19].
4. Y. Nambu, "Quasi-particles and gauge invariance in the theory of superconductivity," *Phys. Rev.* **117** (1960) 648.
5. Y. Nambu, "Axial vector current conservation in weak interactions," *Phys. Rev. Lett.* **4** (1960) 380.
6. Y. Nambu and G. Jona-Lasinio, "Dynamical model of elementary particles based on an analogy with superconductivity I, II," *Phys. Rev.* **122** (1961) 345; **124** (1961) 246.
7. P.W. Anderson, "Random-phase approximation in the theory of superconductivity," *Phys. Rev.* **112** (1958) 1900.
8. P.W. Anderson, "Plasmons, gauge invariance, and mass," *Phys. Rev.* **130** (1963) 439.
9. J. Goldstone, "Field theories with 'superconductor' solutions," *Il Nuovo Cimento* **19** (1961) 154.
10. J. Goldstone, A. Salam and S. Weinberg, "Broken symmetries," *Phys. Rev.* **127** (1962) 965.
11. F. Englert, "Broken symmetry and Yang-Mills theory," in *50 years of Yang-Mills Theory*, ed. by G.'t Hooft, World Scientific (2005), pp. 65–95, hep-th/0406162.
12. S. Elitzur, "Impossibility of spontaneously breaking local symmetries," *Phys. Rev.* **D12** (1975) 3978.
13. P.W. Higgs, "Broken symmetries, massless particles and gauge fields," *Phys. Lett.* **12** (1964) 132.
14. G.S. Guralnik, C.R. Hagen and T.W.B. Kibble, "Global conservation laws and massless particles," *Phys. Rev. Lett.* **13** (1964) 585.
15. F. Englert, R. Brout and M. Thiry, "Vector mesons in presence of broken symmetry," *Il Nuovo Cimento* **43A** (1966) 244; F. Englert, *Proceedings of the 1997 Solvay Conference, Fundamental Problems in Elementary Particle Physics*, Interscience Publishers J. Wiley and Sons, p. 18.
16. G. 't Hooft, "Renormalizable lagrangians for massive Yang-Mills fields," *Nucl. Phys.* **B35** (1971) 167; G. 't Hooft and M. Veltman, Regularization and renormalization of gauge fields, *Nucl. Phys.* **B44** (1972) 189.
17. B.W. Lee and J. Zinn-Justin, "Spontaneously broken gauge symmetries," *Phys. Rev.* **D5** (1972) 3121; 3137; 3155.
18. S.L. Glashow, "Partial-symmetries of weak interactions," *Nucl. Phys.* **22** (1961) 579; S. Weinberg, A model of leptons, *Phys. Rev. Lett.* **19** (1967) 1264; A. Salam,

Proceedings of the 8th Nobel Symposium, Elementary Particle Physics, ed. by N. Svartholm, (Almqvist and Wiksell, Stockholm) p. 367.

19. S.L. Glashow, J. Iliopoulos and L. Maiani, "Weak interactions with lepton-hadron symmetry," *Phys. Rev.* **D2** (1970) 1285.
20. M. Kobayashi and T. Maskawa, "CP Violation in the renormalizable theory of weak interaction," *Prog. Theor. Phys.* **49** (1973) 652.

Peter Ware Higgs. © Nobel Media AB. Photo: A. Mahmoud

Peter Ware Higgs

Peter Higgs was born on 29 May 1929 in the Elswick district of Newcastle upon Tyne, UK. He graduated with First Class Honours in Physics from King's College, University of London, in 1950. A year later, he was awarded an MSc and started research, initially under the supervision of Charles Coulson and, subsequently, Christopher Longuet-Higgins. In 1954, he was awarded a PhD for a thesis entitled 'Some Problems in the Theory of Molecular Vibrations', work which signalled the start of his life-long interest in the application of the ideas of symmetry to physical systems.

In 1954, Peter Higgs moved to the University of Edinburgh for his second year as a Royal Commission for the Exhibition of 1851 Senior Student, and remained for a further year as a Senior Research Fellow. He returned to London in 1956 to take up an ICI Research Fellowship, spending a year at University College and a little over a year at Imperial College, before taking up an appointment as Temporary Lecturer in Mathematics at University College. In October 1960 Peter Higgs returned to Edinburgh, taking up a lectureship in Mathematical Physics at the Tait Institute. He was promoted to Reader in 1970, became a Fellow of the Royal Society of Edinburgh in 1974 and was promoted to a Personal Chair of Theoretical Physics in 1980. He was elected Fellow of the Royal Society in 1983 and Fellow of the Institute of Physics in 1991. He retired in 1996, becoming Professor Emeritus at the University of Edinburgh. He was awarded Fellowship of the University of Swansea in 2008, Honorary Membership of the Saltire Society and Fellowships of the Royal Scottish Society of the Arts and the Science Museum London in 2013.

Peter Higgs' contribution to physics has been recognised by numerous academic honours: the Hughes Medal of the Royal Society (1981, shared with Tom Kibble), the Rutherford Medal of the Institute of Physics (1984, also shared with Tom Kibble), the Saltire Society & Royal Bank of Scotland Scottish Science

Award (1990), the Royal Society of Edinburgh James Scott Prize Lectureship (1993), the Paul Dirac Medal and Prize of the Institute of Physics (1997), and the High Energy and Particle Physics Prize of the European Physical Society (1997, shared with Robert Brout and François Englert), the Royal Medal of the Royal Society of Edinburgh (2000), the Wolf Prize in Physics (2004, shared with Robert Brout and François Englert), the Royal Swedish Academy of Sciences Oskar Klein Memorial Lecture and Medal (2009) and the American Physical Society J. J. Sakurai Prize (2010), shared with Robert Brout, François Englert, Gerry Guralnik, Carl Hagen and Tom Kibble. He received a unique personal Higgs medal from the Royal Society of Edinburgh on 1 October 2012 and the 2013 Nonino Prize 'Man of Our Time'. He shared the award of the 2013 Edinburgh International Science Festival Edinburgh Medal with CERN and the 2013 Prince of Asturias Award for Technical and Scientific Research with François Englert and CERN.

He has received honorary degrees from the Universities of Bristol (1997), Edinburgh (1998), Glasgow (2002), King's College London (2009), University College London (2010), Cambridge (2012), Heriot-Watt (2012), Manchester, (2013), Durham (2013), La Scuola Internazionale Superiore di Studi Avantzi di Trieste (2013), St. Andrews (2014) and the Université Libre de Bruxelles (2014).

In 2011 he was awarded the Edinburgh Award for his outstanding contribution to the city. In the 2013 New Year Honours List he was appointed a Companion of Honour. In 2013 he was granted the Freedom of the City of Bristol. In 2014 he was awarded the Freedom of the City of Newcastle, his birthplace and the Freedom of the City of Edinburgh.

Date and Place of Birth

29 May 1929 at Newcastle upon Tyne, Northumbria, United Kingdom

Childhood

- Birmingham, 1930–1941
- Bristol, 1941–1946

Secondary Education

- Halesowen Grammar School, Worcestershire, 1940–1941
- Cotham Grammar School, Bristol, 1941–1946
- City of London School, 1946–1947

University Education

- King's College, University of London, 1947–1954

Degrees

- BSc (First Class Honours) in Physics, 1950
- MSc, 1951
- PhD, 1954

Professional Career

- Royal Commission for the Exhibition of 1851 Senior Student
  - King's College London, 1953–1954
  - University of Edinburgh, 1954–1955
- Senior Research Fellow, University of Edinburgh, 1955–1956
- ICI Research Fellow, University of London
  - University College, 1956–1957
  - Imperial College, 1957–Dec 1958
- Temporary Lectureship in Mathematics, University College, Jan 1959–1960
- Lecturer in Mathematical Physics, University of Edinburgh, 1960–1970
- (On leave at University of North Carolina, Chapel Hill, 1965–1966)
- Reader in Mathematical Physics, University of Edinburgh, 1970–1980
- (On leave at CERN, Geneva, Oct–Dec 1976)
- Professor of Theoretical Physics, University of Edinburgh, 1980–1996
- Professor Emeritus, University of Edinburgh, 1996–

Fellowships

- Fellow of the Royal Society of Edinburgh (FRSE), 1974
- Fellow of the Royal Society, London (FRS), 1983
- Fellow of the Institute Of Physics (FInstP), 1991
- Fellow of the King's College London, 1998
- Honorary Fellow of the Institute of Physics, 1999
- Fellow of the University of Swansea, 2008
- Honorary Fellow of the Royal Scottish Society of Arts, 2013
- Honorary Member of the Saltire Society, 2013
- Honorary Fellow of the London Science Museum, 2013

Prizes

- Hughes Medal, Royal Society (with T. W. B. Kibble), 1981
- Rutherford Medal, Institute of Physics (with T. W. B. Kibble), 1984
- Scottish Science Award, Saltire Society and Royal Bank of Scotland, 1990
- James Scott Prize Lectureship, Royal Society of Edinburgh (delivered April 1995), 1993
- Paul Dirac Medal and Prize, Institute of Physics, 1997
- High Energy and Particle Physics Prize, European Physical Society (with R. Brout, F. Englert), 1997
- Royal Medal, Royal Society of Edinburgh, 2000
- Wolf Prize in Physics (with R. Brout and F. Englert), 2004
- Oskar Klein Memorial Lecture and Medal, Royal Swedish Academy of Sciences, 2009
- J. J. Sakurai Prize, American Physical Society (with R. Brout, F. Englert, G. S. Guralnik, C. R. Hagen and T. W. B. Kibble), 2010
- Higgs Medal, Royal Society of Edinburgh, 2012
- Nonino 'Man of Our Time' Prize, 2013
- Edinburgh Medal of the Edinburgh International Science Festival (with CERN), 2013
- Prince of Asturias Award for Technical and Scientific Researchl (with F. Englert and CERN), 2013
- Nobel Prize in Physics (with F. Englert), 2013

Honorary Degrees

- DSc University of Bristol, 1997
- DSc University of Edinburgh, 1998
- DSc University of Glasgow, 2002
- DSc King's College London, 2009
- DSc University College London, 2010
- DSc University of Cambridge, 2012
- DSc Heriot-Watt University, 2012
- DSc University of Durham, 2013
- PhD La Scuola Internazionale Superiore di Studi Avanzati (SISSA) di Trieste, 2013
- DSc University of Manchester, 2013
- DSc University of St Andrews, 2014
- DSc Université Libre de Bruxelles, 2014

Other Awards

- Freedom of the City of Bristol, 2013
- Freedom of the City of Newcastle, 2014
- Freedom of the City of Edinburgh, 2014

Publications

"Theoretical Determination of Electron Density in Organic Molecules," (with C.A. Coulson, S.L. Altmann and N. H. March) *Nature* **168**, 1039 (1951).

"Perturbation Method for the Calculation of Molecular Vibration Frequencies I," *J. Chem. Phys.* **21**, 1131 (1953).

"A Method for Computing Zero-Point Energies," *J. Chem. Phys.* **21**, 1330 (1953).

"Vibration Spectra of Helical Molecules," *Proc. Roy. Soc.* **A220**, 472 (1953).

"Vibrational Modifications of the Electron Density in Molecular Crystals I," *Acta. Cryst.* **6**, 232 (1953).

"Perturbation Method for the Calculation of Molecular Vibration Frequencies II," *J. Chem. Phys.* **23**, 1448 (1955).

"Perturbation Method for the Calculation of Molecular Vibration Frequencies III," *J. Chem. Phys.* **23**, 1450 (1955).

"Vibrational Modifications of the Electron Density in Molecular Crystals II," *Acta. Cryst.* **8**, 99 (1955).

"A Method for Calculating Thermal Vibration Amplitudes from Spectroscopic Data" *Acta. Cryst.* **8**, 619 (1955).

"Vacuum Expectation Values as Sums over Histories," *Nuovo Cimento* (10) **4**, 1262 (1956).

"On Four-Dimensional Isobaric Spin Formalisms," *Nuclear Physics* **4**, 1262 (1957).

"Integration of Secondary Constraints in Quantized General Relativity," *Phys. Rev. Lett.* **1**, 373 (1958).

"Integration of Secondary Constraints in Quantized General Relativity," *Phys. Rev. Lett.* **3**, 66 (1959).

"Quadratic Lagrangians and General Relativity," *Nuovo Cimento* (10) **11**, 816 (1959).

"Broken Symmetries, Massless Particles and Gauge Fields," *Physics Letters* **12**, 132 (1964).

"Broken Symmetries and the Masses of Gauge Bosons," *Phys. Rev. Letters.* **13**, 508 (1964).

"Spontaneous Symmetry Breakdown without Massless Bosons," *Phys. Rev.* **145**, 1156 (1966).

"Spontaneous Symmetry Breaking," two lectures at the 14th Scottish Universities Summer School in Physics (1973). Published in *Phenomenology of Particles at High Energy*, R. L. Crawford, R. Jennings (eds.) Academic Press (1974) ISBN 9780121971502.

"Dynamical Symmetries in a Spherical Geometry I," *J. Phys.* **A12**, 309 (1979).

"SBGT and All That," International Conference "50 Years of Weak Interactions from the Fermi Theory to the W" Wingspread, Racine, Wisconsin (29 May–1 June 1984). Published in the conference proceedings by University of Wisconsin at Madison and reproduced in *AIP. Conf. Proc.* **300**,159–163 (1994).

"Inventing an Elementary Particle," INFN Eloisatron Project 9th Workshop "Higgs Particles – Physics Issues and Experimental Searches in High-energy Collisions," Erice,

Italy (15–26 Jul 1989). Published in *Higg(s) Particle(s): Physics Issues and Experimental Searches in High-Energy Collisions*, A. Ali (ed.) Ettore Majorana International Science Series **50**, 1–5 Plenum Press (1990) ISBN 9780306435898.

"Spontaneous Symmetry Breaking 25 Years Ago," 26th International Conference on Subnuclear Physics Physics up to 200 TeV, Erice, Italy (16–24 Jul 1990). Published in *Physics up to 200TeV*, A. Zichichi (ed.) The Subnuclear Series **28**, 439–444 Plenum Press (1991) ISBN 9780306439353.

Panel Session "Spontaneous Breaking of Symmetry," (with L. M. Brown, R. Brout, T. Y. Cao, Y. Nambu) 3rd International Symposium on the History of Particle Physics "The Rise of the Standard Model" (1992): published in *The Rise of the Standard Model*, L. Hoddesdon, L. M. Brown, M. Riordan, M. Dresden (eds.) Cambridge University Press, (1997) ISBN 978052157165.

"My Life as a Boson: The Story of 'The Higgs'," Inaugural Conference of the Michagan Center for Theoretical Physics "2001 A Spacetime Odyssey" Ann Arbor, Michigan (21–25 May 2002). Published in *2001 A Spacetime Odyssey*, M. J. Duff, J. T. Liu (eds.) World Scientific (2002) ISBN 9789810248062 and reproduced in Int. J. Mod. Phys. **A17S1**, 86–88 (2002).

"Prehistory of the Higgs Boson," *Comptes Rendus Physique* **8**, 970–972 (2007).

Evading the Goldstone Theorem

Nobel Lecture, 8 December 2013

by Peter Higgs
University of Edinburgh, Edinburgh, United Kingdom.

My story begins in 1960, when I was appointed Lecturer in Mathematical Physics at the University of Edinburgh. Before I took up my appointment, I was invited to serve on the committee of the first Scottish Universities Summer School in Physics. I was asked to act as Steward at the School in July, my principal duty being to purchase and look after supplies of the wine which was to be served at dinner each evening.

The students at the School included four who stayed up late into the night in the common room of Newbattle Abbey College (the crypt of a former abbey) discussing theoretical physics, and rarely got up in time for the first lecture of the following day. They were Dr. N. Cabibbo (Rome), Dr. S. L. Glashow (CERN), Mr. D. W. Robinson (Oxford) and Mr. M.J.G. Veltman (Utrecht). Many years later, Cabibbo told me that their discussions had been lubricated by bottles of wine collected after dinner and hidden inside the grandfather clock in the crypt.

I did not take part in these discussions, since I had other things to do (such as conserving wine). Consequently, I did not learn about Glashow's paper on electroweak unification, which had already been written.

BROKEN SYMMETRIES

During my first year as a lecturer, I was in search of a worthwhile research programme. In the previous four years in London I had rather lost my way in particle physics and had become interested in quantum gravity. Symmetry had fascinated me since my student days, and I was puzzled by the approximate symmetries (what are now called flavour symmetries) of particle physics.

Then in 1961, I read Nambu's and Goldstone's papers on models of symmetry breaking in particle physics based on an analogy with the theory of superconductivity. (Nambu's models were inspired by the Bardeen, Cooper & Schrieffer theory, based on Bose condensation of Cooper pairs of electrons: Goldstone used scalar fields, with a 'wine bottle' potential to induce Bose condensation, as in the earlier Ginzburg-Landau theory.) What I found very attractive was the concept of a *spontaneously* broken symmetry, one that is exact in the underlying dynamics but appears broken in the observed phenomena as a consequence of an asymmetric ground state ("vacuum" in quantum field theory).

Most particle theorists at the time did not pay much attention to the ideas of Nambu and Goldstone. Quantum field theory was out of fashion, despite its successes in quantum electrodynamics; it was failing to describe either the strong or the weak interactions.

Besides, condensed matter physics was commonly viewed as another country. At a Cornell seminar in 1960, Victor Weisskopf remarked (as recalled by Robert Brout)

> Particle physicists are so desperate these days that they have to borrow from the new things coming up in many body physics—like BCS. Perhaps something will come of it.

THE GOLDSTONE THEOREM

There was an obstacle to the success of the Nambu-Goldstone programme. Nambu had shown how spontaneous breaking of a chiral symmetry could generate the masses of spin-½ particles, such as the proton and neutron, but his model predicted massless spin-0 particles (pions?), contrary to experimental evidence. (As noted by Weinberg, any such particles would dominate the radiation of energy from stars). Goldstone had argued that such massless particles would always be the result of excitations around the trough of the wine bottle potential.

In 1962 a paper entitled "Broken Symmetries" by Goldstone, Salam and Weinberg proved the "Goldstone Theorem," that "In a manifestly Lorentz-invariant quantum field theory, if there is a continuous symmetry under which the Lagrangian is invariant, then *either* the vacuum state is also invariant *or* there must exist spinless particles of zero mass."

This theorem appeared to put an end to Nambu's programme.

CAN ONE EVADE THE GOLDSTONE THEOREM?

In 1963 the condensed matter theorist Phil Anderson pointed out that in a superconductor the Goldstone mode becomes a massive "plasmon" mode due to long-range (Coulomb) forces, and that this mode is just the longitudinal partner of transverse electromagnetic modes, which are also massive. Anderson remarked "The Goldstone zero-mass difficulty is not a serious one, because we can probably cancel it off against an equal Yang-Mills zero-mass problem." However, he did not show that there was a flaw in the Goldstone theorem and he did *not* discuss any relativistic model, so particle theorists such as myself received his remark with scepticism.

In March 1964 Abe Klein and Ben Lee suggested that, even in relativistic theories, a certain equation which was crucial for the proof of the Goldstone theorem could be modified by the addition of an extra term, just as in condensed matter theories. But in June Wally Gilbert (who was in transition from theoretical physics to molecular biology, for which he later won a Nobel Prize for Chemistry) ruled out this term as a violation of Lorentz invariance.

It was at this point that my intervention took place.

HOW TO EVADE THE GOLDSTONE THEOREM

I read Gilbert's paper on 16 July 1964—it had been published a month earlier, but in those days the University of Edinburgh's copies of *Physical Review Letters* came by sea—and I was upset because it implied that there was no way to evade Goldstone's theorem. But over the following weekend I began to recall that I had seen similar apparent violations of Lorentz invariance elsewhere, in no less a theory than quantum electrodynamics, as formulated by Julian Schwinger.

Quantum electrodynamics is invariant under gauge transformations and the gauge must be fixed before well-defined quantum formalism can be set up. The fashionable way to do this was to choose a Lorentz gauge, which was manifestly compatible with relativity. However, such a gauge had unsatisfactory features that led Schwinger to prefer a Coulomb gauge, which introduces an apparent conflict with relativity. Nevertheless, it was well known that this choice did not lead to any conflict between the predicted physics and relativity.

Schwinger had, as recently as 1962, written papers in which he demolished the folklore that it is gauge invariance alone that requires photons to be massless. He had provided examples of some properties of a gauge theory containing *massive* "photons," but without describing explicitly the underlying dynamics.

During the weekend of 18–19 July it occurred to me that Schwinger's way of formulating gauge theories undermined the axioms which had been used to prove the Goldstone theorem. So gauge theories might save Nambu's programme.

During the following week I wrote a short paper about this. It was sent to *Physics Letters* on 24 July and was accepted for publication.

By then I had written down the (classical) field equations of the simplest illustrative model that I could imagine, the result of introducing an electromagnetic interaction into Goldstone's simplest scalar model. It became obvious that in this model the Goldstone massless mode became the longitudinal polarization of a massive spin-1 "photon," just as Anderson had suggested.

My second short paper, consisting of a brief account of this model, was sent to *Physics Letters* on 31 July. It was rejected. The editor (at CERN) suggested that I develop my ideas further and write a full account for *Il Nuovo Cimento.*

I was indignant; it seemed that the referee had not seen the point of my paper. (Later, a colleague who returned from a month's visit to CERN told me that the theorists there did not think it had any relevance to particle physics.) Besides, it seemed odd that the earlier paper had been accepted but the more physical sequel had not.

I decided to augment the paper by some remarks on possible physical consequences, and to send the revised version across the Atlantic to *Physical Review Letters.* Among the additional material was the remark, "It is worth noting that an essential feature of this type of theory is the prediction of incomplete multiplets of scalar and vector bosons."

The revised paper was received by *Physical Review Letters* on 31 August and was accepted. The referee invited me to comment on the relation of my paper to that of Englert and Brout, whose paper (received on 22 June) had been published that day. Until then I had been unaware of their work, but I added a footnote to my paper as soon as I had received a copy of theirs. Twenty years later, at a conference in 1984, I met Nambu, who revealed that he had refereed both papers.

POSTSCRIPT

It took some time for the work of Englert and Brout and myself (and of Guralnik, Hagen and Kibble, who published a little later) to gain acceptance.

My longer (1966) paper was written in autumn 1965 at Chapel Hill, North Carolina, where I was spending a sabbatical year at the invitation of Bryce DeWitt as a consequence of my interest in quantum gravity. A preprint sent to

Freeman Dyson received a positive response; he invited me to give a talk at I.A.S. Princeton. There, in March 1966, I faced an audience including axiomatic quantum field theorists who still believed that there could be no exceptions to the Goldstone theorem.

The next day I gave a talk at Harvard (arranged by Stanley Deser) to another sceptical audience, including Wally Gilbert. I survived this too. After the seminar Shelly Glashow complimented me on having invented 'a nice model,' but he did not recognise its relevance to his electroweak theory—a missed opportunity!

Like Nambu, the six of us who published in 1964 expected to apply our ideas to the broken *flavour* symmetries of the strong interactions, but this did not work. So it was left to Weinberg and Salam in 1967 to find the right application.

Four more years passed before Gerard 't Hooft, in an extension of Veltman's programme, proved the renormalisability of such theories and another two before the discovery of weak neutral currents indicated that Glashow's electroweak unification was the correct one. And in 1976 Ellis, Gaillard and Nanopoulos at CERN encouraged experimentalists to look for the massive spinless boson that the theory predicted.

Physics 2014

Isamu Akasaki, Hiroshi Amano and Shuji Nakamura

"for the invention of efficient blue light-emitting diodes which has enabled bright and energy-saving white light sources"

The Nobel Prize in Physics

Speech by Professor Anne L'Huillier of the Royal Swedish Academy of Sciences.

Your Majesties, Your Royal Highnesses, Ladies and Gentlemen,

In many fairy tales, light is a power that is used to overcome the forces of evil. In *The Lord of the Rings*, written by J.R.R. Tolkien in 1954, the elf queen gives the ring-bearer a glittering crystal vial and says, "May it be a light to you in dark places, when all other lights go out."

When our ancestors learned to master the fire about 300,000 years ago, they used it not only as a source of heat or as a weapon against wild animals. The fire shined in the dark when all other lights had gone out.

In the late 19th century, the American inventor Thomas Edison was able to improve the incandescent light bulb to make it usable. It has been said that it took him around 2,000 experiments to succeed. When people said that he had failed, he replied, "No, I have not failed. I have found two thousand ways not to make a light bulb." As the electrical power grid expanded and the light bulb improved, millions and soon billions of people gained a cheap and sustainable source of lighting. The light bulb shined in the dark when all other lights had gone out.

After the Second World War, electronics based on semiconductor materials was developed. Scientists discovered at an early stage that certain combinations of semiconductor materials, diodes, could emit light. In a light bulb, a filament is heated so that it begins to shine. Light emitting diodes transform electricity directly into light and are therefore more efficient than light bulbs. In Europe, they were called crystal lamps, since they are made of crystals. In the United States, they were called LEDs.

The first LEDs emitted red light. Soon afterward came LEDs that emitted green light. If you mix red, green and blue light, you get white. Creating white light sources also required blue LEDs. From the 1960s, industry and universities invested a lot of effort to create blue LEDs, but the necessary gallium nitride crystals turned into a powder. Many researchers tried, failed and gave up. In the early 1970s, a new technique for growing gallium nitride crystals was developed. This led to new research efforts, new experiments and new failures. It was simply too difficult.

Isamu Akasaki and his PhD student Hiroshi Amano at Nagoya University in Japan, as well as Shuji Nakamura, then researcher at a small company in Tokushima, succeeded where all others had given up. After several years of hard work in the late 1980s – with great persistence, skill and perhaps a bit of luck – they were able to produce fine gallium nitride crystals and give them the properties needed for efficient light emission. The Laureates certainly performed a lot more than 2,000 experiments, and they found many ways not to make a blue LED. But they succeeded in the end.

Thanks to the blue light emitting diode, it became possible to produce white lamps. Today these are everywhere – in our mobile telephones, our bicycles, our cars, our cities and our homes. By using an LED lamp for lighting instead of an incandescent light bulb or a fluorescent lamp, we save energy and thereby protect the environment. An LED lamp also has an incredibly long lifetime: approximately 100,000 hours, or about 11 years. Since it is so efficient, it can shine for many hours using the electricity from a battery, which can be charged using sunlight. LED lamps light up places on earth where there is no electrical power grid. They shine in the dark, when all other lights have gone out.

More than a century ago, Alfred Nobel wrote in his will that the Nobel Prize in Physics should be awarded to those who shall have conferred the greatest benefit to mankind. This year's prize fulfils Alfred Nobel's wish extremely well.

Professor Akasaki, Professor Amano, Professor Nakamura,

You have been awarded the 2014 Nobel Prize in Physics for your invention of efficient blue light emitting diodes, which has enabled bright and energysaving white light sources. On behalf of the Royal Swedish Academy of Sciences, it is my honour and pleasure to convey to you the warmest congratulations for your outstanding work. I now ask you to step forward to receive your Nobel Prizes from the hand of His Majesty the King.

Isamu Akasaki. © Nobel Media AB. Photo: A. Mahmoud

Isamu Akasaki did not submit an autobiography. See https://www.nobelprize.org/prizes/physics/2014/akasaki/biographical/

Fascinated Journeys into Blue Light

Nobel Lecture, December 8, 2014

by Isamu Akasaki
Meijo University, 1-501 Shiogama-guchi, Tempaku-ku, Nagoya 468-8502 Japan, and
Nagoya University Akasaki Research Center, Furo-cho, Chikusa-ku, Nagoya 464-8601 Japan

1. INTRODUCTION

"In the beginning there was light," emphasizes how closely light is tied to our lives. Light is indispensable for mankind and for many other creatures, and humans have pursued light sources since ancient times. Starting with flame, humans have developed electric light bulbs, fluorescent lamps, and then semiconductor light-emitting devices (light-emitting diodes (LEDs) and laser diodes (LDs)) in the second half of the last century. Although these light sources cover a wide wavelength range, the development of high-energy light sources has largely lagged behind. Development of an efficient blue LED had been a long-term dream for researchers worldwide, since it is indispensable for realizing LED-based full-color displays and general lighting applications.

Drastic improvements in the crystal quality of gallium nitride (GaN) [1] and the ability to control the conductivity in both p- and n-type nitride semiconductors [2, 3] in the late 1980s, have enabled the production of high-brightness GaN-based p-n junction blue/ultraviolet (UV) LEDs [2], high-performance blue-violet LDs [4] and many other novel devices. These successes triggered the opening of an entirely new field of electronics.

In this paper, I would like to describe the historical progress that led to the invention of the first p-n junction blue/UV LED and related optical devices.

2. LED RESEARCH IN THE EARLY DAYS

In 1962, a red LED based on gallium arsenide phosphide (GaAsP) alloys was developed by N. Holonyak Jr. and S. F. Bevacqua [5]. This was the first LED in the world to emit visible light. In 1968, a green LED was produced by R. A. Logan and his colleagues based on nitrogen-doped gallium phosphide (GaP: N) [6]. At that time, however, there was no prospect of developing practical blue-light-emitting devices, which operate at the shortest wavelength in the visible spectrum and produce the highest energy.

The energy of photons from light-emitting semiconductor devices, such as LEDs, is approximately equal to the bandgap energy (Eg) for the semiconductor that is being used. The wavelength of blue light is in the range 445–480 nanometers (nm), which is equivalent to a bandgap energy of 2.6–2.8 eV. There were therefore two requirements for creating blue-light-emitting devices.

Requirement [**A**]: it is essential to use semiconductors with an Eg of approximately 2.6 eV or larger, equivalent to a wavelength of 480 nm or shorter (blue light). Semiconductors that have such a large Eg are referred to as "wide-bandgap semiconductors." In contrast, the Eg for the most commonly used semiconductor, silicon, is 1.1 eV.

Requirement [**B**]: it is advantageous to use direct bandgap semiconductors in which the momentum of electrons at the bottom of the conduction band is almost equal to that of holes at the top of the ground state valance band, as shown in Fig. 1, yielding a high radiative recombination probability. In

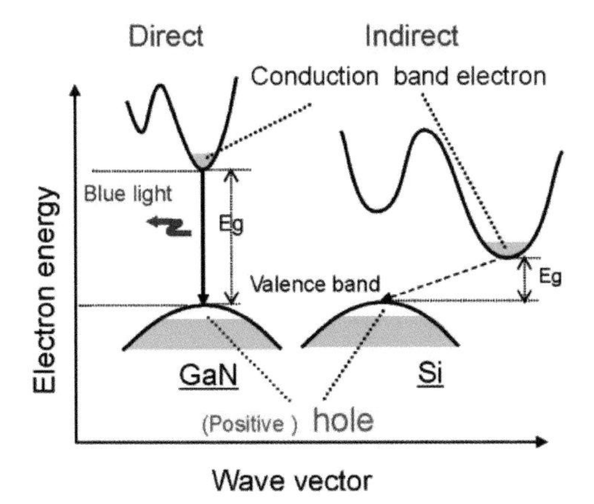

FIGURE 1. Band structures of GaN (direct transition type) and Si (indirect transition type).

contrast, indirect bandgap semiconductors exhibit a lower radiative recombination probability, because the momentum of these electrons and holes is different.

However, requirements **[A]** and **[B]** are not always sufficient conditions. To realize high-performance LEDs, it is essential to **(1)** grow high-quality single crystals, and **(2)** successfully produce p-n junctions (Fig. 2). Semiconductors that have more holes (electron deficiency) than electrons are referred to as p-type semiconductors, whereas those with more electrons than holes are referred to as n-type semiconductors. A p-n junction is an atomically continuous boundary between a p- and n-type semiconductor, and is necessary for the fabrication of devices such as highly-efficient light-emitters, solar cells, and transistors.

It is, however, extremely difficult to achieve **(1)** and **(2)** in wide-bandgap semiconductors, and this prevented the development of high-performance blue-light-emitting devices for many years.

In the late 1960s and the 1970s, candidate materials for blue-light-emitting devices included silicon carbide (6H-SiC) with an Eg of 3.0 eV, zinc selenide (ZnSe) with an Eg of 2.7 eV, and GaN with an Eg of 3.4 eV. Of those, SiC was the only wide-bandgap semiconductor for which p-n junctions could be created in those days, and some researchers attempted to develop blue LEDs based on this material [7]. I had, however, absolutely no interest in this material for photonic device applications because of its indirect band structure which prevented

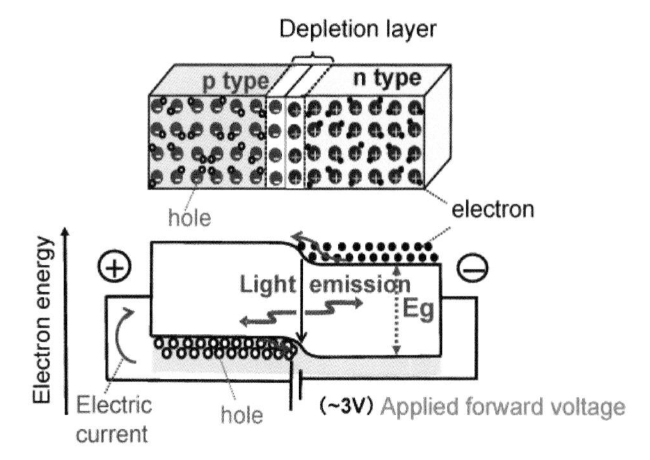

FIGURE 2. A schematic structure of a p-n junction LED.

TABLE 1. A comparison between ZnSe and GaN.

		ZnSe	GaN
[A]	Energy gap (Eg)	**2.7 eV**	**3.4 eV**
[B]	Energy band structure	**direct**	**direct**
[1]	Crystal growth	straightforward	**too difficult**
	Substrate	GsAs	sapphire
	Lattice mismatch	**0.26%**	**16%**
[2]	p-n junction	**not realized at that time**	
	Number of researchers	many	few
	Physical & chemical stabilty	**low**	**high**

efficient light emission. Meanwhile, although ZnSe and GaN as summarized in Table 1 were known to be direct bandgap semiconductors, it was difficult to grow large bulk crystals of these materials, and no p-type crystals had been realized at that time. When it is difficult to produce large bulk crystals, epitaxial growth of single-crystal thin films is generally used. In vapor-phase epitaxial growth, source materials for a crystal growth are provided to the substrate in a form of a gas. Growth then takes place such that there is general alignment between the crystallographic axes of the grown crystals and the substrate. This method has been widely used in the growth of high-quality semiconductor with nanostructures. The terms "homoepitaxy" or "heteroepitaxy" are used when the grown crystal is the same as or different from the substrate crystal, respectively. In the latter case, it is necessary for the lattice constants of the two crystals to be as similar as possible to each other.

ZnSe emits bright light under excitation by an electron beam. Good-quality single-crystal ZnSe film can be grown using vapor-phase epitaxial growth on GaAs single-crystal substrates because the lattice constants are very similar. Thus, many researchers had been working on ZnSe, aiming to develop blue-light-emitting devices.

I myself, however, worried about the instability of ZnSe due to its low cohesive energy (bonding energy), and its poor crystallinity because of the low growth temperature required. In fact a technique for p-type doping of ZnSe was developed later in 1988 [8], and a lasing operation in a zinc cadmium selenide (ZnCdSe)/ZnSe heterostructure was demonstrated in 1991 [9]. However, the lifetime of ZnSe-based optical devices was found to be very short, and the researchers gave up attempting to develop this material further.

3. BRIEF HISTORY OF RESEARCH ON GaN-BASED MATERIALS AND DEVICES

3.1. Early attempts at development of GaN blue LED

During the early stages of group-III nitride semiconductor research, I had an insight into the great potential of this material for blue-light-emitting devices, and yearned to pioneer a new field founded on the unique properties of nitrides, such as their toughness, wider direct energy gaps, and non-toxicity, while I was working from 1964 to 1981 at Matsushita Research Institute Tokyo, Inc. (MRIT).

In 1967, I and Masafumi Hashimoto at MRIT grew aluminum nitride (AlN) crystals by vapor phase reaction, and determined the angular frequencies of longitudinal and transverse optical phonons by fitting the calculated reflectivity to Reststrahlen (residual ray) [10]. It was very difficult, however, to use AlN as an electroluminescent material, because of its excessively large Eg of about 6.2 eV.

In the meantime (in 1969), H. P. Maruska and J. J. Tietjen successfully grew single-crystal GaN films on sapphire substrates using hydride vapor phase epitaxy (HVPE) and found that GaN is a direct bandgap semiconductor with an Eg of 3.34 eV at room temperature (RT) [11]. Then, in 1971, J. I. Pankove et al. developed GaN-based metal-insulator-semiconductor (MIS) type blue LEDs [12]. R. Dingle et al. observed stimulated emission and laser action in single crystal needles of GaN at 2–4 K [13]. These achievements intensified research and development of blue-light-emitters based on GaN (period (A) in Fig. 3).

In the mid-to-late 1970s, however, GaN researchers almost withdrew from the field, and activity on GaN-based devices declined (period (B) in Fig. 3), because they could neither grow high-quality semiconductor-grade GaN single crystal nor control the electrical conductivity of the material (realize p-type conduction in particular), both of which are indispensable for producing high-performance light-emitters based on a semiconductor p-n junction, although some researchers had continued to work on the basic and physical properties of GaN [14, 15].

Besides, at that time, theoretical studies indicated the impossibility of achieving p-type conduction in wide-bandgap semiconductors such as GaN and ZnSe due to the "self-compensation effect" [16].

Despite this stalemate, I started to work on the growth of GaN single crystal film by molecular beam epitaxy (MBE) in 1973, and then by HVPE in 1975, aiming at the development of GaN-based p-n junction LEDs and LDs. In 1978, by utilizing HVPE, my group at MRIT developed a MIS-type GaN blue LED with a unique device structure of as-grown highly n-type (n^+) GaN pillars buried in a thick n-GaN/a thin insulating GaN structure as shown in Fig. 4. The n^+-pillars could be used as cathodes, which greatly simplified the fabrication

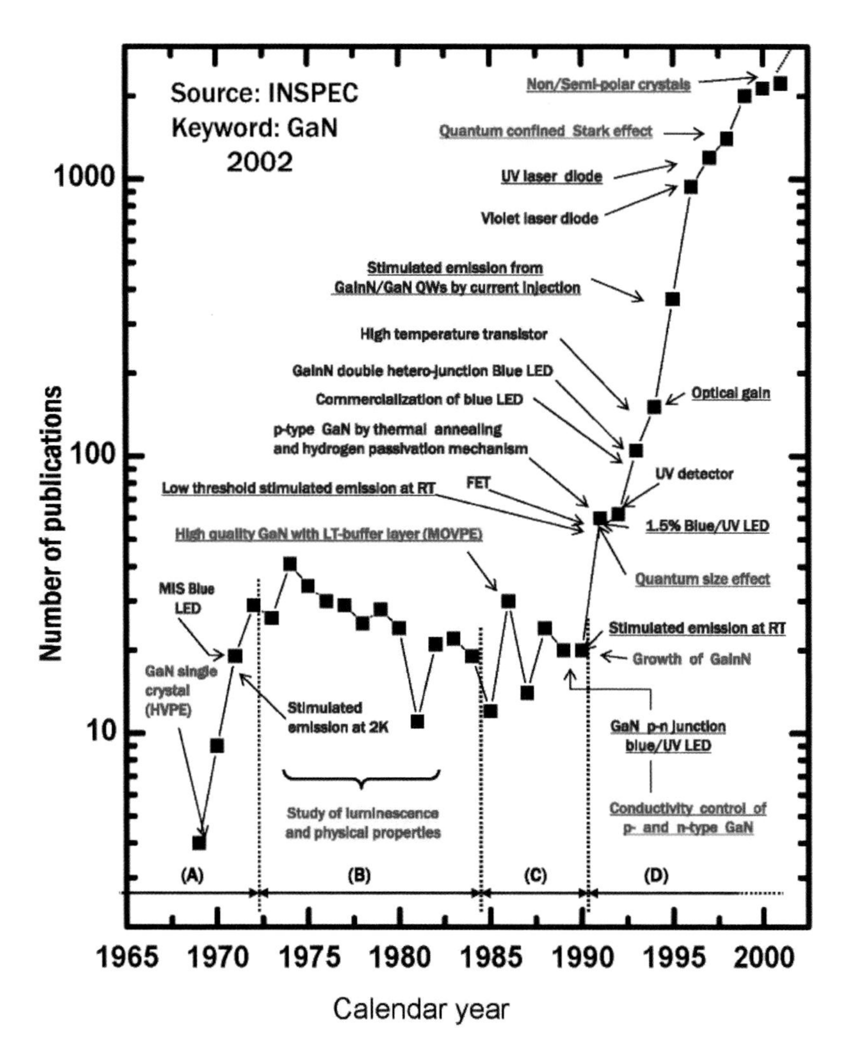

FIGURE 3. Number of publications (INSPEC) and activities related to nitrides from 1969 to 2002. All events are marked in the years when they were first achieved. Most of the important results were achieved by MOVPE using LT-buffer layer after 1986. It is clear that the start of the steep increase in number of publications and accomplishments is due to the key inventions (high-quality GaN, p- and n-type conductivity control, and p-n junction blue/UV LED) in the late 1980s and 1990. Achievements with underlines are works done by Akasaki's group. Green: Crystal Growth, Blue: Devices, Red: Conductivity Control and Physics.

of MIS-type LEDs [17, 18]. The external efficiency was 0.12%, which was the highest ever reported at that time. However, due to the use of a MIS structure, the operating voltage was high and the brightness was low, in contrast to p-n junction LEDs that we invented later [2].

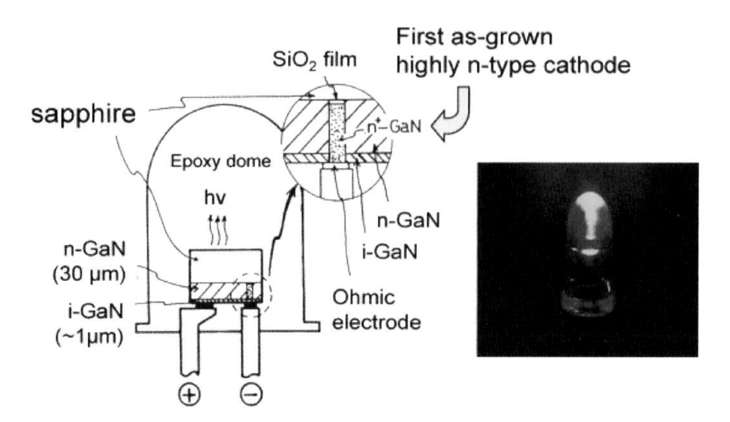

FIGURE 4. A schematic and a photograph of a MIS-type GaN blue LED developed in 1978.

3.2. Reconsideration of growth technology

In parallel with the work described above [18], I also recognized the great potential of GaN as a blue luminescent material, when I found tiny but high-quality crystallites embedded in HVPE-grown crystals containing many cracks and pits in the field of view of microscopes. I was intuitively convinced that it would be possible to achieve conductivity control (even p-type GaN) if this kind of quality could be obtained over an entire wafer.

Thus, in 1978, I made up my mind to go back one more time to the beginning, i.e., "crystal growth," which is an interdisciplinary sciences and essential for the realization of quantum devices with nanostructure.

This decision, I think of, as a major turning point not only in my own GaN research, but also GaN research and development throughout the world, which had been stagnating at that time (Period (B) in Fig. 3).

It is known that the quality of crystal is greatly affected by the nature of the chemical reactions involved in their production, in other words, the growth method and condition. Hence, the choice of growth method was critical for determining the future of the research. Epitaxial GaN can be grown by MBE, HVPE or metalorganic vapor phase epitaxy (MOVPE), the latter of which is also known as metalorganic chemical vapor deposition (MOCVD), as described in Table 2.

On the basis of my crystal growth experience, I realized that MBE was prone to introducing a nitrogen deficiency and the growth rate was very slow at that time. In the case of HVPE, the crystal quality was degraded by appreciable

TABLE 2. Crystal growth methods for GaN.

Method	Reference
Molecular Beam Epitaxy (MBE)	I. Akasaki: (1974) (in Japanese).
Ga (g) + NH_3 (g) = GaN (s) + $\frac{3}{2}$ H_2 (g) **Issues: Prone to nitrogen deficiency, slow growth rate (at that time)**	
Hydride Vapor Phase Epitaxy (HVPE)	H. P. Maruska and J. J. Tietjen: (1969).
GaCl(g) + NH_3 (g) = GaN (s) + HCl (g) + H_2 (g) **Issues: Susceptible to reverse reactions, too fast growth rate**	
Metalorganic Vapor Phase Epitaxy (MOVPE)	H. M. Manasevit et al: (1971).
$Ga(CH_3)_3$ (g) + NH_3 (g) $\rightarrow$ GaN (s) + $3CH_4$ (g) Advantages: • No reverse reactions • Easy to control growth rate, allow composition, and impurity-doping	

reverse reactions, and the growth rate was too high to fabricate devices with layer thicknesses on the order of nanometers. Therefore, these methods were not suitable for producing well-controlled devices based on high-quality GaN crystals.

On the other hand, MOVPE, which was firstly applied to the growth of GaN by H. M. Manasevit et al. in 1971 [19], but almost never employed for this purpose thereafter, seemed to be more suitable, because of the absence of reverse reactions. Furthermore, the composition of alloys such as aluminum gallium nitride (AlGaN) and gallium indium nitride (GaInN) and the level of impurity doping could be readily controlled by varying the flow rates of the source gases in MOVPE. Thus, in 1979, I decided to adopt MOVPE as the optimal crystal growth method for GaN. It was a crucial decision. As for the substrate for GaN growth, I tentatively (until a more suitable substrate would become available) chose the c-face of sapphire as before, because it was stable even under the harsh MOVPE conditions, namely a temperature above 1000 °C and an ammonia (NH_3) atmosphere, and is similar to GaN in terms of crystallographic symmetry.

The fact that, even today, GaN-based crystals and devices are mainly grown on sapphire substrates by MOVPE is a clear indication that my choices were not wrong.

4. CREATION OF GaN SINGLE CRYSTAL WITH EXCELLENT QUALITY

4.1. Development of low-temperature buffer layer technology in MOVPE

After making these crucial decisions, I returned to my old nest, Nagoya University, where I started anew to drastically improve the crystal quality of GaN grown by MOVPE in collaboration with my graduate students: Yasuo Koide and Hiroshi Amano, who put a lot of effort into crystal growth. Even with the MOVPE method, however, it was not easy for us to develop homogeneous GaN films. After many trials and errors, we made drastic innovations and improvements to the reactor tube and growth conditions.

The first improvement was that Koide mixed organometallic compounds such as trimethylgallium (TMGa) (and trimethylaluminum (TMAl) in the case of AlGaN growth) with NH_3 and hydrogen (H_2) gas as a carrier right in front of the opening of the reactor tube, and blew this mixture through a gas delivery tube onto a substrate that was inclined at a 45 degree angle rather than being placed horizontally as in previous attempts as shown in Fig. 5. We also drastically increased the flow velocity of the gases in the reactor tube from only 2 cm per second to approximately 110 cm per second. We were thus able to reduce the formation of adducts of NH_3 and the organometallic sources, and to suppress convective gas flows on the high-temperature substrate, which resulted in a uniform gas flows and the production of homogeneous GaN films.

Even though the film thickness was fairly constant over the entire wafer, this did not mean that there were no pits or cracks. There was also no substantial improvement in the electrical or optical properties, which suggested the presence of lattice defects and unintentionally incorporated impurities.

I suspected that for the most part, this was due to the large interfacial free energy between GaN and sapphire caused by the huge lattice mismatch of 16%

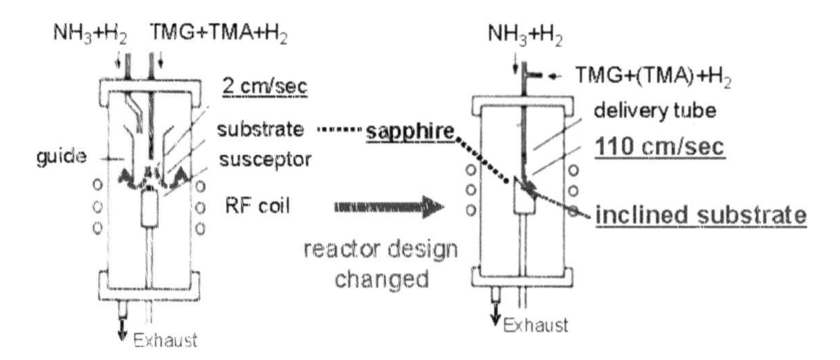

FIGURE 5. Schematic drawings of the reactor part of the MOVPE system before and after the reactor design was changed.

between the two crystals as in Fig. 6. In fact, for epitaxial growth of semiconductor crystals, it was considered to be "gospel" to have complete lattice matching as in the case of GaAs growth on a GaAs substrate. For heteroepitaxial growth, even a mismatch of about 1% would make it difficult to grow good-quality crystals.

To overcome this issue, we developed low-temperature (LT-) buffer layer technology in 1985 [1, 18, 20]. Specifically, this is a method for producing a thin buffer layer from a material with physical properties similar to those of GaN and sapphire with a thickness of 20–50 nm, which is thin enough not to interfere with the transmission of crystallographic information from the substrate to the epitaxial layer as shown in Fig. 7. Temperature for the deposition of the buffer layer might suitably be several hundred °C, which is considerably lower than the typical growth temperature for single crystal GaN. The temperature would then be raised to that required for epitaxy growth of GaN single crystal, which is approximately 1000° C.

This is based on the idea of having a soft or flexible thin layer without a rigid structure like that of a single crystal, inserted between the substrate and the GaN film. The purpose of the buffer layer is to create conditions as close as possible to those for homoepitaxy, where no interfacial free energy exists in principle. For the buffer layer materials, I considered AlN, GaN, zinc oxide (ZnO), and SiC. First, we tried AlN, with which I was already familiar [10].

In addition to the first improvement by Koide, Amano used the LT-AlN buffer layer technology combined with a further accelerated gas flow velocity of about 430 cm per second. By using this approach, in 1985, we eventually succeeded in growing the world's first extremely high-quality (semiconductor

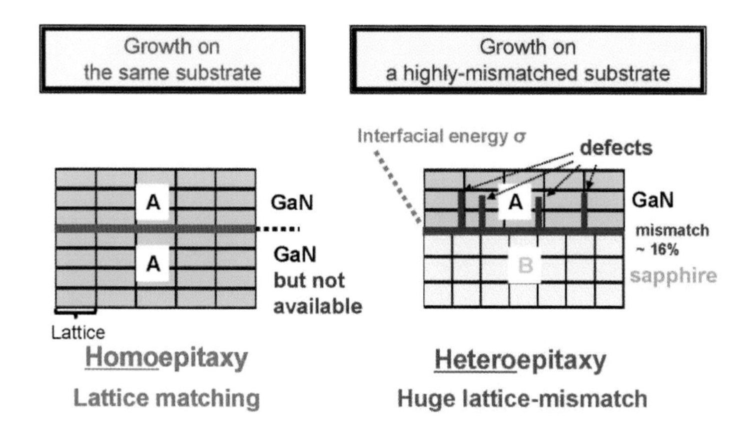

FIGURE 6. Schematics of homoepitaxy (ex. GaN on GaN case) and heteroepitaxy (ex. GaN on sapphire case).

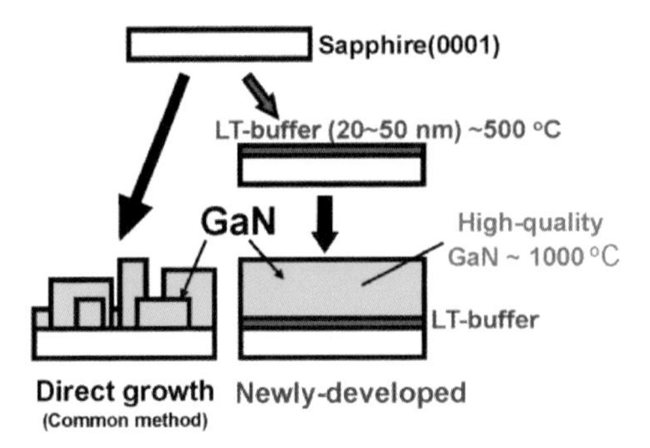

FIGURE 7. A procedure with newly developed low-temperature buffer layer technology for high-quality GaN.

grade) epitaxial GaN film [1], which has been the drastic innovation in the GaN research.

Scanning electron micrographs (SEM) of surfaces of GaN films grown on a sapphire substrate showed the surface morphology of the films to be markedly improved by the LT-AlN buffer layer [1, 22] as shown in Fig. 8. The GaN film had a specular surface with no pits or cracks, and was so transparent that letters written on the underlying paper could be clearly seen as shown in Fig. 8 (c).

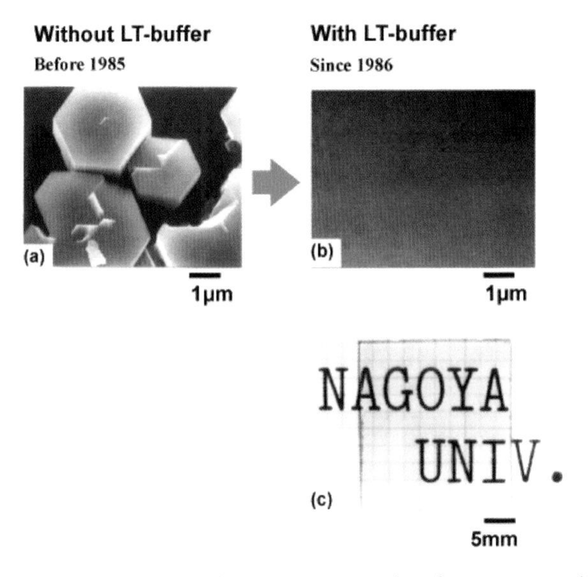

FIGURE 8. Scanning electron micrographs of GaN on sapphire (a) without LT-buffer layer and (b) with LT-buffer layer. (c) a photograph of specular and transparent GaN film grown on sapphire with LT-buffer layer.

Cross-sectional transmission electron microscopy (TEM) showed that the density of crystal defects such as dislocations markedly decreased with the use of the LT-AlN buffer layer [21]. X-ray diffraction profiles also showed that the crystal quality of GaN was significantly improved by this method [1, 22]. The residual donor (electron) concentration for GaN grown with the LT-AlN buffer layer decreased to the order of 10^{17} cm^{-3}, which is more than two orders of magnitude lower than that for GaN grown without the LT-AlN buffer layer [22]. Soon after, the electron concentration was further reduced to less than 10^{15} cm^{-3} [18]. Simultaneously, the electron mobility markedly increased to several hundred cm^2 V^{-1} s^{-1} [22, 23].

Figure 9 shows that near-band-edge emission dominated the photoluminescence (PL) spectrum of the undoped GaN grown with the LT-AlN buffer layer, whereas deep-level-related yellow emission was the major emission from GaN grown without the LT-AlN buffer layer [22]. In the spectrum from another GaN film grown with a LT-AlN buffer layer, free exciton lines (FE^A and FE^B) and impurity-bound excitons (DBE and ABE) were clearly observed showing that the crystal quality was drastically improved [24]. By using the LT-AlN buffer layer deposited under optimum conditions, not only the crystal quality but also the electrical and luminescence properties of GaN were greatly improved [1, 22, 23, 24], as already seen in Figs. 8 and 9.

The thrill I felt when I finally saw transparent and specular GaN crystals is simply unforgettable. It was something that I had dreamed of realizing ever since the early 1970s.

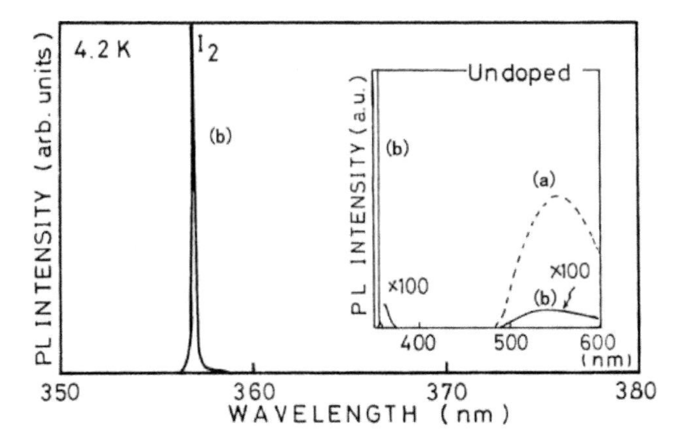

FIGURE 9. 4.2K Photoluminescence spectra of undoped GaN films on sapphire substrates (a) without (dashed line in the inset) and (b) with (solid line) LT-buffer layer.

This technology is an epochal concept that provides control over the growth of GaN, dramatically improving the quality of GaN and its alloys, which in turn has resulted in conductivity control for both p- and n-type nitrides, p-n junction blue-light-emitting device applications and a revival in the materials science of nitrides.

In 1991, S. Nakamura reported that a LT-GaN buffer layer had the same effect as a LT-AlN buffer layer [25].

Today, this LT-buffer layer technology in MOVPE has become the standard approach to growing high-quality GaN and nitride alloys.

4.2. Model for the growth mode

In order to clarify the role of the LT-AIN buffer layer, different stages of GaN growth on the c-face of sapphire were studied by SEM and reflection high-energy electron diffraction (RHEED) [21, 22, 23].

Figure 10 taken by Kazumasa Hiramatsu et al. shows (a) SEM images of the changes in surface morphologies during growth and (b) the corresponding RHEED patterns from the same region. Image (1) in Fig. 10(a) was obtained for the deposited AIN buffer layer, whose thickness was about 50 nm; the layer appears uniform and featureless. However, the image in the inset, which was taken at high magnification, appears to show fine particles with sizes of several nanometers, which seem to be embedded in the featureless structure.

The corresponding RHEED pattern in Fig. 10(b) shows somewhat diffuse spots together with a halo-like pattern, which is in agreement with the SEM results. The AIN buffer layer is thought to consist of fine crystallites in an amorphous-like matrix. After 5 min of growth, as seen in image (2) in Fig. 10(a), many truncated hexagonal pyramidal mesas (trapezoidal crystals) are seen to have formed. Two-dimensional lateral growth then occurs for a certain period, as shown in image (3). Subsequently, quasi-lateral growth dominates when the GaN film reaches a certain thickness, as seen in image (4). Finally, the entire substrate is covered by a flat GaN film, as seen in image (5). In Fig. 10(b), RHEED pattern (2), obtained after 5 min of growth, is spot-like, indicating that the GaN islands are three-dimensional. Subsequently, the pattern becomes streaky, as the surface becomes flatter.

Figure 11 shows a low-magnification cross-sectional bright-field TEM image of a wide region of a GaN film grown using a LT-AlN buffer layer [21]. It can be seen that three zones with different microstructures exist within the GaN layer. The zone nearest to the AlN buffer layer has fine image contrasts due to a

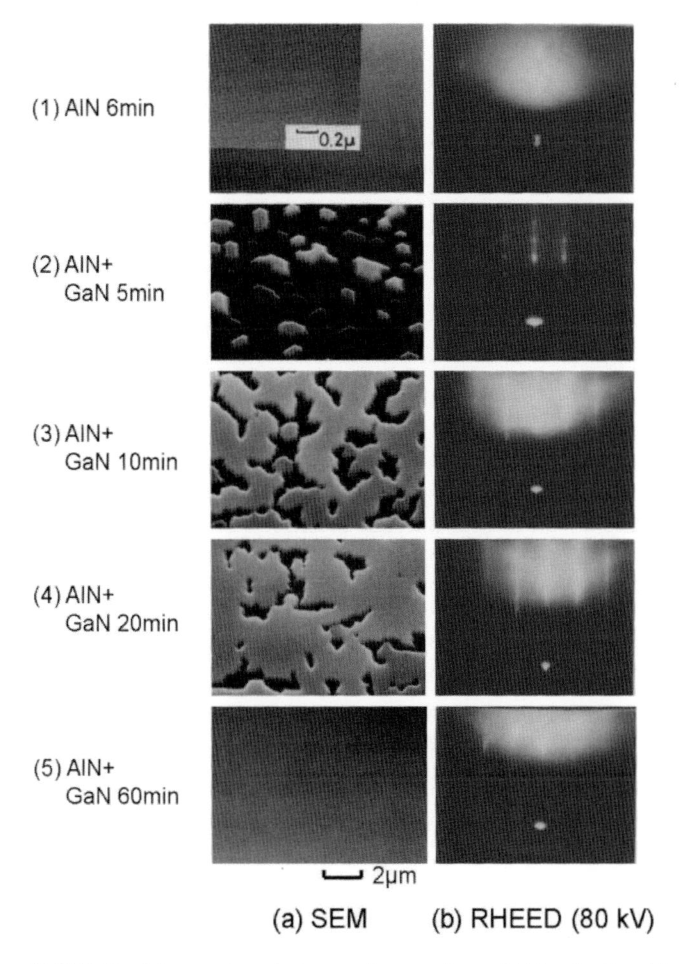

FIGURE 10. (a) scanning electron micrographs and (b) reflection high-energy electron diffraction patterns during the GaN growth on sapphire with LT-buffer.

number of defects. Consequently, this zone is referred to as the "*faulted zone*" (Z_F) and is about 50 nm thick. Above the faulted zone, there is another zone, which contains a number of trapezoid crystals, corresponding to the truncated hexagonal pyramidal GaN islands observed by SEM, as shown in image (3) in Fig. 10(a). Since the defect density in this zone is much lower than that in the faulted zone, it is called here the "*semi-sound zone*" (Z_{ss}), and is about 150 nm thick. The upper zones is referred to as the "*sound zone*" (Z_s), as it contains only a small number of defects. In the sound zone, the defect density decreases abruptly for GaN layers thicker than about 300 nm, resulting in high-quality uniform GaN.

FIGURE 11. A bright field image of GaN on sapphire with LT-AlN buffer by cross-sectional TEM.

Figure 12 shows a dark-field cross-sectional TEM image with a little higher magnification in order to clarify the microstructure of the AlN layer and the different zones in the GaN layer. The image contrasts in the figure reveal that the AlN layer is composed of columnar fine crystals like frost columns. The diameter is of the order of 10 nm, which corresponds to the front of the columnar crystals. The AlN layer has an amorphous-like structure at the deposition

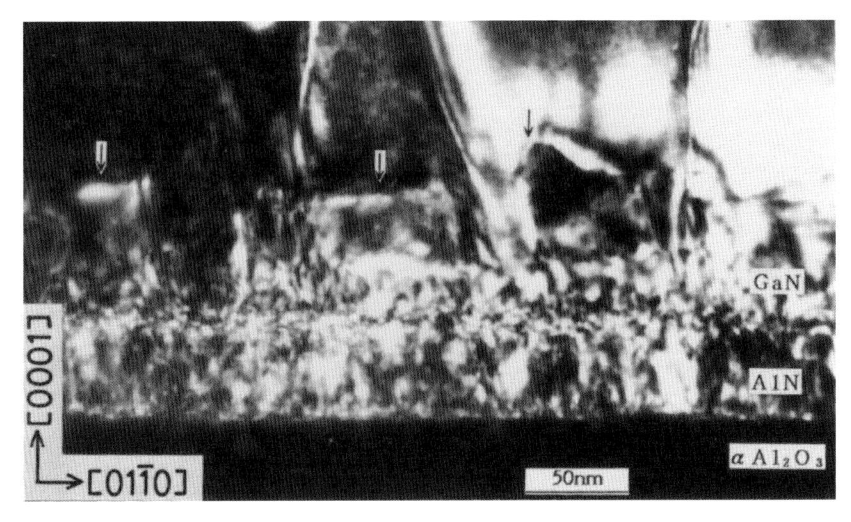

FIGURE 12. A dark field image of GaN on sapphire with LT-AlN buffer by cross-sectional TEM.

temperature, as shown in Fig. 10 (1). However, as the temperature is ramped from the low AlN deposition temperature to the GaN epitaxial temperature of about 1000° C, solid-phase regrowth occurs, and a fine columnar structure forms in the AlN. Similar contrast is seen in the faulted zone of the GaN, which suggests that GaN in the faulted zone is also composed of columnar fine crystals. It is likely that each GaN column from a GaN nucleus has formed on top of a columnar AlN crystal. Therefore, it is thought that *high-density nucleation* of GaN occurs owing to the much higher density of AlN columns as described in Fig. 13 (1)–(3), compared to the case where GaN is grown directly on a sapphire substrate as seen in Fig. 8 (a).

Figure 13 shows the proposed growth model for a GaN film on a c-face sapphire substrate with a LT-AlN buffer layer. Each fine GaN crystal begins to grow along the *c*-axis, forming a columnar structure. Each of the columns is slightly tilted or twisted at this stage, order to absorb the strain due to the large lattice mismatch. This suggests that geometric selection [26] of the GaN fine crystals occurs, and the columnar fine GaN crystals increase accordingly in size during the growth, as shown in Fig. 13 (4). The number of columns emerging at the front gradually decrease with the front area of each column increasing accordingly. Because the only columns that survive are those that grow along the fastest growth directions, which is the *c*-axis, all columns eventually have their *c*-axis oriented normal to the substrate, as indicated by arrows in Fig. 13 (4), which shows the growth model for GaN film on c-face sapphire with the LT-AlN buffer layer. In the faulted zone, trapezoidal islands are formed on top of the columnar crystals, as seen in Fig. 11. At this stage, all of the columnar crystals are *c*-axis oriented, and this orientation is transferred to the trapezoidal islands, as shown in Fig. 13 (5). These islands preferentially grow up to become lager trapezoid crystals, which cover the minor island nearby, and form the semi-sound zone.

Subsequently, "*lateral growth and coalescence*" of the islands occur in the stages shown in SEM images (3) and (4) in Fig. 10(a). The trapezoid crystals grow at a higher rate in the lateral direction, as shown in image (6) in Fig. 13, because the growth rate of the *c*-face is much lower. This lateral growth continues until the islands fully coalesce to form a continuous layer.

Since the crystallographic orientation of all of the islands is almost the same, a smooth well-oriented GaN layer with a low defect density is formed, as seen in Fig. 8(b). Thus, the uniform growth due to layer-by-layer growth occurs creating the sound zone.

On the other hand, in the case of direct growth without an LT-buffer layer, hexagonal GaN columns with many different sizes and heights are formed. These grow three-dimensionally resulting in a rough surface, and with many

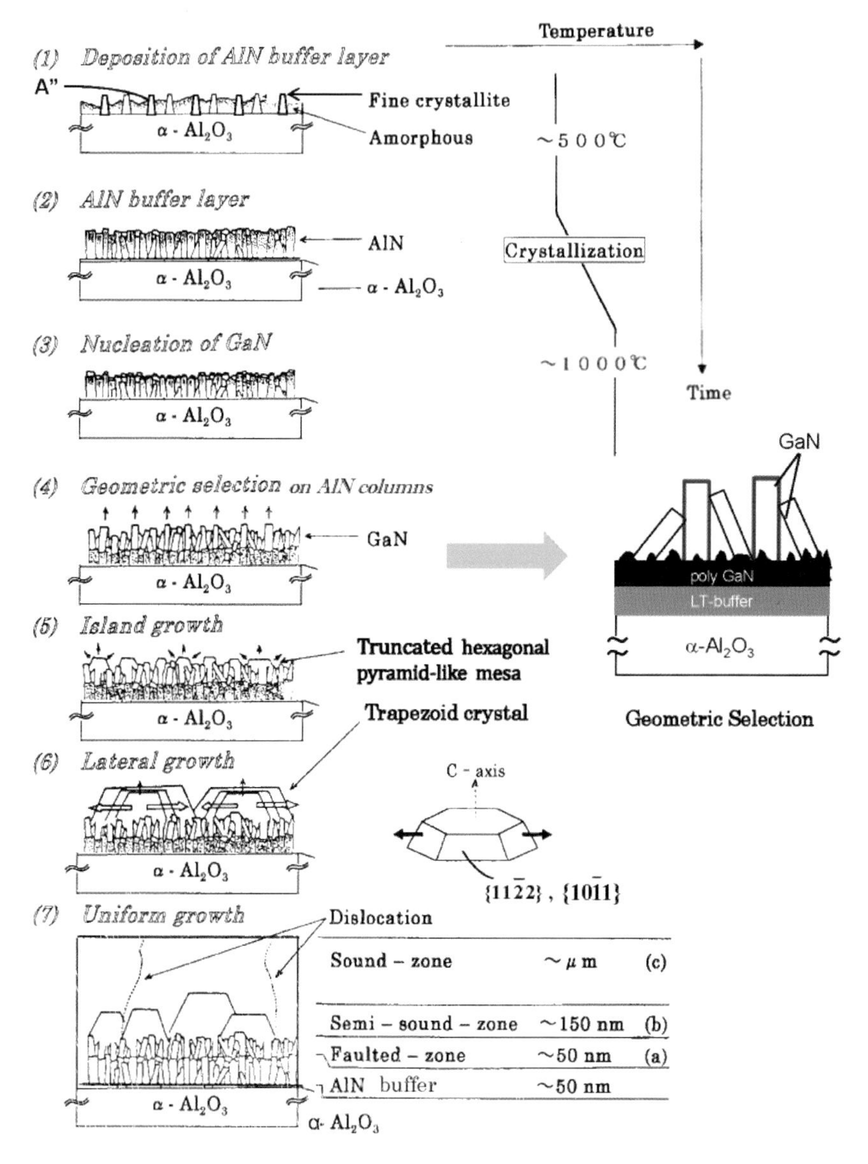

FIGURE 13. A growth model of GaN on sapphire with LT-buffer layer. "Geometric selection" of the GaN fine crystals was suggested.

pits at their boundaries, exposing the bare sapphire substrate, as shown in Fig. 8(a). The optimum thickness of the deposited AIN layer was found to be around 50 nm. If it was too thick (for example, 150 nm), the GaN film became polycrystalline. Also, if the deposition temperature of the AIN buffer layer was close to the temperature at which AlN crystal could be grown, the surface of GaN showed many hexagonal hillocks [23].

5. DISCOVERY OF p-TYPE GaN AND DEMONSTRATION OF p-n JUNCTION BLUE/ UV LED

With the successful control of the quality of GaN, we could start working on p-type doping. Despite repeated efforts on Zn-doping of high-quality GaN films grown with a LT-AlN buffer layer, no successful results were obtained (although the film resistivity was increased). In 1988, Amano found that the Zn-related luminescence intensity was greatly increased with no change in spectral shape, when a high-quality Zn-doped GaN film was irradiated with low-energy electron-beam (LEEBI) at RT or lower [27]. We suspected that the Fermi level of the sample was changed, which potentially led to a shift towards the p-type. However, the samples did not exhibit p-type conduction. We then realized that Mg was a potentially shallower acceptor than Zn, since the difference in electro-negativity between Mg and Ga is smaller than that between Zn and Ga [28]. In early 1989, graduate student Masahiro Kito carried out Mg-doping of high-quality GaN grown by the LT-buffer technology. For Mg-doping, bis-cyclopentadienyl Mg (CP_2Mg) or methyl-CP_2Mg (MCP_2Mg) was used as a dopant precursor [29, 30].

Then we applied the LEEBI treatment to the sample and found that the Mg-related blue-luminescence intensity was markedly enhanced, while the spectral shape remained unchanged, and that the sample had been converted to a low-resistivity p-type crystal, as confirmed by Hall effect measurement [2] and later by electron beam induced current (EBIC) measurement [31].

We immediately produced a high-performance GaN p-n junction blue/UV-LED with encouraging current-voltage (I–V) characteristic. This was demonstrated in 1989 and was the first such device in the world [2] (Fig. 14(a)(b)). This LED exhibited blue light emission with a low current injection (<20 mA), and additional near UV light emission with a high current injection [31] (Fig. 14(c)).

As a possible mechanism for the LEEBI effect, we proposed the "Coulombic explosion" model [32]. In this model, electron-beam excitation generates many free electrons and holes (Coulombic explosion) which stimulate the breaking of Mg-H bonds, and create active Mg as an acceptor impurity. The validity of this mechanism was later confirmed by researchers at the University of Illinois [33], and it was found to occur even at low temperatures [27, 33].

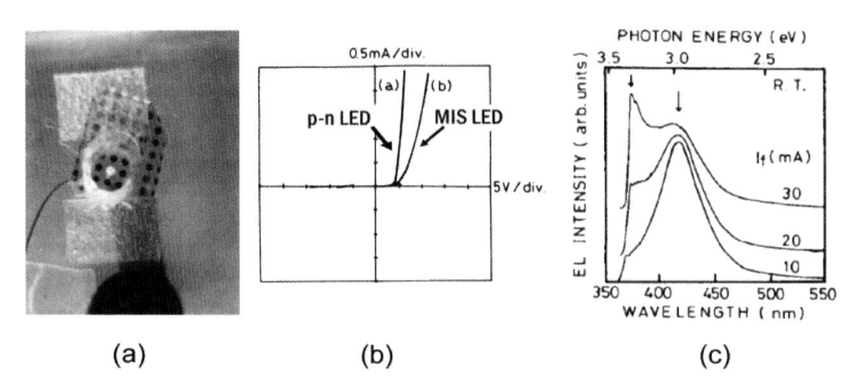

FIGURE 14. (a) GaN p-n junction blue/UV LEDs (black dots). An electric current is only being passed through the centered LED that is emitting blue light. (b) I-V characteristics of p-n junction LED (left) and MIS LED (right). (c) EL spectra of GaN p-n junction blue/UV LED.

In 1992, S. Nakamura and his group obtained p-type GaN by thermal annealing above 400° C in a H_2-free atmosphere, rather than by electron irradiation of Mg-doped GaN using CP_2Mg. It was clarified that the Mg becomes passivated by forming a Mg-H complex [34]. To realize p-type nitrides, it is therefore essential to activate the Mg acceptor by releasing the hydrogen [35]. Indeed, there is a high concentration of hydrogen atoms in MOVPE-grown nitrides. It should be noted, however, that the concentration of residual donors should be drastically reduced before solving problems related to hydrogen passivation of acceptors [36], which is also a well-known issue with regard to both III–V compounds and Si.

These results clearly shows that improving crystal quality by the LT-buffer layer technology, Mg-doping and its activation are the keys for the development of the p-type conduction in wide-bandgap nitrides.

We later went on to develop p-type nitride alloys; p-AlGaN in 1991–92 [37] and p-GaInN in 1995 [38] for the first time.

6. CONDUCTIVITY CONTROL OF n-TYPE GaN AND NITRIDE ALLOYS

Control of electric conductivity of n-type GaN grown without the LT-buffer layer had been quite difficult, because of the high residual donor concentration of greater than 10^{19} cm^{-3}. On the other hand, when an LT-AlN buffer layer was used, the conductivity of n-type GaN became extremely low due to the drastic

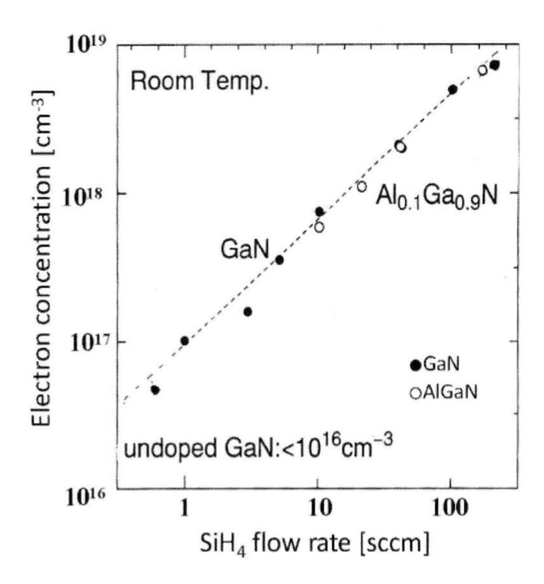

FIGURE 15. Electron concentrations in GaN (●) and AlGaN (○) as a function of SiH_4 flow rate.

reduction in the residual donor concentration. Control of n-type conductivity is extremely important for many types of nitride-based devices. In 1990, we succeeded in controlling the conductivity of n-type GaN [3] (and AlGaN in 1991 [39]), over a range of about two orders of magnitude, by Si doping using SiH_4, as shown in Fig. 15. High crystalline quality was again maintained by the use of the LT-buffer layer technology.

This conductivity-control method now enjoys widespread use around the world.

The conductivity control of high-quality p-type and n-type GaN and nitride alloys has allowed the use of heterostructure and multi-quantum wells (MQWs) for more efficient p-n junction light-emitting structures.

7. BRIEF HISTORY OF NITRIDE-BASED BLUE LEDs

In 1992, we improved the external quantum efficiency of the blue/UV LED to 1.5% (power conversion efficiency of more than 1%) by employing an AlGaN/GaN double heterostructure (DH) [40]. And in 1994, the first commercial AlGaN/GaInN: Zn, Si DH blue LED with the external quantum efficiency of 2.7 % was launched [41]. In 1992, we also developed the first UV LEDs based on p-AlGaN/GaN/n-AlGaN DH with an output power of several mW at RT [42].

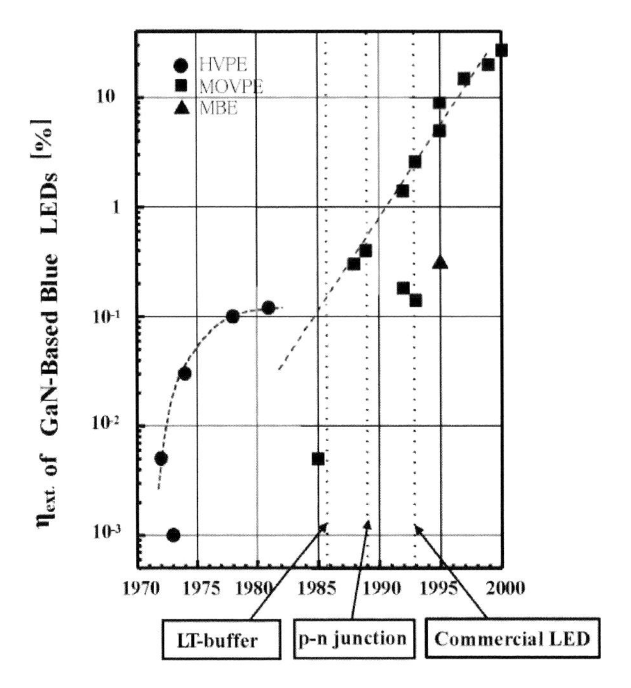

FIGURE 16. Chronological developments in external quantum efficiency, η_{ext} of nitride-based blue LEDs (1971–2000).

Chronological developments in the external quantum efficiency, η_{ext} of nitride-based blue LEDs between 1971 and 2000 are shown in Fig. 16 [43]. It can be seen that η_{ext} was relatively constant at about 0.1%, before the breakthroughs described in this paper. It began to increase steeply soon after success was achieved in growing high-quality nitride crystals, which resulted in the GaN p-n junction LED [2]. To improve the emission efficiency of blue LEDs and the performance of violet LD, the use of GaInN alloy as an active layer is essential. The first successful growth of single-crystal GaInN alloy was reported in 1989–1991 [44, 45]. Although it was soon used as the active layer of blue DH-LEDs, the devices functioned via donor-acceptor-pair emission rather than band-edge emission because of the low quality of GaInN. Growth of high-quality GaInN/GaN quantum wells (QWs) [46] and multiple QWs (MQWs) [47] exhibiting distinct band-edge emission, which is currently used as an active layer in blue LEDs, was first reported in 1995. In 2002, M. Yamada et al. reported that blue LEDs using such high-quality GaInN/GaN MQWs exhibited a η_{ext} of higher than 36% [48]. They have been used as an excitation source for yellow phosphors in white LEDs [49, 50].

8. DEVELOPMENT OF NITRIDE-BASED LDs AND OTHER DEVICES

In 1990, we also succeeded for the first time in achieving RT stimulated emission in the UV range by optical excitation at a power that was one order of magnitude lower than before that previously required [51], showing that GaN is of high-quality as shown previously. Figure 17 [42, 43] shows the threshold power, P_{th} for stimulated emission from nitrides over the years. Before 1986, stimulated emission with optical excitation could only be achieved at low temperatures and P_{th} was very high. We proposed that a significant reduction in P_{th} would be possible with the use of DHs or separate-confinement heterostructures (SCHs). This is due to the fact that the refractive index of the nitride system decreases with increasing bandgap energy [52]. Similar to the increase in blue LED efficiency shown in Fig. 16, P_{th} began to decrease exponentially immediately after

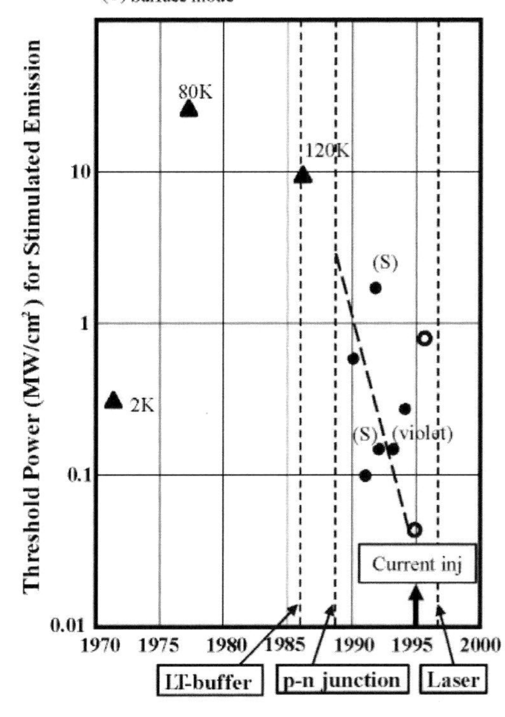

FIGURE 17. Chronological developments in threshold power for stimulated emission by optical pumping (before 1995) and current injection (after late 1995) from nitrides (1971–2000).

the marked improvement in the crystal quality of nitrides, described in the previous section. By optimizing the heterostructure, we obtained a new record of the lowest P_{th} almost every year between 1990 and 1995.

In 1995, we found that band-edge emission for GaInN/GaN MQWs with well widths of less than 3 nm, which roughly corresponds to the Bohr radius for excitons or electrons in GaN, was three orders of magnitude higher than that for a thick GaInN QW as shown in Fig. 18 [53]. In 1997, this phenomenon was qualitatively attributed to suppression of the quantum confined Stark effect (QCSE) [54, 55] in such narrow wells. The QCSE is caused by the presence of a large piezoelectric field [54] in the GaInN well, which was found to be about 1 MV/cm for an InN molar fraction of 0.1. By adopting such thin QWs as the active layer, we succeeded in demonstrating the first stimulated emission at 388

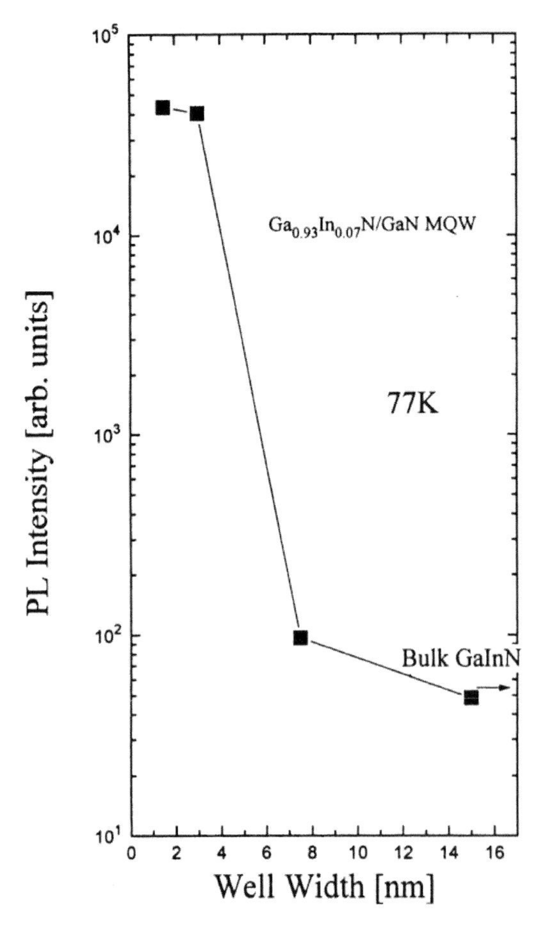

FIGURE 18. 77K PL intensities of GaInN/GaN QWs as a function of GaInN well width.

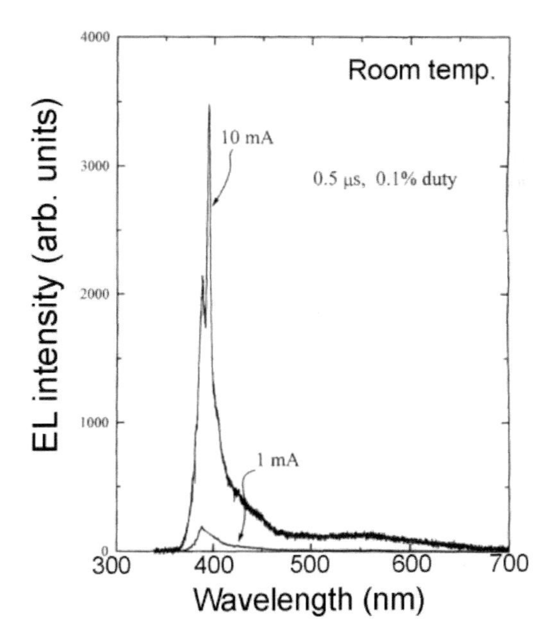

FIGURE 19. A stimulated emission spectrum from a GaInN/GaN QW device under pulsed current injection at room temperature.

nm from a GaInN/GaN QW device with well width of 2.5 nm under pulsed current injection at RT as shown in Fig. 19 [56].

Shortly after that, a group at Nichia Corporation reported a distinct pulsed lasing operation at 405 nm using an active layer composed of thin QWs [57]. We achieved pulsed laser oscillations from a diode with a single quantum well at 376 nm in 1996 [58] and at 350.9 nm in 2004 [59], which were the shortest wavelengths at that time. The first RT continuous-wave operation of a 405 nm blue-violet LD was reported in 1996 [60], and such devices were later adapted as the light source for high density digital disk systems. In 2000, we developed a high-sensitivity solar-blind UV sensor with low dark current [61], and in 2006 a high on-off ratio, low on-resistance, and normally-off mode AlGaN/GaN heterostructure field effect transistors [62].

With regard to quantum effects in the nitride system, we verified quantum size effect in 1991 [63], and the piezoelectric and quantum confined Stark effects in 1997 [54] as described above. In 2000 [64], Tetsuya Takeuchi et al. theoretically determined the orientation dependence of the piezoelectric field and showed the existence of non-/semi-polar nitride crystal planes (see Fig. 20). This has triggered world-wide efforts to grow such crystals in order to produce

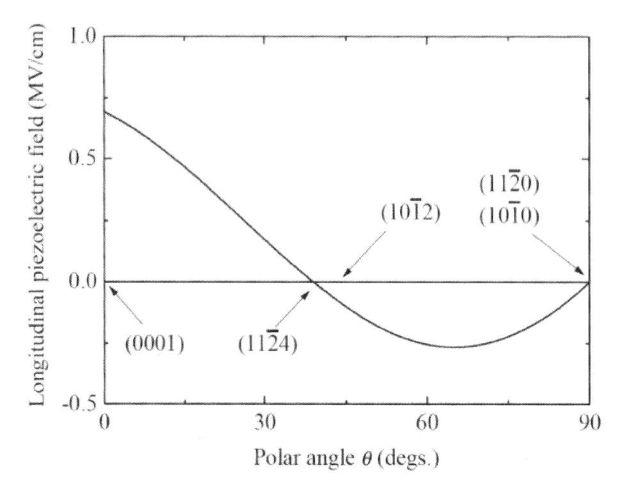

FIGURE 20. Calculated longitudinal piezoelectric field in strained $Ga_{0.9}In_{0.1}N$ on GaN as a function of the polar angle from (0001).

more efficient light-emitters such as high-efficiency green/yellow-green LEDs [65], high-power blue-violet LEDs [66], and high-power green LDs [67].

9. SUMMARY AND FUTURE EXPECTATIONS

The creation of high-quality semiconductor-grade single crystals of GaN and related alloys, realization of low-resistivity p-type GaN, and conductivity control in n-type nitrides are all of great practical importance, and by 1991, these final pieces of the puzzle were in place. Stretching back to the late 1980s to the early 1990s, the remarkable achievements in GaN-based materials and devices have caught the eye of many scientists around the world, who become engaged in research in this field. Many topics relating to GaN-based LEDs and LDs have been published in scientific journals and presented at international conferences. Nitride research has become one of the most exciting themes in the world.

Turning back to Fig. 3, it shows the number of publications per year referred from INSPEC using the keyword GaN; major events in the history of nitride research are indicated. Events are labeled by the year when they were first achieved, anywhere in the world. As can be seen, the number of papers in 1969 is only 4 but it increased to 34 in 1975, inspired by the first GaN single crystal [11], the MIS LED [12], and stimulated emission from GaN at 2K [13] (period (A)).

The number then decreased until the year 1985, indicating that GaN research activity declined (period (B)), since the difficulty in achieving high-quality GaN and p-type conduction had come to be realized. However, after we succeeded in creating GaN single crystal with excellent quality (1986) and creating a GaN p-n junction blue/UV LED (1989), the number of published papers began to increase. The increase was exponential to 60 in 1991 after a short incubation period (period (C)), to 370 in 1995, to 1200 in 1997, and to 2040 in 2000 (period (D)).

At present, the aforementioned innovations of blue LEDs allow us to complete the set of three primary colors of light using semiconductors. Coupled with existing high-brightness red LEDs, blue/green LEDs are leading to the development of completely solid-state full-color displays, traffic lights, signals, signage, and specialized lighting applications. White LEDs composed of nitride-based blue/UV LEDs and yellow phosphor are two times more efficient than fluorescent lamps, and as such, white LEDs are now being used in TVs, mobile phones, computer displays and general lighting equipment. The white and UV LEDs are expected to be used for bio-medical applications, such as sterilization, microscopic level diagnoses, and medical treatments. Blue LEDs, combined with existing red LEDs are also being used for agricultural lighting sources. GaN-based LED lamps can operate for many hours on electricity produced by a solar battery, and can be used in places on Earth where there is no electrical power supply.

Long-lifetime nitride-based violet LDs are being used to read and write data in optical disc systems. Coupled with existing red LDs, GaN-based blue and green LDs are expected to be used in mini-projectors.

GaN-based group-III nitride semiconductors are also promising for high-speed/high-power electronic devices, due to their high electron saturation velocities and high breakdown voltages. Such devices are important in the mobile telecommunications industry, for energy saving in vehicles such as trains and automobiles, and also for home electrical appliances such as air conditioners and refrigerators.

All of these nitride-based devices are robust in harsh environments, allow a significant amount of energy to be saved, and provide a route to avoid the use of hazardous materials. The use of nitride-based devices will be one of our most powerful tools in the fight against global warming.

ACKNOWLEDGEMENTS

The author wishes to express his sincere appreciation to many collaborators: M. Hashimoto, Y. Ohki, Y. Toyoda, M. Ohshima, N Mazda, H. Kobayasi, and M. Matsuki at Matsushita Research Institute Tokyo, Inc., numerous students and co-researchers: N. Sawaki, K. Hiramatsu, Y. Koide, H. Amano, M. Kito, and H. Murakami at Nagoya University and S. Kamiyama, T. Takeuchi, and M. Iwaya at Meijo University, and B. Monemar at Linköping/Lund Universities, for their great contributions throughout this research.

His special thanks are extended to Toyoda Gosei Company Ltd., Toyota Central Research Laboratory, and Japan Science and Technology Agency for their co-operation in the development and production of GaN-based blue LEDs and related materials.

This work received financial supports from the Ministry of International Trade and Industry, Japan (at that time), the Ministry of Education, Culture, Sports, Science and Technology, Japan, and the Japan Society for Promotion of Science.

Finally, the author would like to thank the Nobel Foundation and members of the Royal Swedish Academy of Sciences for honouring our "invention of efficient blue-LEDs which has enabled bright and energy-saving white light sources."

REFERENCES

1. H. Amano, N. Sawaki, I. Akasaki, and Y. Toyoda, "Metalorganic vapor phase epitaxial growth of a high quality GaN film using an AlN buffer layer," *Appl. Phys. Lett.*, **48**, 5, 353 (1986).
2. H. Amano, M. Kito, K. Hiramatsu, and I. Akasaki, "P-Type conduction in Mg-doped GaN treated with low-energy electron beam irradiation (LEEBI)," *Jpn. J. Appl. Phys.*, **28**, 12, L2112 (1989).
3. H. Amano and I. Akasaki, "Fabrication and properties of GaN p-n junction LED," *Ext. Abstr. Mat. Res. Soc.*, (EA-21), 165 (1990).
4. S. Nakamura, M. Senoh, S. Nagahama, N. Iwasa, T. Yamada, T. Matsushita, H. Kiyoku, and Y. Sugimoto, "InGaN-based multi-quantum-well-structure laser diodes," *Jpn. J. Appl. Phys.*, **35**, L74 (1996).
5. N. Holonyak Jr. and S. F. Bevacqua, "Coherent (visible) light emission from Ga(As_1-xPx) junctions," *Appl. Phys. Lett.*, **1**, 82 (1962).
6. R. A. Logan, H. G. White, and W. Wiegman, "Efficient green electroluminescence in nitrogen-doped GaP p-n junctions," *Appl. Phys. Lett.*, **13**, 139 (1968).
7. H. Matsunami, M. Ikeda, A. Suzuki, and T. Tanaka, "SiC blue LED's by liquid phase epitaxy," *IEEE Trans. Electron Devices*, **ED-24**, 958 (1977).
8. R. M. Park, M. B. Troffer, C. M. Rouleau, J. M. DePuydt, and M. A. Haase, "p-type ZnSe by nitrogen atom beam doping during molecular beam epitaxial growth," *Appl. Phys. Lett.*, **57**, 2127 (1990).
9. M. A. Hasse, J. Qiu, J. M. DePuydt, and H. Cheng, "Blue-green laser diodes," *Appl. Phys. Lett.*, 59, 1272 (1991).

10. I. Akasaki and M. Hashimoto, "Infrared lattice vibration of vapor-grown AlN," *Solid State Commun.*, Vol. 5, pp. 851–853, (1967).
11. H. P. Maruska and J. J. Tietjen, "The preparation and properties of vapor-deposited single-crystalline GaN," *Appl. Phys. Lett.*, **15**, 327 (1969).
12. J. I. Pankove, E. A. Miller, and J. E. Berkeyheiser, "GaN electroluminescent diodes," *RCA Review*, **32**, 383 (1971).
13. R. Dingle, K. L. Shaklee, R. F. Leheny, and R. B. Zetterstrom, "Stimulated emission and laser action in gallium nitride," *Appl. Phys. Lett.*, **19**, 5 (1971).
14. e.g. E. Ejder, "Refractive index of GaN," *Physica Status Solidi*, A**6**, 445 (1971).
15. e.g. B. Monemar, "Fundamental energygap of GaN from photoluminescence excitation stectra," *Phys. Rev.*, B**10**, 676 (1974).
16. G. Mandel, "Self-compensation limited conductivity in binary semiconductors. I. Theory," *Phys. Rev.*, 134, A1073 (1964).
17. Y. Ohki, Y. Toyoda, H. Kobayasi and I. Akasaki, "Fabrication and properties of a practical blue-emitting GaN m-i-s diode," *Inst. Phys. Conf. Ser.*, **63**, 479 (1981).
18. I. Akasaki and H. Amano (Invited Review Paper), "Breakthroughs in improving crystal quality of GaN and invention of the p-n junction blue-light-emitting diode," *Jpn. J. Appl. Phys.*, **45**, 9001 (2006), I. Akasaki and H. Amano: Erratum, "[Jpn. J. Appl. Phys. **45**, 9001]," *Jpn. J. Appl. Phys.*, **47**, 3781 (2008).
19. H. M. Manasevit, F. M. Erdmann and W. I. Simpson, "The use of metalorganics in the preparation of semiconductor materials: IV. The nitrides of aluminum and gallium," *J. Electrochem. Soc.*, **118**, 1864 (1971).
20. I. Akasaki and N. Sawaki, Jpn. Patent 1,708,203 (appl:1985.11.18), U.S. Patent 4,855,249, and later K. Manabe, H. Kato, I. Akasaki, K. Hiramatsu and H. Amano, Jpn. Patent 3,026,087 (appl:1989.3.1), U.S. Patent 5,122,845 etc.
21. K. Hiramatsu, S. Itoh, H. Amano, I. Akasaki, N. Kuwano, T. Shiraishi and K. Oki, "Growth mechanism of GaN grown on sapphire with A1N buffer layer by MOVPE," *J. Crystal Growth*, **115**, 628 (1991).
22. I. Akasaki, H. Amano, Y. Koide, K. Hiramatsu and N. Sawaki, "Effects of AlN buffer layer on crystallographic structure and on electrical and optical properties of GaN and $Ga_{1-x}Al_xN$ ($0 < x \leqq 0.4$) films grown on sapphire substrate by MOVPE," *J. Crystal Growth*, **98**, 209 (1989).
23. H. Amano, I. Akasaki, K. Hiramatsu, N. Koide and N. Sawaki, "Effects of the buffer layer in metalorganic vapor phase epitaxy of GaN on sapphire substrate," *Thin Solid Films*, **163**, 415 (1988).
24. B. Monemar, J. P. Bergman, I. A. Buyanova, W. Li, H. Amano and I. Akasaki, "Free excitons in GaN," *MRS Internet Journal of Nitride Semiconductor Research*, **1**, Art. 2 (1996).
25. S. Nakamura, "GaN growth using GaN buffer layer," *Jpn. J. Appl. Phys.*, **30**, L1705 (1991).
26. A. A. Chernov, *Modern Crystallography III, Crystal Growth* (Springer series in Solid-State Science, vol. 6, Springer-Verlag, Berlin, 1984) p. 64.
27. H. Amano, I. Akasaki, T. Kozawa, K. Hiramatsu, N. Sawaki, K. Ikeda and Y. Ishii, "Electron beam effects on blue luminescence of zinc-doped GaN," *Journal of Luminescence*, **40/41**, 121 (1988).

28. J. C. Phillips, *Bonds and bands in semiconductors*, Academic Press, New York and London (1973).
29. I. Akasaki, H. Amano, M. Kito, K. Hiramatsu, and N. Sawaki, "Pure-blue electroluminescence from Mg-doped GaN grown by MOVPE," *Electrochem. Soc. 175th Meeting* (SOTAPOCS-X) 673-SOA, 1989.
30. H. Amano, M. Kitoh, K. Hiramatsu and I. Akasaki, "Growth and luminescence properties of Mg-doped GaN prepared by MOVPE," *J. Electrochem. Soc.*, **137**, 1639 (1990).
31. I. Akasaki, H. Amano, M. Kito and K. Hiramatsu, "Photoluminescence of Mg-doped p-type GaN and electroluminescence of GaN p-n junction LED," *J. Luminescence*, **48/49**, 666 (1991).
32. T. A. Carlson and R. M. White, "Measurement of the relative abundances and recoil-energy spectra of fragment ions produced as the initial consequences of x-ray interaction with CH_3I, HI, and DI," *J. Chem. Phys.*, **44**, 4510 (1966).
33. X. Li and J. J. Coleman, "Time-dependent study of low energy electron beam irradiation of Mg-doped GaN grown by metalorganic chemical vapor deposition," *Appl. Phys. Lett.*, **69**, 1605 (1996).
34. S. Nakamura, T. Mukai, M. Senoh and N. Iwasa, "Thermal annealing effects on p-type Mg-doped GaN films," *Jpn. J. Appl. Phys.*, **31**, L139 (1992).
35. J. A. Van Vechten, J. D. Horning and B. Goldenberg, "Defeating compensation in wide gap semiconductors by growing in H that is removed by low temperature de-ionizing radiation," *Jpn. J. Appl. Phys.*, **31**, 3662 (1992).
36. "Hydrogen in Semiconductors II," ed. by N. H. Nickel in *Semiconductors and Semimetals*, Vol. 61, Academic Press, San Diego, USA and others (1999).
37. I. Akasaki and H. Amano, "Conductivity Control of AlGaN, fabrication of AlGaN/GaN multi-heterostructure and their application to UV/blue light emitting devices," *Mat. Res. Soc. Symp. Proc.*, **242**, 383 (1992) (Symposium held December 2–6,1991,Boston).
38. S. Yamasaki, S. Asami, N. Shibata, M. Koike, K. Manabe, T. Tanaka, H. Amano and I. Akasaki, "*p*-type conduction in Mg-doped $Ga_{0.91}In_{0.09}N$ grown by metalorganic vapor-phase epitaxy," *Appl. Phys. Lett.*, **66**, 9, 1112 (1995).
39. H. Murakami, T. Asahi, H. Amano, K. Hiramatsu, N. Sawaki and I. Akasaki, "Growth of Si-doped $Al_xGa_{1-x}N$ on (0001) sapphire substrate by metalorganic vapor phase epitaxy," *J. Crystal Growth*, **115**, 648 (1991).
40. I. Akasaki, H. Amano, K. Itoh, N. Koide and K. Manabe: "GaN-based UV/blue light emitting devices," *Inst. Phys. Conf. Ser.*, **129**, 851 (1992)
41. S. Nakamura, T. Mukai and M. Senoh, "Candelaclass highbrightness InGaN/AlGaN double heterostructure blue light emitting diodes," *Appl. Phys. Lett.*, **64**, 1687 (1994).
42. I. Akasaki and H. Amano, "Room temperature ultraviolet/blue light emitting devices based on AlGaN/GaN multi-layered structure," *Ext. Abstr. 24th Int. Conf. Solid State Devices and Mater.*, p. 327 (1992).
43. I. Akasaki and H. Amano, "Crystal growth and conductivity control of group III nitride semiconductors and their application to short wavelength light emitters," *Jpn. J. Appl. Phys.*, **36**, 5393 (1997).

44. T. Nagatomo, T. Kuboyama, H. Minamino and O. Otomo, "Properties of $Ga_{1-x}In_xN$ films prepared by MOVPE," *Jpn. J. Appl. Phys.*, **28**, L1334 (1989).
45. N. Yoshimoto, T. Matsuoka, T. Sasaki and A. Katsumi, "Photoluminescence of InGaN films grown at high temperature by metalorganic vapor phase epitaxy," *Appl. Phys. Lett.*, **59**, 2251 (1991).
46. S. Nakamura, M. Senoh, N. Iwasa, S. Nagahama, T. Yamada and T. Mukai, "Superbright green InGaN single-quantum-well-structure light-emitting diodes," *Jpn. J. Appl. Phys.*, **34**, L1332 (1995).
47. H. Amano and I. Akasaki, "Fabrication and properties of GaN-based quantum well structure for short wavelength light emitter," *Ext. Abstr. Int. Conf. Solid State Devices and Materials*, **V-7**, 683(1995).
48. M. Yamada, T. Mitani, Y. Narukawa, S. Shioji, I. Niki, S. Sonobe, K. Deguchi, M. Sano and T. Mukai, "InGaN-based near-ultraviolet and blue-light-emitting diodes with high external quantum efficiency using a patterned sapphire substrate and a mesh electrode," *Jpn. J. Appl. Phys.*, **41**, L1431 (2002).
49. K. Bando, K. Sakano, Y. Nobuchi, and Y. Shimizu, "Development of high-bright and pure-white LED lamp," *Journal of Light and Visual Environment*, **22**, 2 (1998).
50. Y. Shimizu, A. Banno, Jpn. Pat, 2,927,279 (1999) "Light-emitting diode" [in Japanese].
51. H. Amano, T. Asahi and I. Akasaki, "Stimulated emission near ultraviolet at room temperature from a GaN film grown on sapphire by MOVPE using an AlN buffer layer," *Jpn. J. Appl. Phys.*, **29**, 2, L205 (1990).
52. I. Akasaki and H. Amano, "MOVPE growth of high quality $Al_xGa_{1-x}N/Ga_yIn_{1-y}N$ ($x \geq 0$, $y \leq 1$) heterostructure for short wavelength light emitter," *Mat. Res. Soc. Symp. Proc.*, **339**, 443 (1994).
53. I. Akasaki, H. Amano, and I. Suemune, *Inst. Phys. Conf. Ser.*, **142**, 7 (1996).
54. T. Takeuchi, S. Sota, M. Katsuragawa, M. Komori, H. Takeuchi, H. Amano and I. Akasaki, "Quantum-confined stark effect due to piezoelectric fields in GaInN strained quantum wells," *Jpn. J. Appl. Phys.*, **36**, 2, 4A, L382 (1997).
55. T. Takeuchi, C. Wetzel, S. Yamaguchi, H. Sakai, H. Amano and I. Akasaki, "Determination of piezoelectric fields in strained GaInN quantum wells using the quantum-confined Stark effect," *Appl. Phys. Lett.*, **73**, 1691 (1998).
56. I. Akasaki, H. Amano, S. Sota, H. Sakai, T. Tanaka and M. Koike, "Stimulated Emission by Current Injection from an AlGaN/GaN/GaInN Quantum Well Device," *Jpn. J. Appl. Phys.*, **34**, 11B, L1517 (1995).
57. S. Nakamura, M. Senoh, S. Nagahama, N. Iwasa, T. Yamada, T. Matsushita, H. Kiyoku, Y. Sugimoto, "InGaN-Based Multi-QuantumWell-Structure Laser Diodes," *Jpn. J. Appl. Phys.*, **35**, 1B, L74 (1996).
58. I. Akasaki, S. Sota, H. Sakai, T. Tanaka, M. Koike and H. Amano, "Shortest wavelength semiconductor laser diode," *Electronics Letters*, **32**, 12, 1105 (1996).
59. K. Iida, T. Kawashima, A. Miyazaki, H. Kasugai, S. Mishima, A. Honshio, Y. Miyake, M. Iwaya, H. Amano and I. Akasaki, "350.9nm UV laser diode grown on low-dislocation-density AlGaN," *Jpn. J. Appl. Phys.*, **43**, 4A, L499 (2004).
60. S. Nakamura, M. Senoh, S. Nagahama, N. Iwasa, T. Yamada, T. Matsushita, Y. Sugimoto and H. Kiyoku, "Room-temperature continuous-wave operation of

InGaN multi-quantum-well structure laser diodes," *Appl. Phys. Lett.*, **69**, 4056 (1996)
61. C. Pernot, A. Hirano, M. Iwaya, T. Detchprohm, H. Amano and I. Akasaki, "Solar-blind UV photodetectors based on GaN/AlGaN p-i-n photodiodes," *Jpn. J. Appl. Phys.*, **39**, 2, 5A, L387 (2000).
62. T. Fujii, N. Tsuyukuchi, M. Iwaya, S. Kamiyama, H. Amano and I. Akasaki, "High on/off ratio in enhancement-mode $Al_xGa_{1-x}N$/GaN junction heterostructure field-effect transistors with p-Type GaN gate contact," *Jpn. J. Appl. Phys.*, **45**, 39, L1048 (2006).
63. K. Itoh, T. Kawamoto, H. Amano, K. Hiramatsu and I. Akasaki, "Metalorganic vapor phase epitaxial growth and properties of GaN/$Al_{0.1}Ga_{0.9}N$ layered structures," *Jpn. J. Appl. Phys.*, **30**, 9A, 1924 (1991).
64. T. Takeuchi, H. Amano and I. Akasaki, "Theoretical study of orientation dependence of piezoelectric effects in wurtzite strained GaInN/GaN heterostructures and quantum wells," *Jpn. J. Appl. Phys.*, **39**, 413 (2000).
65. S. Yamamoto, Y. Zhao, C. C. Pan, R. B. Chung, K. Fujito, J. Sonoda, S. P. DenBaars, and S. Nakamura, "High-efficiency single-quantum-well green and yellow-green light-emitting diodes on semipolar ($20\bar{2}1$) GaN substrates," *Appl. Phys. Exp.*, **3**, 122102 (2010).
66. Y. Zhao, S. Tanaka, C.-C. Pan, K. Fujito, D. Feezell, J. S. Speck, S. P. DenBaars, and S. Nakamura, "High-Power Blue-Violet Semipolar ($20\bar{2}\bar{1}$) InGaN/GaN Light-Emitting Diodes with Low Efficiency Droop at 200 A/cm^2," *Appl. Phys. Exp.*, **4**, 082104 (2011).
67. S. Takagi, Y. Enya, T. Kyono, M. Adachi, Y. Yoshizumi, T. Sumitomo, Y. Yamanaka, T. Kumano, S. Tokuyama, K. Sumiyoshi, N. Saga, M. Ueno, K. Katayama, T. Ikegami, T. Nakamura, K. Yanashima, H. Nakajima, K. Tasai, K. Naganuma, N. Fuutagawa, Y. Takiguchi, T. Hamaguchi, and M. Ikeda, "High-power (over 100 mW) green laser diodes on semipolar {$20\bar{2}1$} GaN substrates operating at wavelengths beyond 530nm," *Appl. Phys. Exp.*, **5**, 082102 (2012).

Hiroshi Amano. © Nobel Media AB. Photo: A. Mahmoud

Hiroshi Amano

I was born in Hamamatsu, Shizuoka Prefecture, Japan, on September 11, 1960, to my father Tatsuji and mother Yoshiko, and I grew up with younger brother Takashi. Hamamatsu is famous as the birthplace of Professor Kenjiro Takayanagi, a Japanese pioneer in the development of television, who succeeded in transmitting the first Katakana character "イ" by wireless transfer using a Braun tube in 1926, and also Soichiro Honda, who established Honda Motor Co., Ltd., in 1946. In addition, several engineering and manufacturing companies, such as motorbike companies, musical instrument companies, and optoelectronics companies, are based in Hamamatsu. A possible reason why I considered a career in engineering may have been the influence of growing up in such an industrial city.

I spent elementary school, junior high school, and high school at Hamamatsu. I was a weak child who was often ill. My grandmother Ken always nursed me, during which she often told me of her wretched experiences during World War II, so I learned a lot about the war. At elementary school, I concentrated on sports such as baseball and football. During elementary school and junior high school, I did not enjoy studying because the only reason to study seems to be to pass the entrance examination for high school. At high school, I had the same mathematics teacher for three years who taught me the importance of logical thinking and how to approach difficult problems in mathematics. I found that I could solve difficult problems if I thought logically. Although I became very interested in solving mathematics problems, I still did not have a good reason for studying.

I moved to Nagoya in 1979 to enter Nagoya University as a student of the Department of Electrical Engineering. In the introductory class to engineering, I heard a very profound interpretation of the meaning of the Kanji character "工," which means engineering. The lecturer explained that the meaning of "工"

is the connection of people with people, which means that the ultimate goal of engineering is to enrich the lives of people. I was astonished with this explanation and felt that my view of study had suddenly opened through recognizing that the meaning of study is to benefit the people. As a result, I became interested in all fields of study offered by my department, particularly computer science. This was soon after Bill Gates and Paul Allen established Microsoft (1975) and Steve Jobs and Stephen Wozniak started Apple Computers (1976). After the establishment of these now giant companies, the development of personal computer (PC) systems proceeded rapidly, and I hoped to contribute to the further development of PC systems.

In 1982, when I was in my third year of university, I had to choose a dissertation research topic. Unfortunately, there were no topics concerning computer science, especially the design of central processing units. But when I found that GaN-based blue LEDs could be researched in Professor Isamu Akasaki's laboratory, I decided to pursue this topic as my dissertation topic. At that time, Braun tubes were used as the monitors of PCs and also in television systems. Because Braun tubes were so large, I thought that if I could develop blue LEDs, I could change the world by improving people's lives by providing the means to develop more smart PC and TV systems. At that time, I did not know how difficult it would be to develop blue LEDs.

At that time, funding of our laboratory was very limited. Therefore, the students at Nagoya University had to develop their own system to enable the growth of nitride crystals by metal-organic vapor phase epitaxy (MOVPE). Through the efforts of a master's degree student followed by myself and Mr Koide, a student in the year above me, we succeeded in developing an MOVPE system. I then used our laboratory-built MOVPE system to try to grow high-quality GaN, and Mr. Koide focused on growing AlN and AlGaN. However, the growth of GaN on a foreign substrate such as sapphire was so difficult that I spent three years in vain trying to grow high-quality GaN.

In February 1985, almost at the end of my master's course, I was still carrying out lonely fruitless experiments. When I compared my GaN and AlN grown by Mr Koide, I found that the surface morphology of his AlN was slightly better that that of my GaN, so I decided to deposit AlN just before growing GaN. I noticed that a very high temperature was necessary to grow AlN. However, the heating system of our MOVPE system was so old that it could not reach the required temperature. I then remembered a discussion with Dr. Sawaki, at that time Associate Professor at Akasaki Laboratory. He said that in the growth of boron phosphide (BP) on Si for which the lattice mismatch is 16%, almost the same as that of GaN on sapphire, the predeposition of P was effective for growing BP with a flat surface. He also mentioned that a P cluster should act as a nucleation center causing BP to grow laterally in the initial stage. I decided to deposit a very small amount of AlN at a low temperature, which I anticipated would act as a nucleation center. When I took the resulting sample out from the MOVPE reactor, I first thought that I had forgotten to supply the Ga source. But when I checked the surface morphology using an optical microscope, I finally recognized that I had succeeded in growing GaN with an atomically flat surface. It took almost one year for this result to be published in Applied Physics Letters because I had to check not only the surface morphology but also the crystalline quality, electrical properties and optical properties. I found that all the properties of the GaN film were far superior to those reported previously. This process is called low-temperature deposited buffer layer technology for growing GaN on a sapphire substrate by MOVPE.

Our next challenge for us was to realize p-type GaN. I unsuccessfully tried to grow p-type GaN using Zn as an acceptor dopant from 1985 to 1988. At NTT, where I spent the internship of my PhD, I found that blue luminescence increased irreversibly when Zn-doped GaN was irradiated with electrons. I called this low-energy electron beam irradiation (LEEBI) treatment. However, even after LEEBI treatment, the Zn-doped samples did not show p-type conduction. In 1989, I became a Research Associate of Akasaki laboratory before completing my PhD. When I read the book "Bonds and Bands in Semiconductors" written by J. C. Phillips, I found that Mg is a better acceptor impurity than Zn in GaP. Then, myself and Mr. Kito, a master's student, started to investigate the Mg doping of GaN. We found that grown Mg-doped GaN showed high resistivity, but after LEEBI treatment, it showed distinct p-type conduction. We also succeeded in fabricating the world's first pn-junction-type UV/blue LED. In 1991 Nichia Chemicals team led by Dr. Nakamura found that p-type GaN could be grown by simple thermal annealing, which became the de fact standard method for the growth of p-type GaN.

In 1992, I moved to the School of Science and Technology, Meijo University, to work with Professor Akasaki, where I became an Assistant Professor. In 1998 and 2002, I became an Associate Professor and a Professor, respectively. In 2010, I moved to the Engineering Department, Nagoya University, where I continued to work as a Professor. I have been Director of Akasaki Research Center, Nagoya University, since 2011.

I was awarded several honors such as the IEEE/LEOS Engineering Achievement Award in 1996; the Rank Prize, Rank Prize Foundation, UK, in 1998; the Marubun Academic Award, Marubun Research Promotion Foundation, Japan, in 2002; the Takeda Award, Takeda Foundation, Japan, in 2002; the Japanese

Association for Crystal Growth JACG Award, Japan, in 2008; the NISTEP Award, National Institute of Science and Technology Policy, Japan, in 2009; the Order of Culture from the Japanese Emperor in 2014; and the Nobel Prize in Physics, Nobel Foundation, Sweden, in 2014.

I have been a fellow of the Japan Society of Applied Physics (JSAP) since 2009 and a fellow of the Institute of Physics (IOP), UK, since 2011. I am also a member of several academic organizations.

Selected activities include Sub Chair of the Program Committee of the International Symposium on Compound Semiconductors in 2007, Program Committee Chair of the Second International Symposium on the Growth of Nitride semiconductors in 2010, Program Committee Chair of the Third International Symposium on Growth of Nitride Semiconductors in 2012, and Organizing Committee Chair of the International Workshop on Nitride Semiconductors in 2012.

Growth of GaN on Sapphire via Low-Temperature Deposited Buffer Layer and Realization of p-Type GaN by Mg Doping Followed by Low-Energy Electron Beam Irradiation

Nobel Lecture, December 8, 2014

by Hiroshi Amano

Department of Electrical Engineering and Computer Science, Venture Business Laboratory, Akasaki Research Center, Nagoya University, Japan.

ABSTRACT

This is a personal history of one of the Japanese researchers engaged in developing a method for growing GaN on a sapphire substrate, paving the way for the realization of smart television and display systems using blue LEDs. The most important work was done in the mid- to late 80s. The background to the author's work and the process by which the technology enabling the growth of GaN and the realization of p-type GaN was established are reviewed.

1. MOTIVATION FOR STARTING BLUE LED RESEARCH

To explain blue light-emitting diodes (LEDs), it is worth showing an example of how they have changed our lives. Portable games machines and cellular or smart phones are very familiar items, especially to young people. The world's first portable games machine was released in 1979 [1] and cellular phones first became commercially available in 1984 [2]. But until the end of the 90s, all the displays of portable games machines and cellular phones were monochrome.

So, it should be emphasized that the younger generation can now enjoy full-color portable games and cellular/smart phones because of the emergence of blue LEDs. Today, the applications of blue LEDs are not limited to displays. In combination with phosphors, blue LEDs can act as a white light source [3] and are also used in general lighting.

In this introduction, let me briefly explain why I became interested in the development of blue LEDs. The two giant computer-related companies, Microsoft and Apple, were established by Bill Gates and Paul Allen in 1975 [4] and by Steve Jobs and Stephen Wozniak in 1976 [5], respectively. Since then, the market for computers, especially personal computer (PC) systems has expanded enormously [6]. When these companies were first established, Braun tubes were used in almost all displays as well as in television systems, and Braun tubes were too big to use in laptop PCs. Also, the use of Braun tubes in televisions meant that they were too bulky to be comfortably used in small Japanese houses. So, when I found nitride-based blue LEDs listed as an undergraduate dissertation topic at Akasaki Laboratory, Nagoya University, in 1982, I was so excited. The reason why I chose this laboratory was that as a naive undergraduate student, I thought that the subject of nitride-based blue LEDs would be easy to understand. I thought if I could achieve blue LEDs, I would contribute to improving the quality of life of people by helping to realize wall-mounted television systems and elegant PC systems, meaning that I would change the world. Of course, I was not aware at that time of the difficulty of this subject.

2. DIFFICULTY OF REALIZING HIGH-PERFORMANCE BLUE LEDs BASED ON GaN

If we try to grow bulk GaN crystals from a solution, we need a very high pressure and high temperature, similar to those needed for diamond growth, or even higher [7, 8]. So, we have to use a chemical reaction to reduce the pressure and temperature required for the growth of GaN. Also, we have to use foreign substrates. For the synthesis of GaN, we used ammonia as the nitrogen source [9] because nitrogen molecules are inert and do not actively react with metallic Ga. Ammonia is very active at temperatures of around 1000 °C at which GaN can be synthesized, therefore the range of materials that could be used as the substrate was limited.

Sapphire was one of the most promising substrate materials because it is stable at high temperatures and does not react with ammonia so strongly [10]. But the most serious problem with sapphire is its large mismatch with GaN of up to 16% for each (0001) plane. In general, for heteroepitaxial growth, some people think that the lattice mismatch should be less than a few percent [11], so a mismatch of 16% should make it almost impossible.

In 1971, Professor Jacques Pankove developed the first GaN-based blue LEDs, which were a metal-insulator-semiconductor (MIS)-type fabricated by hydride vapor phase epitaxy (HVPE), which involved the chemical reaction of Ga and hydrogen chloride to form GaCl and ammonia [12]. At that time, it was believed to be impossible to grow p-type GaN because of self-compensation [13]. Self-compensation means that if we dope acceptors as an impurity, the same number of intrinsic donors such as nitrogen vacancies are generated to compensate for the doped acceptors.

Another reason why bright blue LEDs are so difficult to achieve is related to the sensitivity of the human eye. The responsivity of the human eye to pure blue light is only 3% of that to 555 nm yellow-green light [14].

3. FUNDING SITUATION OF OUR LABORATORY IN THE MID-1980S AND THE DIFFICULTY OF GROWING GaN ON A SAPPHIRE SUBSTRATE

Let me go back to the early 80s. Professor Isamu Akasaki started his research on nitrides in 1967 [15] at Matsushita Research Institute Tokyo (MRIT), now Panasonic, first investigating powdered AlN. Then, his group started to grow GaN by molecular beam epitaxy (MBE) and observed its cathodoluminescence. His group subsequently switched to HVPE and succeeded in fabricating MIS-type blue LEDs with a flip-chip configuration in the late 70s [16]. Unfortunately, however, MRIT decided to abandon its project on GaN-based blue LEDs, so Professor Akasaki moved from MRIT to Nagoya University in 1981. I joined his laboratory in 1982 as an undergraduate student.

The problem of fabricating MIS-type blue LEDs using HVPE was that the growth rate was so high that it was difficult to control the thickness of the insulating layer in the MIS-type structure. Therefore, the operating voltage could not be controlled. Also, Professor Akasaki noticed the difficulty of growing GaN by MBE. He thus decided to use metalorganic vapor phase epitaxy (MOVPE) for the growth of GaN. At that time, funding for research at our laboratory was insufficient [17]. Also, there was no commercially available MOVPE system especially designed for the growth of GaN. Consequently, it was impossible to buy an MOVPE system. So in 1982, a master's degree student two years older than I developed the first vertical type MOVPE reactor [18]. At that time, the flow rate was so low that we could not grow GaN using hydrogen as the carrier gas. I tried to visualize the flow pattern by using the reaction between $TiCl_4$ and H_2O

to form TiO_2 powder and found that the flow rate would be insufficient if I used hydrogen as the carrier gas.

In 1984, a PhD student, now Dr. Yasuo Koide, joined Professor Akasaki's laboratory and started research on AlGaN and AlN, while I focused on growing GaN. From experience, I knew that the flow rate would be insufficient if I used the old configuration of gas supply tubes in the reactor, so I merged all the gas lines into one line and increased the flow rate from a few cm/s to more than 4 m/s [19]. Then, I successfully grew GaN on a sapphire substrate even though I used hydrogen as the carrier gas, although the surface was quite rough and the quality was very poor.

I tried to grow GaN many times while varying the growth temperature, the flow rate of the source and carrier gases, the configuration of the linear tubes, the susceptor shape, and other parameters. But I could not grow high-quality GaN with a smooth surface. The problem of the large lattice mismatch of 16% was too great for a master's student to overcome. So, almost two years passed without any success.

4. LOW-TEMPERATURE DEPOSITED BUFFER LAYER

In February 1985, I was almost at the end of my master's course. A foreign student and I had decided to start a PhD program from April. While all the other Japanese students went on a graduation trip, I carried out lonely experiments. At that time, Dr. Koide was growing Al-containing nitrides such as AlN and AlGaN and I was growing GaN. When we compared his Al-containing crystals and my GaN, the surface of his crystals seemed to be smoother. Therefore, I thought that AlN could be used to effectively grow GaN with a better surface morphology. So, I tried to grow a thin AlN layer on a sapphire substrate just before the growth of GaN. At that time, I knew that the epitaxial temperature of AlN should be higher than 1200°C. Because the old oscillator did not work well, I could not get the temperature to reach 1200°C. However, I suddenly remembered a discussion in the laboratory. Dr. Sawaki, an associate professor at that time, explained the growth process of boron phosphide (BP) on Si [20], for which the lattice mismatch is as large as 16%. He explained the effectiveness of a preflow of phosphorus as a source gas just before the growth of BP and mentioned that the phosphor atoms appear to act as nucleation centers. So, I imagined that if I supplied a small amount of AlN at a low temperature, it should provide nucleation centers. The temperature sequence in the growth process is shown in Fig. 1. Usually, I looked inside the reactor during growth to see whether there was an interference pattern on the substrate, by which I could check that the source gas

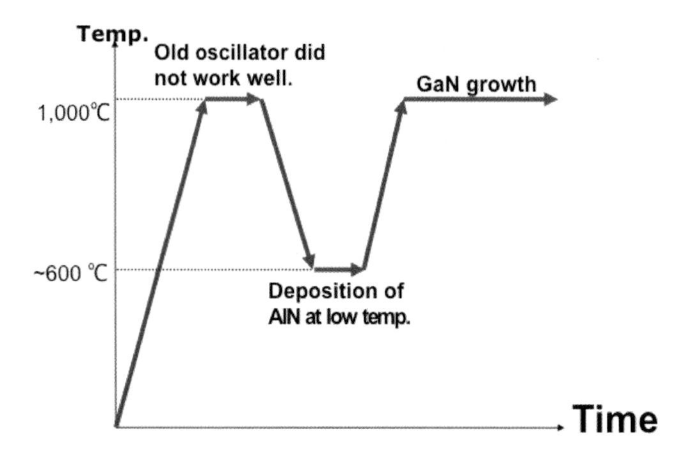

FIGURE 1. Susceptor temperature sequence in the growth of GaN on a sapphire substrate using a low-temperature-deposited AlN buffer layer.

had been properly supplied. But at that time, I was tired and forgot to check the interference pattern. When I took the sample out from the reactor and saw that it had a perfectly smooth surface and was perfectly transparent, I thought, "Oh, I've made a mistake! I forgot to supply trimethylgallium!".

But after rethinking, I recognized that I had not made a mistake.

So, I checked the surface using a Nomarski-type microscope and found that I had succeeded in growing atomically flat GaN as shown in Fig. 2. Following the suggestion of Professor Akasaki, I evaluated other qualities such as the crystalline, optical, and electrical qualities, all of which were superior to those in previous reports. This process is known as "low-temperature deposited buffer layer technology" and has been used by many researchers worldwide [21–37].

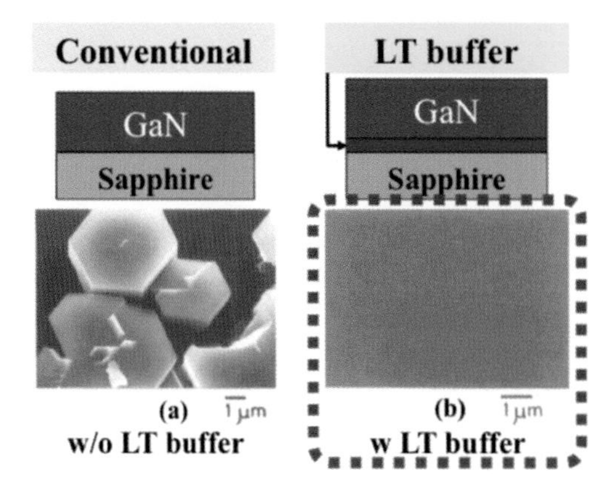

FIGURE 2. Scanning electron microscopic images of GaN on a sapphire (0001) substrate (a) without and (b) with a low-temperature-deposited AlN buffer layer [19].

5. REALIZATION OF p-TYPE GaN

The next task for us was to realize p-type GaN. I grew Zn-doped GaN many times, but all the samples were highly resistive or n-type. In 1987, during my PhD program, I observed very sharp exciton emission from Zn-doped GaN grown on c-plane and a-plane sapphire at a cryogenic temperature [22]. I also measured the deformation potential of the GaN. I was excited by these results and tried to present them at the Japan Society for Applied Physics annual fall meeting held at Nagoya University. However, I was surprised to see that there were only four people in the room for my presentation, the chairman, Prof. Akasaki, one other guy and me. At that time, other researchers were interested in other compound semiconductors such as GaAs and ZnSe, and GaN researchers were in the minority. Also in 1988, during my internship as part of my PhD program, I found that Zn-related blue emission was enhanced irreversibly during cathodoluminescence measurement as shown in Fig. 3 [38]. So, I called this process low-energy electron beam irradiation (LEEBI) treatment. But even after the LEEBI treatment, Zn-doped GaN did not show p-type conductivity. This phenomena was already published by Russian scientist. [39]

In 1989, I became a research associate of the Akasaki Laboratory of Nagoya University. When I read the textbook "Bonds and Bands in Semiconductors," written by Dr. J.C. Phillips [40], I found one graph particularly interesting. It shows that Mg is better than Zn for the activation of acceptors. However, the Mg source, bis-Cp_2Mg, was too expensive. So, I begged Professor Akasaki to let me buy some. He kindly gave permission, and after waiting several months for it to arrive, I was able to grow many Mg-doped samples with my laboratory partner Masahiro Kito, at that time a master's student.

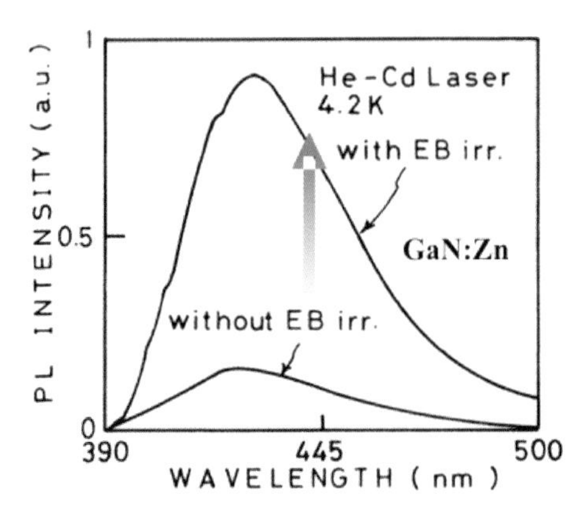

FIGURE 3. Change in blue PL intensity upon electron beam (EB) irradiation of Zn-doped GaN [38].

Here, I would like to mention the pioneering work in 1972 of Dr. H.P. Maruska [41], who at that time was a student at Stanford University. He succeeded in fabricating the world's first MIS-type violet LED using Mg-doped GaN.

All our Mg-doped GaN samples were highly resistive when they were as-grown. But after LEEBI treatment, some samples showed p-type behavior when subjected to hot probe measurement. I knew that hot probes are not so reliable and that no one would believe that p-type conduction had been achieved. So, Mr. Kito subjected the samples to Hall effect measurement and we finally recognized that we had achieved p-type GaN for the first time in the world. We also fabricated pn-junction ultraviolet LEDs as shown in Fig. 4 [42–45]. Soon after that, Dr. Shuji Nakamura's group also used LEEBI treatment [46, 47]. In 1992, Dr. Nakamura claimed that p-type GaN could be obtained by simple thermal annealing [48]. Today, almost all LED companies use thermal annealing.

The mechanism of p-type conduction involves the desorption of hydrogen near Mg acceptors as shown in Fig. 4, as first pointed out by Professor J.A. Van Vechten [49], which was confirmed experimentally by Dr. Nakamura [48].

6. ATTEMPTS TO GROW InGaN

For us, another important task was to realize true blue emission using a band-to-band transition. So, we tried to grow InGaN. However, this was also very difficult and we only succeeded in growing InGaN with an In composition of less than 1.7% [50].

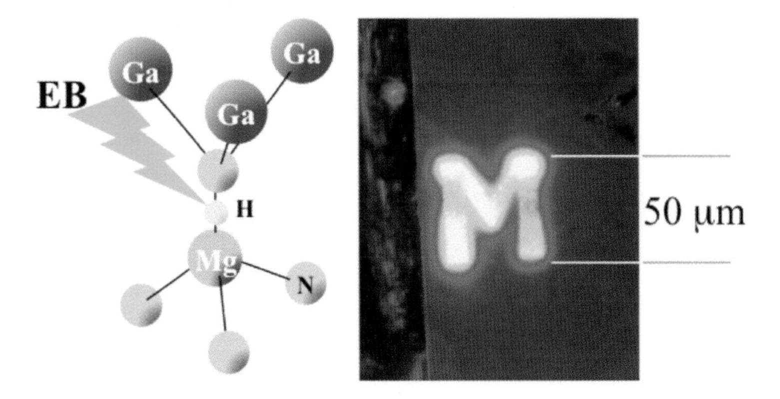

FIGURE 4. Schematic GaN drawing of the activation of hydrogen-passivated Mg in GaN [48] and electroluminescence pattern of a LED in which only the area of the "M" was irradiated with an electron beam [41].

In 1989, Dr. Takashi Matsuoka's group at NTT reported the successful growth of InGaN under an extremely high ammonia supply while also using nitrogen as a carrier gas [51]. They also reported blue-violet photoluminescence (PL) at 77 K, indicating the incorporation of In. At room temperature, deep-level-related yellow emission could be observed. The mechanism of In incorporation in InGaN has been clarified by thermodynamic analysis by Professor Akinori Koukitu et al. [53, 54].

Finally, by combining high-quality-crystal growth technology using a low-temperature-deposited buffer layer with p-type growth technology and InGaN growth technology, Nichia Corporation succeeded in commercializing double-heterostructure-type InGaN blue LEDs for the first time in the world in 1993 [55]. They also fabricated single-quantum-well LEDs in 1995 [56], which are also a very important technology for enhancing the efficiency of nitride LEDs because a very narrow quantum well suppresses the quantum-confined Stark effect [57], thus increasing the transition probability [58].

7. CONTRIBUTION OF InGaN-BASED BLUE LEDs TO ENERGY SAVING

To conclude, let me explain how InGaN LEDs can contribute to improving the electricity situation, especially in Japan. Many people remember the great earthquake of east Japan and the meltdown of the nuclear power plants in 2011. Currently, none of the 48 nuclear electricity generators in Japan are in operation [59]. Before 2011, about 30% of Japan's electricity was generated by nuclear reactors. So, we have to find a way of adapting to the loss of 30% of Japan's generating capacity. The US Department of Energy predicted that more than 70% of lighting will have been replaced with LED lighting systems by the year 2030 in the United States, resulting in a 7% reduction in electricity use [60]. In the case of Japan, the penetration of LED lighting systems into the market is expected to be much faster. A research company in Japan has predicted that by 2020 more than 70% of general lighting systems will have been replaced with LED lighting [61].

More importantly, we can develop and supply compact lighting systems to the younger generation, especially children in remote areas without access to electricity. Figure 5 shows an image of the Earth at night provided by NASA [62]. Using an LED lighting system with a solar cell panel and a battery, children can read books and study at night as shown in the inset images of Fig. 5.

Finally, I would like to address younger researchers. When we achieved the LT buffer, I was a 24-year-old master's student, and when we first realized p-type GaN, I was 28 years old. Of course I was very lucky to have carried out research under the excellent supervision of Prof. Akasaki and many distinguished

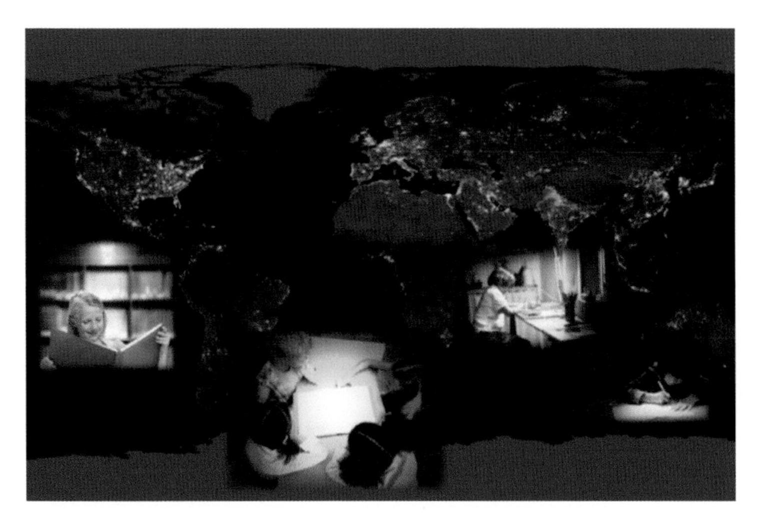

FIGURE 5. Image of the Earth at night provided by NASA [61].

colleagues. These days, facilities and funding are much better than in the 80s. So, I would like to see the younger generation attempting to tackle subjects which will greatly contribute to improving the quality of human lives. By doing so, the younger generation can develop a much better world for themselves.

ACKNOWLEDGEMENTS

I would like to thank the following people: Isamu Akasaki, Nobuhiko Sawaki, Kazumasa Hiramatsu, Shigeru Tamura, Atsushi Shimizu, Yasuo Koide, Kenji Itoh, Takahiro Kozawa, Masahiro Kito, Kouichi Naniwae; the previous students of Akasaki Laboratory at Nagoya University, Satoshi Kamiyama, Tetsuya Takeuchi, and Motoaki Iwaya; the previous students of the Akasaki and Amano Laboratory at Meijo University, Masahito Yamaguchi, Yoshio Honda, Guangju Ju, Kaddour Lekhal, and Siyoung Bae; the students of Amano, Yamaguchi and Honda Laboratory at Nagoya University, Aki Eguchi, Masako Yasui, Yoko Tatsumi, Tomoko Hosoe, Michinari Hamaguchi, Hideyo Kunieda, Yoshihito Watanabe, Yasuo Suzuoki, and Seiichi Matsuo; the staff of Nagoya University, Koichi Ota, Naoki Shibata, Nobuo Okazaki, Katsuhide Manabe, Michinari Sassa, Hisaki Kato, Masahiro Kotaki, and Tadashi Arashima; the staff of Toyoda Gosei, Masafumi Hashimoto, Akira Hirano, Masamichi Ipponmatsu, Cyril Pernot, Hidemasa Tomosawa, and Toshihiko Kai; and the staff of UVCR and Nikkiso.

Finally, I would like to express my sincere gratitude to my parents Yoshiko and Tatsuji Amano, my brother Takashi Amano, and my family Kasumi, Aya, and Mitsuru Amano.

REFERENCES

1. http://gaming.wikia.com/wiki/History_of_handheld_game_consoles
2. http://en.wikipedia.org/wiki/Mobile_phone
3. P. Schlotter, R. Schmidt and J. Schneider, *Appl. Phys.*, A**64** (1997) 417.
4. http://news.bbc.co.uk/2/hi/business/5085630.stm
5. http://www.notablebiographies.com/Ho-Jo/Jobs-Steve.html
6. http://www.maximizingprogress.org/2012/03/platform-wars-history-of-emergent.html
7. S. Porowski and I. Grzegory, *J. Cryst. Growth*, **178** (1997) 174.
8. F. P. Bundy, H. T. Hall, H. M. Strong and R. H. Wentorf, *Nature*, **176** (1955) 51.
9. W. C. Johnson, J. B. Parsons and M. C. Crew, *J. Phys. Chem.*, **36** (1932) 7.
10. H. P. Maruska and J. J. Tietjen, *Appl. Phys. Lett.*, **15** (1969) 327.
11. For example, F. C. Frank and J. H. van der Merwe, *Proc. R. Soc. London*, Ser. A **198** (1949) 205.
12. J. I. Pankove, E. A. Miller, D. Richman and J. E. Berkeyheiser, *J. Lumin.*, **4** (1971) 63.
13. For example, G. Mandel, *Phys. Rev.* A, **134** (1964) 1073.
14. For example, photopic V(λ) modified by Vos (1978) http://www.cvrl.org/ Copyright © 1995–2015 Color and Vision Research Labs.
15. I. Akasaki and M. Hashimoto, *Solid State Commun.*, **5** (1967) 851.
16. Y. Ohki, Y. Toyoda, H. Kobayashi and I. Akasaki, *Inst. Phys. Conf. Ser.*, **63** (1982) 479.
17. For example, http://www.mext.go.jp/english/whitepaper/1302651.htm
18. M. Hashimoto, H. Amano, N. Sawaki and I. Akasaki, *J. Cryst. Growth*, **68** (1984) 163.
19. H. Amano, N. Sawaki, I. Akasaki and Y. Toyoda, *Appl. Phys. Lett.*, **48** (1986) 353.
20. T. Nishinaga and T. Mizutani, *Jpn. J. Appl. Phys.*, **14** (1975) 753.
21. H. Amano, I. Akasaki, K. Hiramatsu, N. Koide and N. Sawaki, *Thin Solid Films*, **163** (1988) 415.
22. H. Amano, K. Hiramatsu and I. Akasaki, *Jpn. J. Appl. Phys.*, **27** (1988) L1384.
23. I. Akasaki, H. Amano, Y. Koide, K. Hiramatsu and N. Sawaki, *J. Cryst. Growth*, **98** (1989) 209.
24. H. Amano, T. Asahi and I. Akasaki, *Jpn. J. Appl. Phys.*, **29** (1990) L205.
25. K. Hiramatsu, H. Amano, I. Akasaki, H. Kato, N. Koide and K. Manabe, *J. Cryst. Growth*, **107** (1991) 509.
26. K. Hiramatsu, S. Itoh, H. Amano, I. Akasaki, N. Kuwano, T. Shiraishi and K. Oki, *J. Cryst. Growth*, **115** (1991) 628.
27. N. Kuwano, T. Shiraishi, A. Koga, K. Oki, K. Hiramatsu, H. Amano, K. Itoh and I. Akasaki, *J. Cryst. Growth*, **115** (1991) 381.
28. S. Nakamura, *Jpn. J. Appl. Phys.*, **30** (1991) 1620.
29. H. Murakami, T. Asahi, H. Amano, K. Hiramatsu, N. Sawaki and I. Akasaki, *J. Cryst. Growth*, **115** (1991) 648.
30. J. N. Kuznia, M. A. Khan, D. T. Olson, R. Kaplan and J. Freitas, *J. Appl. Phys.*, **73** (1993) 4700.
31. S. T. Kim, H. Amano, I. Akasaki and N. Koide, *Appl. Phys. Lett.*, **64** (1994) 1535.
32. T. Sasaoka and T. Matsuoka, *J. Appl. Phys.*, **77** (1995) 192.
33. Y. M. Le Vaillant, R. Bisaro, J. Oliver, O. Durand, J. Y. Duboz, S. Ruffenach-Clur, O. Briot, B. Gil and R. L. Aulombard, *Mater. Sci. Eng.*, B**50** (1997) 32.

34. M. Iwaya, T. Takeuchi, S. Yamaguchi, C. Wetzel, H. Amano and I. Akasaki, *Jpn. J. Appl. Phys.*, **37** (1998) L316.
35. Y. M. Le Vaillant, R. Bisaro, J. Olivier, O. Durand, J-Y Duboz, S. Ruffenach-Clur, O. Briot, B. Gil, and R. L. Aulombard, *J. Cryst. Growth*, **189/190** (1998) 282.
36. Y. Kobayashi, T. Akasaki and N. Kobayashi, *Jpn. J. Appl. Phys.*, **37** (1998) L1208.
37. T. Ito, K. Phtsuka, K. Kuwahara, M. Sumiya, Y. Takano and S. Fuke, *J. Cryst. Growth*, **205** (1999) 20.
38. H. Amano, I. Akasaki, T. Kozawa, K. Hiramatsu, N. Sawaki, K. Ikeda and Y. Ishii, *J. Lumin.*, **40–41** (1988) 121.
39. G.V. Saparin, S.K. Obyden, M.V. Chukichev, S.J. Popov, *J. Lumin.* **31** & **32** (1984), 684.
40. J. C. Phillips, *Bonds and Bands in Semiconductors*, 1st edition, Academic Press 1973.
41. H. P. Maruska, W. C. Rhines and D. A. Stevenson, *Mater. Res. Bull.*, **7** (1972) 777.
42. H. Amano, M. Kito, K. Hiramatsu and I. Akasaki, *Jpn. J. Appl. Phys.*, **28** (1989) L2112.
43. H. Amano, M. Kitoh, K. Hiramatsu and I. Akasaki, *J. Electrochem. Soc.*, **137** (1990) 1639.
44. I. Akasaki, H. Amano, M. Kito and K. Hiramatsu, *J. Lumin.*, **48** & **49** (1991) 666.
45. I. Akasaki, H. Amano, H. Murakami, M. Sassa, H. Kato and K. Manabe, *J. Cryst. Growth*, **128** (1993) 379.
46. S. Nakamura, M. Senoh and T. Mukai, *Jpn. J. Appl. Phys.*, **30** (1991) L1708.
47. S. Nakamura, N. Iwasa, M. Senoh and T. Mukai, *Jpn. J. Appl. Phys.*, **31** (1992) 1258.
48. S. Nakamura, T. Mukai, M. Senoh and N. Iwasa, *Jpn. J. Appl. Phys.*, **31** (1992) L139.
49. J. A. Van Vechten, J. D. Zook, R. D. Horning and B. Goldenberg, *Jpn. J. Appl. Phys.*, **31** (1992) 3662.
50. T. Kozawa, Master's Thesis, Nagoya University, 1987.
51. T. Matsuoka, H. Tanaka, T. Sasaki and A. Katsui, *Inst. Phys. Conf. Ser.*, **106** (1990) 141.
52. N. Yoshimoto, T. Matsuoka, T. Sasaki and A. Katsui, *Appl. Phys. Lett.*, **59** (1991) 2251.
53. A. Koukitu, N. Takahashi, T. Taki and H. Seki, *Jpn. J. Appl. Phys.*, **35** (1996) L673.
54. A. Koukitu, T. Taki, N. Takahashi and H. Seki, *J. Cryst. Growth*, **197** (1999) 99.
55. S. Nakamura, M. Senoh and T. Mukai, *Jpn. J. Appl. Phys.*, **32** (1993) L8.
56. S. Nakamura, M. Senoh, N. Iwasa and S. Nagahama, *Jpn. J. Appl. Phys.*, **34** (1995) L797.
57. H. Amano and I. Akasaki, Ext. Abst. Int. Conf. Solid State Devices and Materials, V-7 (1995) 683.
58. T. Takeuchi, S. Sota, M. Katsuragawa, M. Komori, H. Takeuchi, H. Amano and I. Akasaki, *Jpn. J. Appl. Phys.*, **36** (1997) L382.
59. http://www.enecho.meti.go.jp/category/electricity_and_gas/nuclear/001/pdf/001_02_001.pdf (in Japanese)
60. U.S. Department of Energy, *Energy Savings Potential of Solid-State Lighting in General Illumination Applications*, Jan. 2012, (2012) 4. (http://apps1.eere.energy.gov/buildings/publications/pdfs/ssl/ssl_energy-savings-report_jan-2012.pdf)
61. Fuji Chimera Research Institute, Inc., *2014 LED Related Market Survey*, (2014) 41.
62. http://earthobservatory.nasa.gov/Features/NightLights/page3.php

Shuji Nakamura. © Nobel Media AB. Photo: A. Mahmoud

Shuji Nakamura

Shuji Nakamura was born on May 22, 1954 in Oku, a tiny fishing village on the Pacific coast of Shikoku, the smallest of Japan's four main islands. Farming is the principle occupation in Oku. Local farmers grow yams on steps cut into steep hillsides. Shuji's maternal grandparents owned such a farm. To get to the nearest town, the villagers relied on a ferry. Inconvenient perhaps, but the village was an idyllic place to grow up.

Shuji's father, Tomokichi, worked as a maintenance man for Shikoku Electric Power. From him, Shuji learned how to make wooden toys, like catapults and bamboo propellers. He liked making things and became good at it, a skill that would stand him in good stead.

At school Shuji was not academically gifted. His boyhood was typical. He fought constantly with his elder brother. Smaller than his sibling, Shuji always lost. Though physically defeated, mentally he would never give in. His mother was forever chiding her boys to do their homework. But for the most part, they ignored her admonishments.

Throughout primary and high school, Shuji's passion was volleyball. There was no gym at the school, so his team had to practice outside in the mud. They tried hard, but rarely won. Fiercely competitive from an early age, Shuji always hated to lose. Volleyball left Shuji little time to study for his high school entrance exams. He was bad at rote learning, but good at math and science. Somehow, he managed to scrape into an academically-oriented school.

Here, too, volleyball remained his priority. His classroom teacher told him that to improve his scores, he would have to quit playing. It was time to concentrate on studying for the all-important university entrance exams. But he could not let down his team. Shuji was the only student in the A-stream to continue playing sports until graduation.

FIGURE 1. Shuji and his family in 1979.

Shuji paid a price for his dedication to volleyball. His university entrance exam results were not good enough to win him a place at a prestigious school. Shuji's dream was to become a theoretical physicist or a mathematician. But his teacher told him that he could not make a living from physics—he had better choose a course like engineering so that he could find a job.

Shuji picked electrical engineering because it seemed close to physics. In 1973, aged 19, he entered Tokushima University, a local state school. Many professors there were former high-school teachers. The textbooks were out of date.

His first two years consisted of general studies, including arts courses, which Shuji hated. He couldn't understand why he had to take such irrelevant subjects. Soon he stopped attending classes. All day long he would read books, mostly on physics. But there was a limit to the amount he could absorb by reading alone. Finally, in his third year at Tokushima, Shuji attended a lecture on semiconductors. Fascinated by the physics of solid-state materials, he decided to stay on at university for a further two years and do a master's degree under Professor Osamu Tada.

As his thesis topic, Shuji chose the conductivity mechanism of barium titanium oxide. His focus was theoretical. But Professor Tada was a dyed-in-the-wool experimentalist. He would catch his student reading papers and tell him that knowing theory was no use if he couldn't make actual devices.

Tada's lab was known as "the junk room." It was crammed with broken televisions and old radios which could be cannibalized for spare parts. To build what they needed, students had to acquire manual skills—soldering, cutting and joining glass, beating and welding sheet metal, fashioning parts on a lathe.

Shuji remembered his days as a master's student as like being a sheet-metal worker in a small factory. What he wanted was to study theory. But most of his time was taken up jury-rigging equipment for experiments. In fact, Shuji was gaining precisely the kind of skills that he would later need in his quest to develop a bright blue LED. As a corporate researcher, he would be forced to make or modify much of his own equipment. Ultimately it would largely be this technical mastery that would give him the edge on his rivals.

As a 25-year-old graduate student with a master's degree in electrical engineering, Shuji expected one of Japan's consumer appliance manufacturers would hire him. But the likes of Sony tend not to recruit graduates from local universities.

In his interview at Matsushita, Shuji made the mistake of discussing the theoretical aspects of his thesis research. We don't need theoreticians, the firm's recruiters told him. At Kyocera, he did better, emphasizing the practical applications of his work. The company offered him a job. But as the day on which Shuji was due to report for work in Kyoto neared, he had second thoughts. Prior to his job-hunting trips, he had rarely left Shikoku.

At the same time, he really wanted to work in a proper research laboratory at a major company. Torn between two paths, Shuji asked Professor Tada what he should do. Tada pointed out that in Tokushima, there were no jobs for electrical engineers. If Shuji elected to stay on the island, he would have to give up a career in his chosen field. Eventually Shuji decided to remain in Tokushima.

His professor introduced Shuji to Nobuo Ogawa, the founder-president of an obscure local chemical firm called Nichia. The company was initially reluctant to hire him, but Shuji refused to take no for an answer.

When Shuji joined in April 1979, Nichia had fewer than 200 employees. The firm made phosphors for color televisions and fluorescent lamps. These were mature markets. If Nichia was to grow, it needed new products.

Shuji was assigned to the company's two-man development section. His first job was to refine high-purity gallium metal. This turned out to be a dead end. The company ordered him to produce gallium phosphide, a material used to make red and green LEDs. Knowing next to nothing about LED materials, Shuji had to start from scratch.

There was no budget for equipment. He had to scavenge, fixing broken parts by hand. To build his reactor, Shuji scrounged heat-proof bricks, cables,

FIGURE 2. Shuji's first day at Nichia in April 1979.

a vacuum pump, and an old electric furnace. He had to order quartz tubes. To seal the open-ended tubes so they could be evacuated, he had to learn how to weld quartz.

To make gallium phosphide you heat phosphorus in a tube. If the tube gets too hot, the phosphorus vapor expands, causing the quartz to crack. This lets in oxygen, which reacts with the phosphorus causing an explosion. Such explosions became a feature of Shuji's time at Nichia. His lab would fill with white smoke. Ignited phosphorus would fly everywhere, along with shards of broken quartz. Shuji would run around pouring water over the burning phosphorus, desperately trying to douse the flames.

The blasts happened several times a month, often in the evening. The shock wave would hit his fellow-workers as they were heading for their cars in the parking lot. The first few times it happened, they dashed into his lab to see if he was alright. By the fifth or sixth time, however, they had become so used to the bangs they no longer came to check.

Eventually, Shuji succeeded in developing commercial-grade gallium phosphide. More satisfying than producing the material was how he felt when the company's salesmen told him they had made a sale. It gave him pleasure to think that he had finally managed to contribute to the company's bottom line. But only a little: the market for gallium phosphide was already crowded. A late entrant, Nichia was only able to win a sliver of the pie.

Shuji's next assignment was to produce gallium arsenide, which is also used to make LEDs, typically infrared ones such as those found in television remote controls. But GaAs also has other applications, like the semiconductor lasers used in optical fiber communications. Thus, the potential market for the material was much larger.

Happily, unlike phosphorus, arsenic is not inflammable. Unhappily, the material is poisonous, releasing lethal arsenic oxide gas every time the furnace blew up. Shuji had to wear a home-made "space-suit" and breathe through a respirator. Miraculously he was never adversely affected by having to work in such a toxic environment.

By 1985, Shuji was producing gallium arsenide in bulk. But when it came to selling the product, the market's response was the same. There were plenty of existing suppliers, so why buy from an untried latecomer like Nichia? The next idea the salesmen brought back was, instead of making the starting materials for LEDs, why not make the devices themselves? To fabricate a simple LED required mastering a technique known as liquid phase epitaxy.

Shuji performed countless experiments. Small differences in thickness, he discovered, could make a big difference in brightness and lifetime. As usual, the company pressured him to produce a saleable product quickly; as usual, there was no budget for equipment. Eventually he managed to fabricate some prototype LEDs. Samples were delivered to a client for evaluation. Not having his own measuring equipment meant Shuji was dependent on such external evaluations. He had to wait months to get data back before he could start making improvements.

Shuji felt strongly that if the company was going to enter the LED business, then he should be able to conduct his own evaluations. He presented the case to his boss, but was told no budget, so not possible. Previously, Shuji would have accepted this answer and given up. By now, he had realized that Nichia was run on the say-so of its president and founder, Nobuo Ogawa. He went directly to Ogawa to ask for the equipment he needed. To his surprise, the old man immediately agreed to his request.

Shuji had made many friends among Nichia's employees. When work finished they would often ask him to make up the numbers for a game of softball. Afterwards, they would drop by a local bar. There, his workmates would implore him to develop products that would make the company grow. Knowing that he had yet to produce anything that had a significant impact on the company's bottom line, Shuji would hang his head. Others, especially older employees, were critical. They asked him what he had been doing for the past five years. In their opinion, he was just wasting the company's money.

The only way a corporate researcher can contribute directly to the bottom line is through patent royalties. But fearful of losing trade secrets, Nichia did not permit patent applications. Thus Shuji's apparent sales were zero. In ten years, Shuji had not published a single scientific paper, because of Nichia's policy of keeping its technical know-how secret. From a professional point of view, he had no achievements.

Finally, in desperation, he approached the president with an audacious proposal: to develop the world's first bright blue LED. To do this he would need around five hundred million yen (then worth about US $4 million). This was equivalent to two percent of the company's sales that year, an unbelievably large amount. Nonetheless, Ogawa gave Shuji his blessing.

Two thirds of the money would go to equipment, together with the laboratory and clean-room facilities to house it. Of the remaining third, the largest item was mastering metal-organic chemical vapor deposition, the crystal growth technology needed to make bright blue LEDs.

Shuji selected MOCVD because it could be applied to the factory floor. Shiro Sakai, an expert on the technique, was an old acquaintance from Tokushima University. Now a professor at Tokushima, Sakai was on sabbatical at the University of Florida. Shuji invited him to visit Nichia. There he outlined the significance of MOCVD. Blue LEDs were not mentioned. Sakai recommended that Nichia should send Shuji to Florida for a year to learn the technique.

In March 1988 Shuji flew to Gainesville. It was the first time the country boy had boarded an airplane. Like many first-time fliers, he feared it might fall from the sky. It was also his first trip abroad. He worried that his rudimentary English would not enable him to communicate with Americans.

Shuji was 34 years old, rather long in the tooth for a student. His fellow researchers at the University of Florida were mostly in their mid-twenties. All of them were PhD students. Shuji's status was ambiguous. Since he was not studying for a degree, he was obviously not a student. Nor, since he did not have a PhD, could he be a post-doctoral fellow. As a compromise, he was designated a "guest research associate."

Initially, his fellows treated Shuji as an equal or even, because he was older, as a senior. However, once they discovered that he only had a master's degree and, worse, that he had not published a single paper, their attitude changed. Henceforth they looked down on Shuji, treating him as little more than a technician. It was particularly galling because, from his perspective, these PhDs were mere novices whereas he had years of hands-on experience. They could not do the simplest experiment. Something would go wrong and they would come running to him for help. Their condescending attitude provided Shuji with further motivation. "I feel resentful when people look down on me," he wrote. "At that time, I developed more fighting spirit—I would not allow myself to be beaten by such people."

When Shuji arrived at Gainsville, the MOCVD system at the lab to which he had been assigned had not yet been built. He had to spend ten months of his precious year in the US with his sleeves rolled up, connecting pipes and welding quartz, just like back at Nichia. Here again, adversity in the short term would turn out to be priceless experience in his quest to develop the first bright blue LEDs. He gained an intimate familiarity with the workings of MOCVD equipment. Having managed to assemble the system, Shuji was only able to do a few device-growing runs. Then it was time to go home.

Shuji returned to Nichia in March 1989. While in the US he had ordered his own MOCVD equipment, keeping his goal a secret from the supplier. The reactor had arrived. The question was, what material to grow in it? There were three candidates. One, silicon carbide, despite the fact that it was in limited commercial production, he had already rejected. Silicon carbide had an indirect bandgap, meaning that the material would never be able to emit bright blue.

The other two materials, zinc selenide and gallium nitride, both suffered from the same deficiencies. One was that, to make a proper LED, you need to fabricate both negative- and positive-type material. Thus far, however, it had proved impossible to produce either p-type zinc selenide or p-type gallium nitride. When making his choice Shuji could not have known it, but this was about to change: in 1989, researchers would succeed in fabricating p-type gallium nitride; the following year would see the first p-type zinc selenide.

The second, more serious, drawback was the lack of a suitable base material on which to fabricate an LED. Gallium arsenide LEDs could be grown on gallium arsenide wafers. But nobody had been able to grow bulk zinc selenide or gallium nitride. That meant employing wafers of some "foreign" material as the substrate, which in turn mean a mismatch between substrate and light emitting layers. The result was defects, which are undesirable because they cause LEDs to dissipate energy in the form of heat instead of light.

With zinc selenide, a soft material, the problem seemed much less severe. You could grow ZnSe on a gallium arsenide substrate and the mismatch was only 0.3 percent, not far off the ideal value of 0.01 percent. This translated into a defect density of around one thousand per square centimeter. With gallium nitride, a rock-hard material, the best available substrate was sapphire. But even sapphire produced a huge mismatch, of sixteen percent. That translated into in a defect density of ten billion per square centimeter. It was plausible to imagine that imperfections in crystal ZnSe could be reduced by an order of magnitude. But *ten billion* defects? It seemed unlikely that that figure was going to be significantly reduced during any researcher's working lifetime.

Gallium nitride had been thoroughly investigated by RCA, Bell Labs, and Matsushita. It was almost universally perceived to be a dead end. Few groups were still active in the GaN field. The overwhelming consensus was that zinc selenide was the way to go. Yet zinc selenide devices tended to fall apart when zapped with current. ZnSe simply wasn't strong enough to cope with the stress of giving birth to photons. What nobody could have foreseen in 1989 was that gallium nitride would turn out to behave very differently than previous light emitting materials. Any other semiconductor with that density of defects simply would not function. Much to everyone's surprise, however, with gallium nitride, defects just didn't seem to matter.

Having arrived at what he described as this "fateful fork in the road," Shuji chose gallium nitride. His reason for placing this apparently reckless bet was not because he was confident that he could do what no-one else had done. Rather, it was because he had repeatedly had the bitter experience of developing products only to find that his company could not sell them. If he chose zinc selenide, since big companies had several years' head start, history would likely repeat itself. With gallium nitride, in the unlikely event that he did succeed, there would be no competition, because no other companies were working on GaN.

Another motivation, following his unhappy experience at the University of Florida, was that Shuji wanted to get a PhD. In Japan it was possible to obtain a PhD by publishing a minimum of five scientific papers. If he had selected zinc selenide, it would have been difficult to publish papers because a huge number of papers on ZnSe had already been published. If he selected gallium nitride, it would be easier to publish because only a few papers had been published.

Shuji was able to make this seemingly foolhardy decision by himself without reference to Nichia's senior management because none of them knew anything about semiconductors. All they knew was that his goal was to develop a bright blue LED. The choice of methodology to adopt and material to work on was his alone. Had he been working at a large company, his proposal to work on a known-loser material would undoubtedly have been shot down. But as he himself would later say, "breakthroughs are born out of unusual circumstances."

The quest began. Shuji had no colleagues with whom he could discuss his work. Other than New Year's Day, he never took time off. His solitary routine seldom varied. He would get into work around 7AM, leaving around 7PM. He would go home, eat dinner with his family, have a bath, then go to bed. All the while he would be musing about his work.

Modifying the MOCVD equipment was the key to his success. Shuji took the reactor apart, then put it back together exactly how he wanted. He bent the steel pipe, changing the height and the angle at which it was attached to the reaction chamber. He welded quartz tube, cut high-purity carbon, re-did the wiring. He even altered the shape of the gas nozzles.

Shuji's motto was "remodel in the morning, experiment in the afternoon." Such urgency was not because he was worried that other researchers might overtake him. Rather, impatient by nature, he was eager to see the results of the changes he had made.

To grow high-quality films of gallium nitride, one major problem had to be solved. Nitrides are vulnerable to parasitic reactions. The gases react with each other spontaneously to form an adduct—in this case, a white powder that researchers call "snow." Flakes of snow fall on the wafer, ruining the film. Much ingenuity therefore goes into designing reactors so that the gases are injected separately, keeping them apart as they flow down to the wafer. Shuji conceived a novel way of doing this, which he dubbed "two-flow" MOCVD.

Still, failure followed failure. Nichia kept demanding to know when he could develop a product. But as time went by and no results emerged, his boss stopped bothering him. Even Shuji's friends at the company left him alone. Then, one winter's day, the clouds finally lifted.

Everything was as usual: Shuji arrived at work and grew a thin film of gallium nitride crystal. He hooked up his sample to measure its electron mobility. The figure was surprisingly high: the best result to that point, achieved at Nagoya University by [fellow Nobel Laureates] Professor Isamu Akasaki and his student Hiroshi Amano, was less than half as much. Shuji had succeeded in making the world's best gallium nitride. "It was the most exciting day of my life," he recalled.

Further breakthroughs followed. Akasaki and Amano had blazed the trail, with their buffer layer (1986) and positive-type gallium nitride (1989). The buffer layer was necessary to mitigate the effect of the mismatch between the sapphire substrate and the gallium nitride layers deposited on top. Interposing a buffer enabled the growth of smoother films. For their buffer layer Akasaki and Amano had used aluminum nitride. Shuji was determined not to copy his rivals, knowing patent problems would result if he did. He used gallium nitride as the material for his buffer layer. He was able to achieve a smooth, mirror-like surface that had better electrical characteristics than aluminum nitride.

But to build a blue LED, Shuji needed first to make positive-type gallium nitride. His rivals had produced p-type GaN by irradiating the material with an electron beam. This was a wonderful scientific discovery, but impractical technologically because the method was too slow for LED manufacturing.

Akasaki and Amano announced their discovery at a conference in 1989, just after Shuji got back from Florida. He asked them what the hole concentration of their material was. The answer told him the quality of their material was not high. But at least they had demonstrated it was possible to make p-type GaN. Shuji would thus start his research on GaN just as the hitherto most intractable problem in the field had been shown solvable. It was an incredible stroke of luck.

In July 1991, armed with both negative- and positive-type materials, Shuji was able to proceed to the next stage, making a simple LED. The device lit up with a violet-blue light. Though not very bright, it was fifty percent brighter than conventional, silicon carbide LEDs. The outstanding question was longevity. How long would a fragile thin film with ten billion defects per square centimeter continue to emit light? He went home that night, leaving his LED switched on. Next morning, he returned to the lab, his heart thumping, to find that . . . it was still lit! He measured the output and was elated to discover that it had barely dropped. In fact, on testing, the lifetime turned out longer than 1,000 hours.

Next, he focused on making high-quality p-type gallium nitride. His rivals had not figured out why e-beams caused the transformation. Shuji speculated that it was merely heat that turned the material p-type. In December 1991, he tried annealing magnesium-doped films. The resultant material was p-type. Thermal annealing was simpler and much faster than the e-beams, hence applicable to the production line. It also produced much better quality gallium nitride. E-beam penetration was very shallow, with only a very thin surface layer of the material becoming p-type. Thermal annealing converted the material to p-type all the way through. This was a major breakthrough.

Shuji also clarified the mechanism of hole compensation, which had been a mystery for twenty years. Atomic hydrogen produced from the dissociation of ammonia gas forms Mg-H complexes. This formation prevents magnesium acceptors from behaving as acceptors. Using thermal annealing removes atomic hydrogen from the Mg-H complexes activating the Mg acceptors. The material then becomes p-type gallium nitride.

The world's first conference on nitrides was held in St Louis in 1992. Shuji gave a talk on his prototype blue LED. He revealed that its lifetime was more than 1,000 hours. The audience reacted by giving him a standing ovation. Encouraged by the response, on his return to Japan, Shuji embarked on the final stage of what he called his "climb to the summit of Mount Fuji."

To make a bright blue LED, he had to take two further steps. First, to make the light bright, he had to build a more complex device, called a double heterostructure. Second, in order to make the light pure blue, as opposed to violet-blue, he had to prepare alloys that incorporated indium, whose slightly narrower bandwidth would produce longer-wavelength light.

Thus far, no one had been able to make indium gallium nitride of sufficiently high quality for practical use. The difficulty was that the InGaN layer has to be grown at a much lower temperature than the confining layers of GaN. The bonds between indium and nitrogen are weak. Increase the temperature too quickly and the indium atoms disassociate themselves from their nitrogen neighbors. How to move on to grow the next layer, upping the temperature without destroying the thin layer of InGaN in the process? It was at this final hurdle that Akasaki and Amano fell.

Shuji solved the disassociation problem in two ways. First, by brute force, turning the indium tap on his system all the way open, using ten times as much indium as would turn out to be needed, attempting to get least some of the stuff to stick. Second, by guile, adding an extra "blocking" layer to cap the InGaN layer, preventing the material from disassociating.

In September 1992, he succeeded in fabricating a double heterostructure LED. Its wavelength was still too short to qualify as true blue. By the end of the year, he had adjusted the growth program, increasing the amount of indium and reducing the thickness of the active layer. This time, there was no doubting the result. Thus far, the output of blue LEDs had been given in milli-candelas, or thousandths of a candle. Now, for the first time, Shuji's device crossed into the candela class. It shone with a dazzling sky-blue light, a hundred times brighter than silicon carbide blue LEDs, bright enough to be clearly seen in broad daylight. Shuji felt like he had reached the top of Mount Fuji.

On November 29, 1993, at a press conference in Tokyo, Nichia announced the world's first bright blue LED. The initial reaction was incredulity. Once the disbelief subsided, however, orders for Nichia's LEDs started pouring in.

Shuji continued making breakthroughs. In May 1994, he demonstrated blue and blue-green LEDs capable of emitting two candelas, double the brightness of his original devices. Next year Nichia commercialized bright emerald green light emitters, the first true green LEDs. In September 1995, Shuji announced the first quantum-well-based blue and green LEDs. These featured a brightness of up to ten candelas. Also in 1995, at Shuji's suggestion, the company developed white LEDs. They worked by placing a yellow phosphor in front of a bright blue LED, converting its light to white. Wavelength conversion opened up huge new markets, in particular general illumination.

Perhaps Shuji's biggest coup was developing a blue laser diode. Many people thought such devices would be impossible given that GaN crystal was riddled with micro-cracks. To amplify light, a laser needs a more complicated structure than an LED; it also has to be pumped with more current. The structural defects in the material should have scattered the light, preventing optical amplification. Under high current, the defect-ridden layers should have caused instantaneous catastrophic failure.

In the mid 1990s, blue lasers were seen as more significant than blue LEDs. The reason is that, whereas it was hard to imagine all the applications that would emerge for bright blue LEDs, it was clear what the big application for a blue laser would be: data storage. That was why consumer electronics and disk drive companies were pouring resources into blue laser development. In 1996 Shuji unveiled a prototype violet-blue laser at a conference in Berlin, using the laser as a pointer in his presentation. By the end of the year, he and his group at Nichia announced an improved blue laser that operated for 1,000 hours.

In December 1999, Shuji left Nichia to join the University of California at Santa Barbara (UCSB) as a professor of materials and electrical & computer engineering. Since then he has continued to push the boundaries in solid-state lighting and associated crystal growth methods with his colleagues Professors James Speck, Umesh Mishra and Steven DenBaars.

FIGURE 3. Shuji with his colleagues (from left to right) James Speck, Umesh Mishra and Steven DenBaars at UCSB after the announcement of the 2014 Nobel Prize in Physics.

FIGURE 4. Group photograph of the Solid State Lighting and Energy Electronics Center (SSLEEC) in 2014.

As the Research Director of the Solid State Lighting & Energy Electronics Center (SSLEEC) and the Cree Chair in Solid State Lighting & Displays, he is overseeing the research enabling the next generation opto-electronic devices. Of the various research topics, he is a strong advocate of developing and using native gallium nitride substrates, which offer significant improvements towards efficient operation at high current. Furthermore, it opens the door to investigating the use of laser based solid-state lighting due to the superior performance of lasers over LEDs at very high current densities, and hence light output. To enable this future, he is heavily invested in pursuing bulk single crystal growth of GaN boules using the ammonothermal method—a method which grows single crystals from a supercritical ammonia solution under extreme conditions (thousands of atmospheres pressure and hundreds of degrees Celsius).

Shuji currently holds more than 200 US patents, over 300 Japanese patents and has published more than 550 papers in his field. Since coming to UCSB, Shuji has become a fellow of the National Academy of Engineering (NAE) and the National Academy of Inventors (NAI) and has won numerous awards. They

FIGURE 5. Shuji with his researchers Paul Von Dollen (middle) and Siddha Pimputkar (right) in one of the bulk single crystal GaN growth laboratories.

include the Charles Stark Draper Prize (2015), the Order of Culture Award (2014), the Inventor of the Year Award from the Silicon Valley Intellectual Property Law Association (2012), the Technical and Engineering Emmy Award (2011), the Harvey Prize from Technion, the Israel Institute of Technology (2009), the Japan Science of Applied Physics Outstanding Paper Award (2008), the Prince of Asturias Award for Technical Scientific Research (2008), the Czochralski Award (2007), the Santa Barbara Region Chamber of Commerce Innovator of the Year Award (2007), Finland's Millennium Technology Prize (2006) and the Global Leader Award, Optical Media Global Industry Awards (2006).

Background Story of the Invention of Efficient Blue InGaN Light Emitting Diodes

Nobel Lecture, December 8, 2014

by Shuji Nakamura
University of California, Santa Barbara, CA, USA.

1. INTRODUCTION AND IMPACT OF WHITE LIGHT EMITTING DIODES

The basic structure of an efficient double-heterostructure (DH) light emitting diode (LED) is summarized in Figure 1. This optoelectronic device is composed of semiconductor materials and is fabricated by sandwiching an active, emitting layer between an n-type and p-type layer. The n-type semiconductor layer has an abundance of high-energy electrons, whereas the p-type semiconductor has an abundance of available, empty sites in which the electron may reside at a lower energy level. These sites are also referred to as holes, are positively charged, and are mobile. The energy difference between the high and low energy electron state is referred to as the bandgap of the material. For DH LEDs, the bandgap of the active layer is smaller than that of the n-type and p-type layers.

When forward biasing a DH LED using a battery (or any other direct current source), electrons and holes are injected into the active layer from n-type and p-type layer, respectively. The electrons and holes recombine radiatively in the active layer, thereby emitting photons. This act is very efficient for DH LEDs as the electrons and holes are confined to the active layer due to the smaller bandgap of the active layer with respect to the n-type and p-type cladding layers (see also Figure 7). The resulting photon has an energy approximately equal to the bandgap of the active layer material. Modifying the bandgap of the active layer creates photons of different energies.

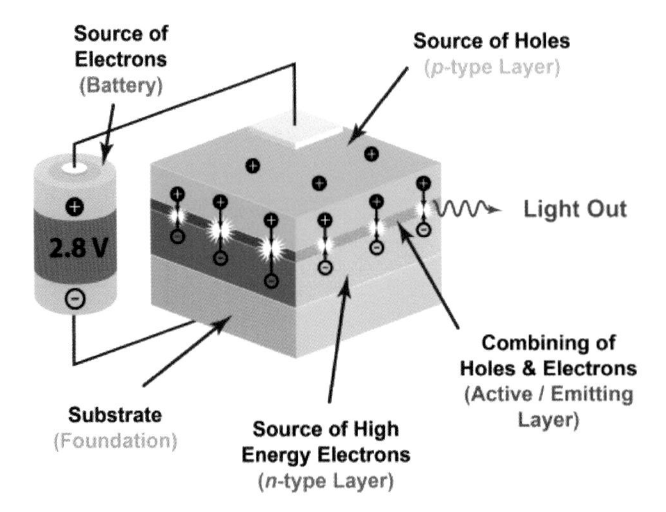

FIGURE 1. Schematic depiction of a double heterostructure (DH) light emitting diode (LED) in operation while being powered by a 2.8 V battery. Within the active, emitting layer, electrons and holes recombine and emit light equal to the bandgap of said layer. High-energy electrons are sourced from the negative terminal of the battery and return to the positive terminal after losing their energy to a photon in the active layer.

In the 1980s, all known material systems possessing the necessary material properties for blue light emission had shortcomings negating the possibility of creating an efficient blue LED. Gallium nitride (GaN) was one possible candidate, though, at the time, no *p*-type or active layer could be created. These challenges were ultimately overcome, leading to the first efficient blue LED using GaN in 1993 by Nakamura *et al.* [1]. Figure 2 shows a close-up image of a bare and packaged blue GaN LED.

Using blue LEDs, highly efficient white light sources become possible. This can be achieved by converting part of the blue light emitted from an LED to yellow using a phosphor [2]. To the human eye, the combination of blue and yellow light is perceived as white. A white LED can be created by embedding phosphors in a plastic cap which surrounds a blue LED (see Figure 3). Higher quality white light can also be created by mixing blue light with other colors as well, including red and green [3].

With the availability of white LEDs, a variety of applications can be significantly improved, if not enabled all together. But arguably, the most important impact of the white LED is its ability to generate white light at an efficiency that was previously impossible. The efficacy, a measure of perceived light power relative to the provided electrical power, of white light has improved over the

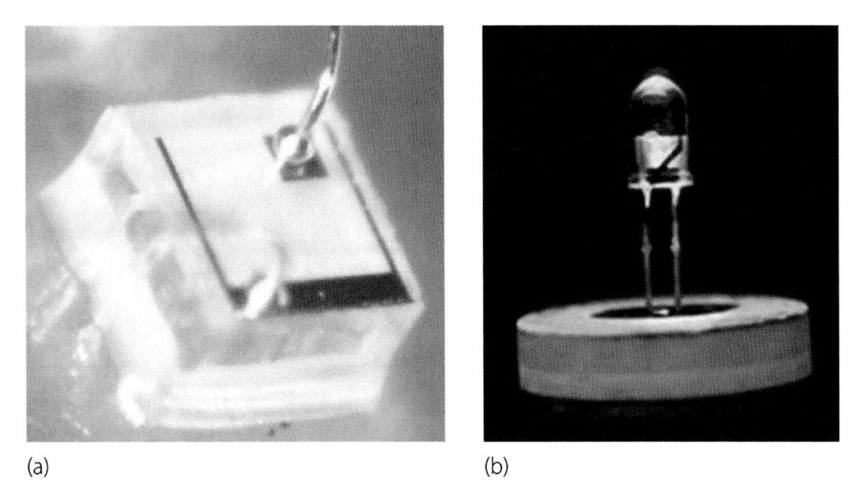

FIGURE 2. (a) Image of a blue GaN LED with attached gold wire contacts (size of diode: 0.4 mm × 0.4 mm) and (b) the same LED packaged as a commercial product [1].

centuries, starting with oil lamps (0.1 lm/W) in the 15,000s B.C., incandescent bulbs (16 lm/W) in the 19th century, fluorescent lamps (70 lm/W) in the 20th century, and LEDs (300 lm/W) in the 21st century (see also Figure 15).

With this significant improvement, substantial energy savings are now possible. It is currently estimated that in 2030 approximately 261 TWh of electrical energy will be saved due to widespread use of white LEDs [4]. This corresponds to an electricity savings of approximately 40% in 2030. Furthermore, this reduction in energy usage eliminates the need for at least 30 1-GW power plants by 2030 and avoids generating 185 million tons of CO_2.

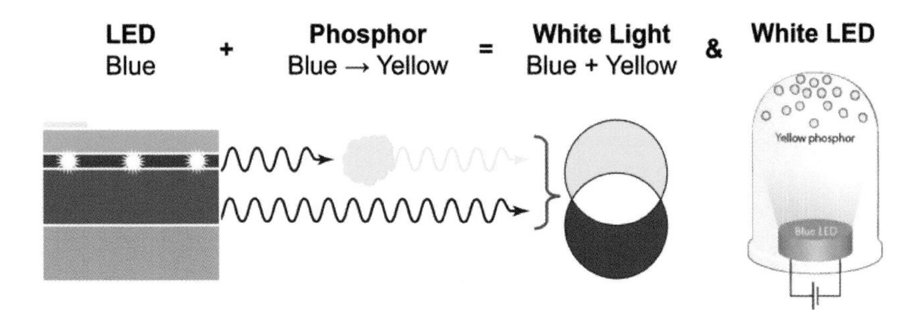

FIGURE 3. From blue LED to white LED. Part of the blue light emitted from a blue LED is converted to lower energy colors, such as yellow, using a phosphor. The combination of blue and yellow light is perceived as white to the human eye. Combining a blue LED with embedded phosphors in the plastic cap creates a white LED. [3]

2. MATERIAL OF CHOICE: ZnSe VS. GaN

In the 1980s, there were two materials considered as possible candidates for efficient blue LEDs: zinc selenide (ZnSe) and GaN [5].

ZnSe could be grown on single crystal gallium arsenide (GaAs) substrates, yielding high structural quality material given the very small lattice mismatch of 0.3% between ZnSe and GaAs. For GaN, on the other hand, no lattice-matched substrate was available and researchers were forced to grow it on sapphire. The large lattice mismatch (~ 16%) resulted in heavily defected material with a high density of dislocations.

When I joined the field in 1989, ZnSe was grown on GaAs with dislocation densities less than 10^3 cm^{-2}. It was very popular among scientists, given the high crystal quality and the prevailing notion that a dislocation density below 10^3 cm^{-2} is needed to achieve optically functional LEDs with a high efficiency and a long lifetime [5]. Most researchers worked in this field. GaN, however, was grown on sapphire, yielding dislocation densities on the order of 10^9 cm^{-2}. Unsurprisingly, few researchers were working in this field except, most notably, fellow Nobel Laureates Professor Isamu Akasaki and his graduate student at the time, Hiroshi Amano.

A striking example to highlight the popularity of ZnSe, as compared to GaN, is provided by looking at the attendance of researchers at the most popular conference for applied physics in Japan. At the Japan Society of Applied Physics (JSAP) conference in 1992, there were approximately 500 individuals attending the ZnSe sessions, whereas for GaN, there were around 5, including the chair Professor Isamu Akasaki, speaker Hiroshi Amano and myself, as a member of the audience. Not only was ZnSe more popular at the time, GaN was actively discouraged with researchers stating "GaN has no future" and "GaN people have to move to ZnSe material."

3. DEVELOPMENT OF GaN

My entry into the field started in April of 1988, when I went to the University of Florida as a visiting researcher. The main purpose of my visit was to learn how to use a MOCVD (Metal Organic Chemical Vapor Deposition) system to growth GaAs crystals on a silicon substrate, as I had no experience in how to use a MOCVD. During my stay there, I worked together with graduate students and they all asked me if I had a Ph.D. I said no. At the time, I only had a Master's. Next, they asked me if I had published any scientific papers. Again, I said no, I had never published a single paper. Consequently, they treated me as a technician. In the U.S., this meant one has to help the researcher and one's name would not appear on papers or patents. Gradually, I became very frustrated with this arrangement.

One year later, in March of 1989, I came back to Japan. It was my dream to get a Ph.D. degree. In Japan, at the time, it was possible to be awarded a Ph.D. if one published five scientific papers. This type of degree was called a paper degree and one did not need to go to the university to get the degree. It was therefore my ultimate dream to publish at least five papers and get a Ph.D.

With this in mind, I noted that the ZnSe field was publishing lots of papers. As I had never published a paper, I had no confidence in publishing a paper. In the GaN field, only very few papers had been published, mainly from Professor Isamu Akasaki and Hiroshi Amano. I was therefore confident that I could publish lots of papers, though had no confidence that I could actually invent the blue LED. My only objective was to get a Ph.D. That's it.

So, after returning to Japan in March of 1989, I wanted to grow GaN using a MOCVD reactor. I purchased a commercially available MOCVD reactor for 2 million U.S. dollars. But this MOCVD reactor was designed for growth of GaAs. At the time, Professor Akasaki and his student Amano had developed a novel, research-scale MOCVD reactor for growth of GaN [6]. Their design required exceptionally high carrier gas velocities (around 4.25 m/s) yielding GaN, though the high carrier velocities presented challenges pertaining to uniformity, scalability and reproducibility. Furthermore, their reactor design could only be used for small area growths, thereby lacking the necessary properties for commercialization. Since I was working for a company, I had to find a way to grow high quality GaN on large area, 2-inch diameter sapphire substrates.

Another challenge related to growing high quality GaN was the use of high concentrations of aluminum in the MOCVD reactor. While the development of the aluminum nitride (AlN) buffer layer by Akasaki and Amano was a major breakthrough providing high quality GaN film growth with a mirror-like surface morphology [6], the use of aluminum caused significant problems to the MOCVD reactor resulting in poor reproducibility in subsequent GaN growths. Eliminating the use of high concentrations of aluminum during growth was strongly desired.

After my purchase of a MOCVD reactor, I attempted a significant number of growths over the course of a few months, but consistently failed. Either no growth of GaN occurred or the grown layer was black. GaN should be transparent. I realized this was a big problem, especially considering the substantial investment in the tool. That is when I decided I had to modify the reactor.

For the next 1.5 years, I modified the reactor design. In the morning, I would go to work and modify the reactor. In the afternoon, I would perform a couple of growths and analyze the results. I would repeat this pattern for 1.5 years until I invented a novel MOCVD reactor design with a low carrier gas flow which I called a two-flow MOCVD (Figure 4 a) [7]. Using this reactor, I was able to get very uniform and high quality 2-inch GaN growth. The main breakthrough of this reactor was the introduction of a subflow (Figure 4 b) which gently pushed the carrier gases down to the substrate, thereby also improving the thermal boundary layer.

This was the most important breakthrough in my life and was instrumental toward all future breakthroughs in GaN research. One significant advancement

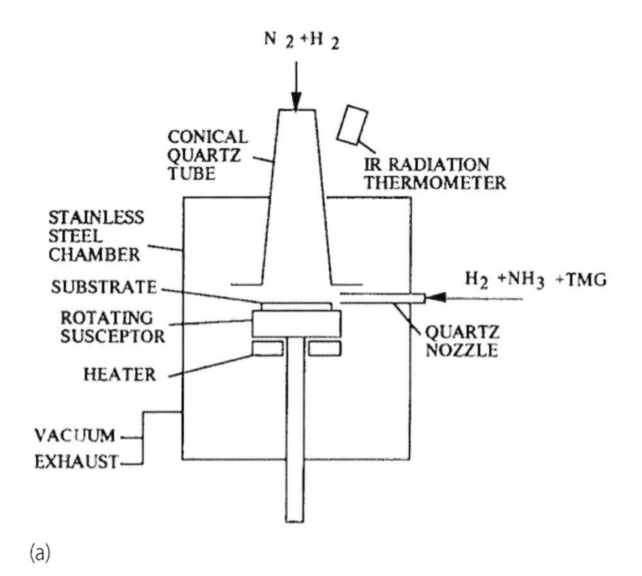

(a)

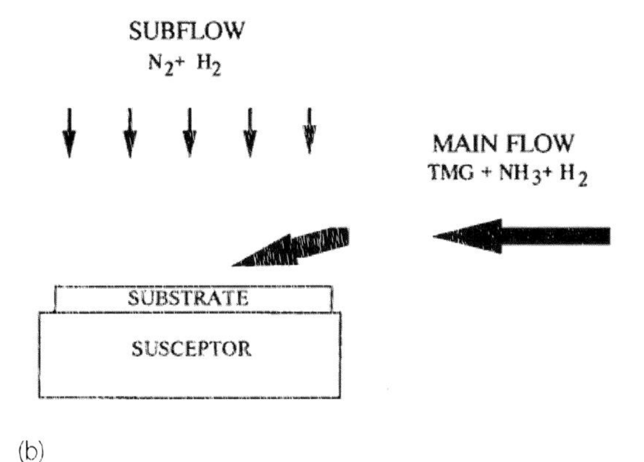

(b)

FIGURE 4. (a) Schematic of a two-flow MOCVD for GaN growth and (b) schematic of the effect of the newly introduced subflow on the carrier gases. [7] (Reprinted with permission. Copyright 1991, AIP Publishing LLC.)

this tool immediately enabled was the development of a GaN buffer layer which was superior to the AlN buffer layer, in part due to the elimination of aluminum from the growth system. With the invention of the two-flow MOCVD and the GaN buffer layer, it was possible to achieve the highest quality GaN material in the world. One measure for crystal quality is the value of the electron mobility in a crystal. Fewer defects result in fewer scattering events, which enhances overall mobility of the electrons. Mobilities for GaN grown directly on sapphire (no buffer layer) by Akasaki and Amano resulted in values around 50 cm^2/Vs [6], whereas use of the two-flow MOCVD yielded 200 cm^2/Vs [7]. Use of an AlN buffer layer improved the mobility to values as high as 450 cm^2/Vs for Akasaki and Amano [8]. Use of a GaN buffer layer and the two-flow MOCVD values as high as 600 cm^2/Vs were measured at room temperature (see Figure 5) [9]. This was a clear sign that the two-flow MOCVD was producing GaN material of higher quality on larger area substrates, a key step towards commercialization of GaN based devices.

The next significant development in creating an efficient blue LED occurred in 1992 when I was able to clarify why *p*-type GaN had remained so elusive for 20 years. While Akasaki and Amano achieved a major breakthrough in 1989 by demonstrating local *p*-type GaN after treating magnesium doped GaN (GaN:Mg) with low-energy electron beam irradiation (LEEBI) [10], its origin

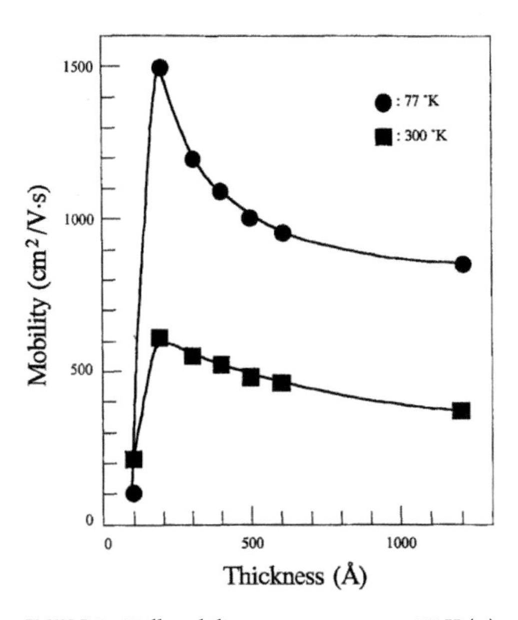

FIGURE 5. Hall mobility measurements at 77 K (•) and 300 K (■) for a 4 μm thick GaN film grown on sapphire as a function of the GaN buffer layer thickness. [9] (Reprinted with permission. Copyright 1991, The Japan Society of Applied Physics)

was not understood for another three years. In 1992, I clarified that hydrogen was the source of passivating *p*-type GaN [11]. A few years later, theoretical computations by Jörg Neugebauer and Chris Van de Walle confirmed hydrogen passivation in Mg-doped GaN [12].

For MOCVD growth of GaN, ammonia (NH_3) is used as the nitrogen source. Ammonia dissociates during growth and atomic hydrogen is introduced into the GaN crystal. If Mg is present in the crystal, the hydrogen atom forms a magnesium hydrogen complex (Mg-H), thereby preventing Mg from acting as an acceptor [11]. Thermal annealing of the GaN:Mg sample in a hydrogen-free environment above approximately 400 °C permits hydrogen to diffuse out of the crystal, thereby breaking up the Mg-H complex [13]. As thermal annealing can be performed quickly and simultaneously on multiple substrates of any size in parallel (an act not achievable using LEEBI), it has become the industrial standard process for *p*-type activation of GaN. The formation of local *p*-type GaN using LEEBI treatments can be explained by local heating of the GaN:Mg by the electron beam, causing the hydrogen to locally diffuse out of the crystal and permitting the affected Mg atoms to act as acceptors yielding *p*-type GaN.

With the ability to grow *n*-type and *p*-type GaN, *p*-*n* homojunction LEDs (DH LEDs lacking an active layer) can be formed. The first *p*-*n* homojunction LED was demonstrated by Amano *et al.* in 1989 using an AlN buffer layer and

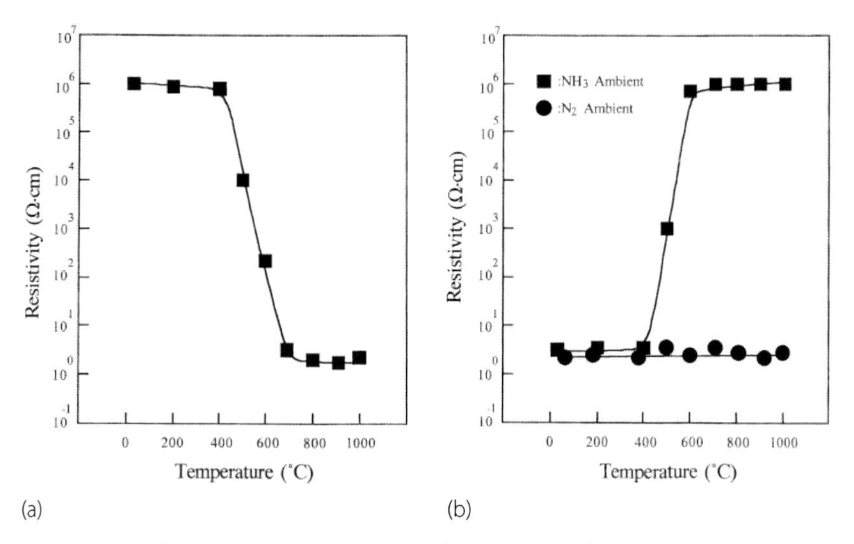

FIGURE 6. (a) Change in electrical resistivity of as grown Mg-doped GaN films as a function of annealing temperature in a nitrogen environment [13]. (b) Change in resistivity of LEEBI-treated Mg-doped GaN films as a function of annealing temperature in a nitrogen (●) or ammonia (■) environment [11]. Annealing time was 20 min for all samples. (Reprinted with permission. Copyright 1992, The Japan Society of Applied Physics)

their newly developed LEEBI treatment process to obtain *p*-type GaN [10]. They reported on the observed current-voltage (I–V) relationship and electroluminescence (EL) of the manufactured LEDs, but did not mention the output power or the efficiency of the LEDs. In 1991, Nakamura *et al.* demonstrated a *p-n* homojunction GaN LED using a low-temperature GaN buffer and the LEEBI treatment [14]. The output power at 20 mA with a forward voltage of 4 V was 42 μW. The external quantum efficiency (EQE) and peak emission wavelength were 0.18% and 430 nm, respectively.

For LEDs to be useful for real world applications, the light output power needs to be well in excess of 1 mW. *P-n* homojunction LEDs cannot reach those levels of output power without generating substantial amounts of waste heat, in large part due to the inefficient device structure. Additionally, *p-n* homojunction LEDs produce light of a fixed wavelength given the exclusive use of GaN, which has a fixed bandgap. In spite of the achievement by Akasaki and Amano to produce optically active GaN *p-n* homojunction LEDs, Toyoda Gosei Co., Ltd. issued a press release for production of Metal-Insulator-Semiconductor (MIS) GaN LEDs with an output power of 70 μW on October 20, 1993 [15]. MIS type LEDs use a semi-insulating layer instead of a *p*-type layer and are also a relatively inefficient device design. The complete omission of *p*-type GaN is advantageous though, as it sidesteps various challenges associated with large area *p*-type activation using the LEEBI process.

Arguably the most efficient LED designs make use of a DH. The concept and invention of the heterostructure in semiconductor materials was so significant that it earned the Nobel Prize in Physics for Zhores Ivanovich Alferov and Herbert Kroemer in 2000. The energy band diagram is shown in Figure 7 for the *p-n* homojunction LED and the DH LED.

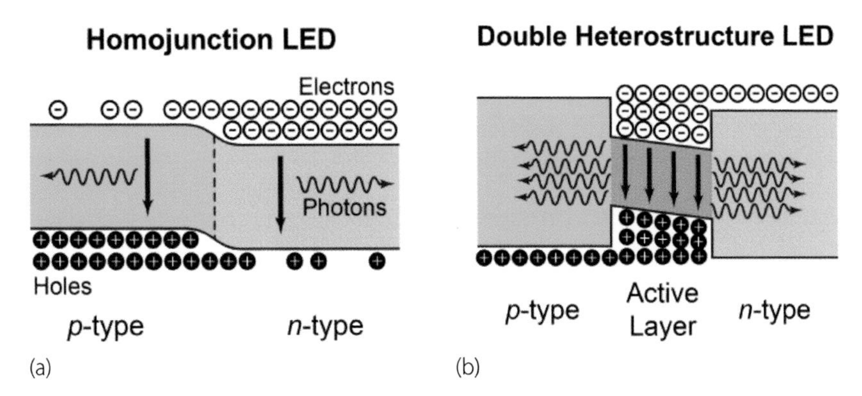

FIGURE 7. Schematic energy band diagram structure for a (a) homojunction LED and (b) double heterostructure LED.

In order to understand the advantage a DH LED provides over a *p-n* homojunction LED, one needs to look at the internal quantum efficiency (IQE) (see Equation (1)).

$$IQE = \frac{\text{Light generated}}{\text{Electrons injected}} = \frac{R_{radiative}}{R_{radiative} + R_{non-radiative}} = \frac{Bn^2}{An + Bn^2 + Cn^3} \quad (1)$$

The IQE is a measure of efficiency and is related to how many electrons are converted into photons within the active region. There are three mechanisms by which a high-energy electron may decay to a lower energy state. One of them is a radiative process ($R_{radiative}$, emission of a photon, desired), while the other two are non-radiative ($R_{non\text{-}radiative}$, emission of phonons, i.e. heat, undesired) and include the Shockley-Read-Hall (SRH) process and Auger recombination process. All of these processes are dependent on the minority carrier concentration (electrons in a *p*-type layer, holes in a *n*-type layer), n. The SRH process increases linearly ($A \cdot n$), the radiative recombination process quadratically ($B \cdot n^2$), whereas the Auger recombination process increases as the cube ($C \cdot n^3$) of n. The coefficients A, B and C are constants.

For low carrier concentrations, the SRH term ($A \cdot n$) dominates, leading to poor efficiencies and significant non-radiative recombination. This is the case for *p-n* homojunction LEDs. Due to the *p-n* structure of the LED, electrons diffuse into the *p*-type layer and holes diffuse into the *n*-type layer. The diffusion length of minority carriers in GaN is approximately 1 μm [16]. This causes carriers to be spread out over a large region reducing their concentration. The DH LED, on the other hand, confines the carriers to within the active layer, which is typically around 3–200 nm thick. This confinement significantly increases their concentration under the same current density and enhances the probability of radiative recombination ($B \cdot n^2$), thereby increasing the efficiency of the LED.

4. DEVELOPMENT OF InGaN

4.1 InGaN based devices

Achieving a high quality active layer with the necessary properties to form a DH LED is the last and, arguably, the most critical, step towards achieving a commercializable and efficient LED. Indium gallium nitride (InGaN) was identified as the ideal candidate for the active layer. Through addition of indium into the GaN crystal, the electrical bandgap of the material shrinks, thereby providing both the ability to confine carrier in a DH arrangement, but also provide the ability to tune the color of the light by changing the amount of indium in the InGaN alloy.

Despite this realization, high quality layers of InGaN could not be realized in the 1970s and 1980s. Room temperature (RT) band-to-band emission, essential for the active layer of a DH LED, could not be achieved given the defective nature of the material.

The reason for this can be traced back to the many challenges associated with InGaN growth, which is in many ways more challenging to grow than *p*-type or *n*-type GaN. Indium has such a high vapor pressure, that at typical growth temperatures of GaN (~ 1000 °C), it would boil off the surface and not incorporate into the crystal. Growth at lower temperatures yielded poor crystal quality along with numerous defects and impurity incorporation. Growths at intermediate temperatures required uniform and stable growth temperatures, as the incorporation of indium is strongly dependent on the temperature, with a few degrees differences across the substrates resulting in noticeable variations in output color across the wafer.

Furthermore, in order to effectively use InGaN in a DH LED, excellent control over the various growth parameters is required, as the interface between GaN and InGaN needs to be smooth on an atomic level. In addition to a smooth surface morphology, the MOCVD reactor requires precise control over all growth parameters and superior uniformity across the entire surface to achieve exceptionally thin layers of high quality (a layer is typically composed of a few 10s to 100s of atom layer thicknesses in DH LEDs). To make matters even more challenging, introduction of indium into the GaN lattice results in significant strain since indium is roughly 20% bigger in size than gallium. Managing this strain and preventing the formation of defects within the layer is important.

The first recorded InGaN alloy growth was performed by electron beam plasma in 1972 and 1975 by Osamura *et al.* on sapphire and quartz substrates [17,18]. In 1989, Nagatomo *et al.* grew InGaN on a sapphire substrate using MOCVD at a growth temperature of 500 °C [19]. In 1991, Yoshimoto *et al.* demonstrated growth of InGaN layers at a growth temperature of around 800 °C using MOCVD [20]. Despite this achievement, their crystal quality was poor as evidenced by their RT photoluminescence (PL) exhibiting only deep level emission (no band-to-band emission) (Figure 8) and by a full width at half maximum (FWHM) of the double crystal X-ray rocking curve (XRC) of 30 arcmin.

In 1992, Mukai and myself succeeded in growing a high quality InGaN layer using the two-flow MOCVD on a GaN template grown on a sapphire substrate [21]. RT PL showed a strong band-to-band emission from violet to blue depending on the indium composition of the InGaN layers. This was the first report of band-to-band emission of InGaN layers at RT. The FWHM of the double crystal XRC was around 8 arcmin. Figure 9 shows the first observation

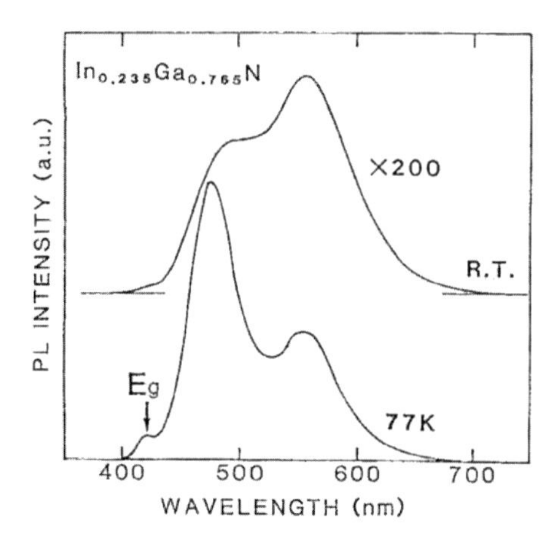

FIGURE 8. Photoluminescence spectra of $In_{0.235}Ga_{0.765}N$ grown at 800 °C using MOCVD on a sapphire substrate by Yoshimoto *et al.* [20]. (Reprinted with permission. Copyright 1991, AIP Publishing LLC.)

of band-to-band emission of the InGaN layers at RT described in reference [21]. With this demonstration, the last remaining barrier for efficient blue LEDs was overcome opening the doors for rapid development of high brightness, high power, high efficiency blue LEDs using the DH structure.

Building on this success, I immediately investigated embedding these InGaN layers within a DH LED structure. The first demonstration of a blue DH LED occurred in 1993 with a *p*-GaN/*n*-InGaN/*n*-GaN structure [22]. The active layer was a Si-doped InGaN layer with a thickness of 20 nm. The LED showed strong band-edge emission in the InGaN layer yielding blue light with a wavelength of 440 nm under forward bias conditions. The output power and the EQE were 125 μW and 0.22%, respectively, at a forward current of 20 mA.

Further improvements to the device led me to demonstrate in 1994 the first commercially available blue LED with an output power of 1.5 mW, an EQE of 2.7% and the emission wavelength of 450nm (Figure 10) [23]. The structure of this DH LED was *p*-GaN/*p*-AlGaN/Zn-doped InGaN/*n*-GaN. The Zn-doped InGaN active layer had a thickness of 45 nm and was used due to an observed increase in luminous efficiency. For the first time, an electron-blocking layer was introduced into the structure. The *p*-AlGaN prevented electrons from overflowing the electron confinement provided by the active layer, further enhancing radiative recombination.

Simultaneously with my publication, Nichia Chemical Corporation released a press statement mentioning production of high brightness blue DH LEDs with *p*-type layers, an output power of 1.5 mW and a brightness of more than 1000

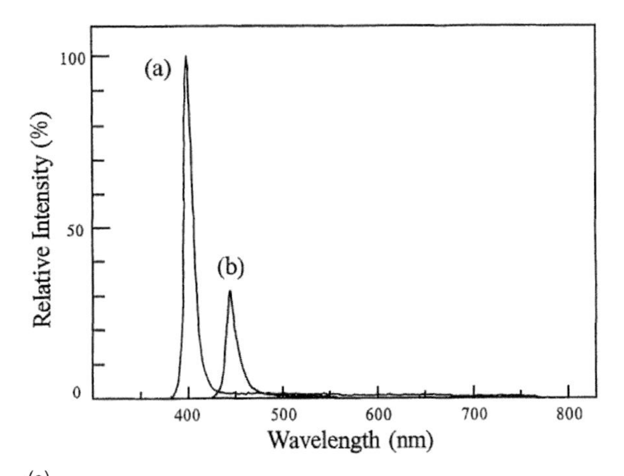

(a)

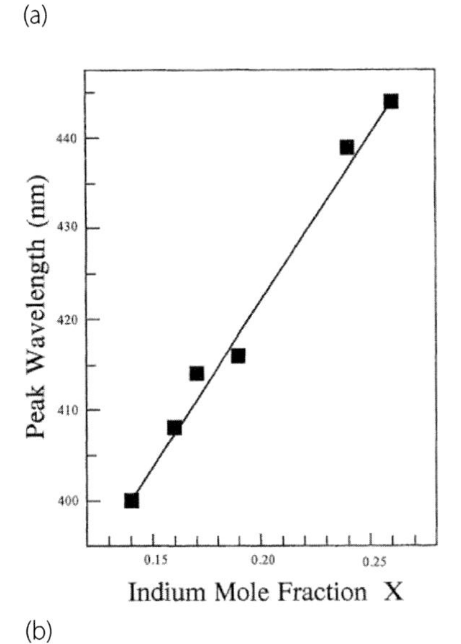

(b)

FIGURE 9. (a) RT PL spectra of InGaN films grown on GaN films under identical growth conditions, except for the InGaN growth temperatures of 830 °C (trace (a)) and 780 °C (trace (b)). (b) Change in peak wavelength of PL spectra as a function of indium mole fraction (x) in $In_xGa_{1-x}N$ films. Indium fraction was determined by X-ray diffraction measurements. [21] (Reprinted with permission. Copyright 1992, The Japan Society of Applied Physics).

mcd (roughly equal to the luminous intensity of 1 candle) on November 30, 1993 [24]—just one month after the press release about MIS type LEDs with an output power of 70 μW by Toyoda Gosei Co. Ltd on October 20, 1993 [15].

Further improvements to the growth conditions allowed me to demonstrate high brightness blue, green and yellow LEDs with InGaN quantum well (QW)

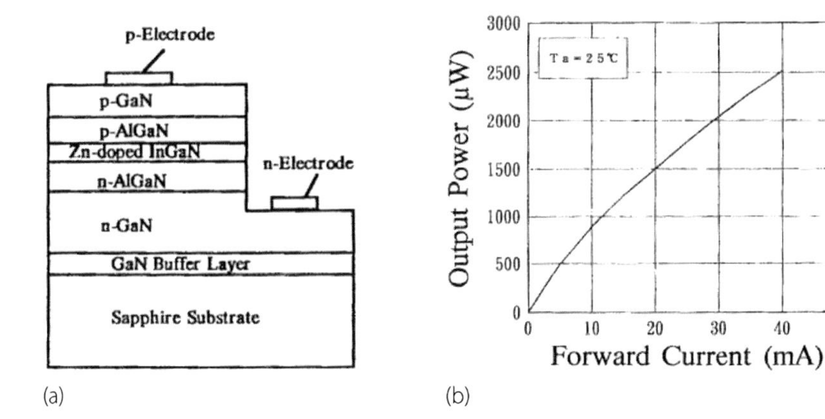

FIGURE 10. (a) Structure of InGaN/AlGaN double-heterostructure blue LED with (b) resulting output power as a function of forward current. [23] (Reprinted with permission. Copyright 1994, AIP Publishing LLC.)

structures in 1995 (see Figure 11) [25]. QW structures are DH structures with a very thin active layer, so thin that quantum confinement effects need to be considered. Layer thicknesses for QW structures are on the order of a few nanometers (a few atom layer thicknesses) versus ~ 100 nm for typical DH structures up to this point. This thinning further improves the IQE due to higher carrier concentrations, but also requires even more stringent controls on uniformity and temperature during MOCVD growth.

Further improvements to growth conditions of the InGaN layers led to the availability of higher power blue and green single QW (SQW) LEDs. At a 20 mA current, the output power and the EQE of the blue SQW LEDs were 5 mW and 9.1%, respectively. Those of green SQW LEDs were 3 mW and 6.3%, respectively. The structure of the green SQW LED and output powers of the blue, green and yellow SQW LEDs are shown in Figure 11 [26]. This LED epitaxial structure is still the basic foundation for all currently commercially available blue and green LEDs.

With the success of the developed high efficiency, high power blue LED, Nichia Chemical Corporation commercialized the first white LEDs by combining the blue InGaN QW DH LED [25, 26] with a yellow yttrium aluminum garnet (YAG, $Y_3Al_5O_{12}$) based phosphor [27].

Having excelled at achieving high efficiency blue LEDs, the next step for me was to demonstrate the first InGaN-based laser diode. I achieved this in 1996 under pulsed [28] and continuous wave (CW) [29] operations. Figure 12 shows the device structure and light output power versus current (L–I) curve for the

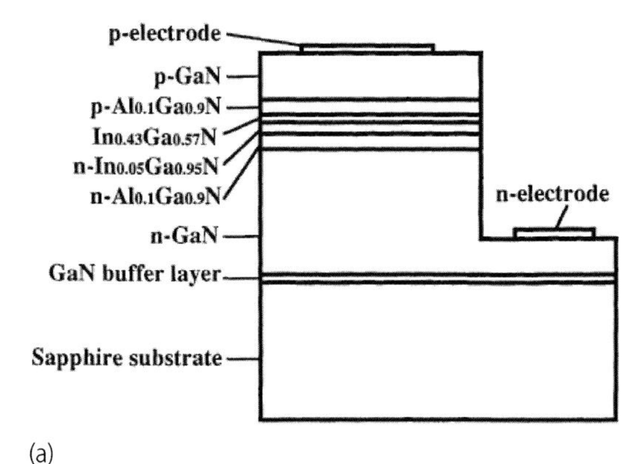

(a)

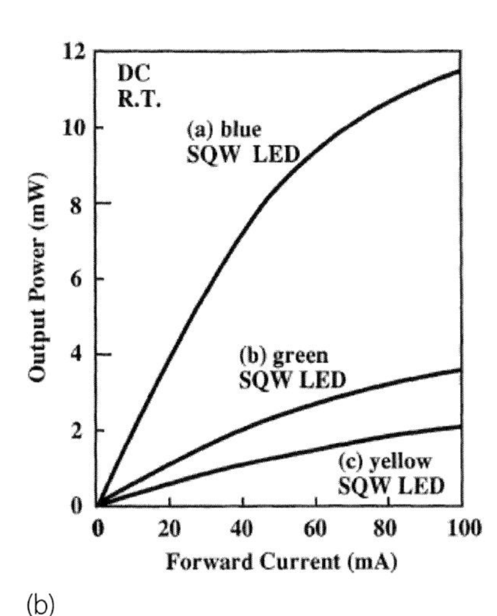

(b)

FIGURE 11. (a) Green single quantum well (SQW) LED structure. (b) Output power of a blue, green, and yellow SQW LED at RT as a function of forward current. [25, 26] (Reprinted with permission. Copyright 1995, The Japan Society of Applied Physics)

first InGaN-based laser diode [28]. The structure was composed of an InGaN multi-quantum wells (MQW) active layer, GaN waveguide layers and AlGaN cladding layers.

4.2 Material Properties of InGaN

One mystery still remains to date, namely why InGaN materials are atypical and luminesce with such high efficiency despite the high density of dislocations. It was deemed common sense in the 1980s that highly efficient LEDs with a long lifetime required dislocation densities less than 10^3 cm^{-2}. Despite all the

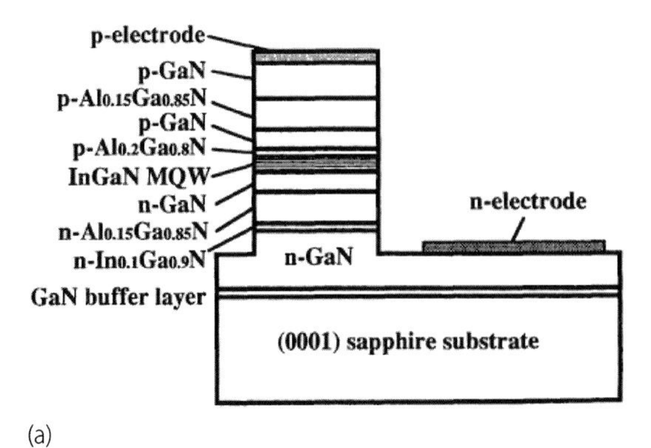

(a)

(b)

FIGURE 12. (a) Structure of a violet InGaN MQW laser diode with (b) corresponding L–I characteristics indicating the onset of lasing (30 μm wide, 1500 μm long) [28] (Reprinted with permission. Copyright 1996, The Japan Society of Applied Physics).

improvements in MOCVD growth of GaN, the fundamental lattice mismatch between sapphire and GaN remains, yielding 10^9 cm^{-2} dislocations in the GaN and InGaN layers, even for the high efficiency devices demonstrated in the 1990s. To highlight the stark difference between InGaN and other semiconductors, Figure 13 depicts the approximate dependence of the LED efficiency on the dislocation density for various semiconductor materials. As can be seen, highly efficient arsenide and phosphide based LEDs could only be achieved for dislocation densities below 10^3–10^5 cm^{-2}. This observation led to the erroneous statements that dislocation densities below 10^3 cm^{-2} are needed for efficient LED operation. Pure GaN LEDs behave comparably to the other arsenide and phosphide LEDs, as demonstrated by dim *p-n* GaN homojunction LEDs developed by Akasaki and Amano in 1989 [10]. Interestingly, InGaN materials

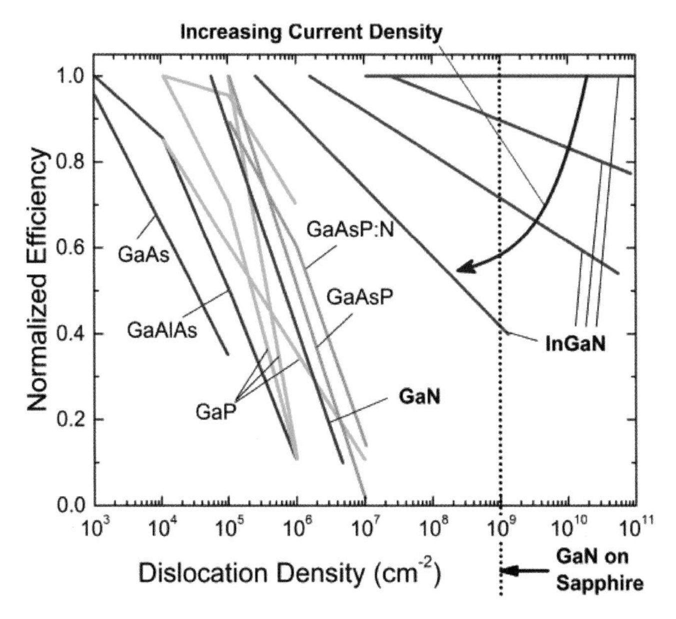

FIGURE 13. Dependence of LED efficiency on dislocation density for various semiconductor materials. (After [30,31,32])

behave quite differently, exhibiting high efficiencies despite high dislocation densities [30].

One explanation for this phenomenon, as proposed by Professor Chichibu from Tohoku University, is the presence of localized states in the InGaN layer [31, 32]. When electrons and holes are injected into the active layer, they are captured by localized states within the layer and radiatively recombine before they are captured by crystal defects (dislocations), which would provide non-radiative recombination pathways. The localized centers can be thought of as enhanced emission centers of light.

Localized states may emerge due to the natural fluctuation of indium within the InGaN layers. Atom probe tomography measurements of the InGaN layers have provided atomic level resolution of the chemical and spatial distribution of atoms within the layer (see Figure 14) [33]. The statistical analysis of the indium distribution in the layer yields a random binomial distribution on the Group III site about some average indium concentration. This naturally occurring random binomial distribution provides fluctuations resulting in regions of high and low indium content. The high indium content regions have a smaller bandgap and hence could act as localized sites. As fluctuations occur on the nanometer scale, there are a significant number of localized sites, comparable, if not greater, to

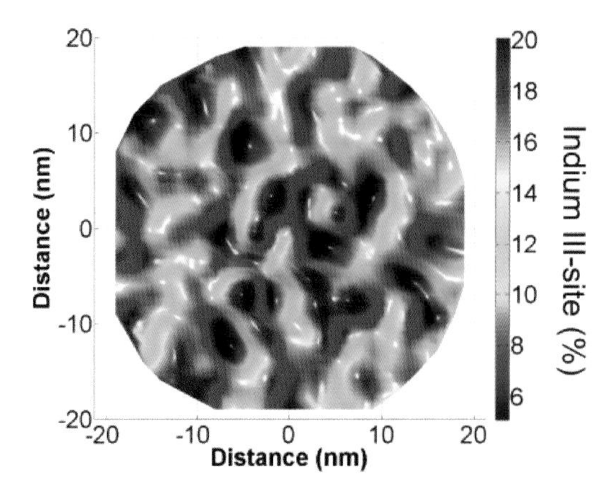

FIGURE 14. Atom Probe Tomography analysis provided as a 2D plot of the lateral compositional variations within a 3 nm InGaN quantum well at nominal 14% indium composition showing local indium fluctuations [33].

the number of defects present. It is important to keep in mind that this is just one possible explanation for localized sites. We currently still do not understand their origin and if they are truly the reason for the highly efficient nature of InGaN.

5. HISTORICAL DEVELOPMENT OF InGaN BASED LEDS

5.1 Luminous Efficacy

To provide an overview of the historical development of the luminous efficacy for the red, green, and blue LEDs, Figure 15 is provided [34]. This figure has been used by many scientists at numerous conferences to explain the historic developments of red, green and blue LEDs. The first visible light LEDs were based on GaP and developed in the late 1950s through early 1960s [35–39]. Over time, the red LED gradually improved in efficacy as shown in the figure.

The first violet/blue LEDs, the MIS LED developed by Maruska *et al.* in 1973 and the *p-n* homojunction LED developed by Akasaki and Amano in 1989, have typically not been included in the figure, probably due to their poor luminous efficacy, though are included in this depiction. It wasn't until 1992, when high quality InGaN became available and could be incorporated into a DH LED structure, did rapid progress ensue, leading to the first efficient yellow, green and blue LEDs in 1992–1995 [23, 24, 25, 26]. The rapid progress that occurred could not have been possible without the atypical properties of high quality InGaN. Without InGaN, I argue, it would not have been possible to make efficient

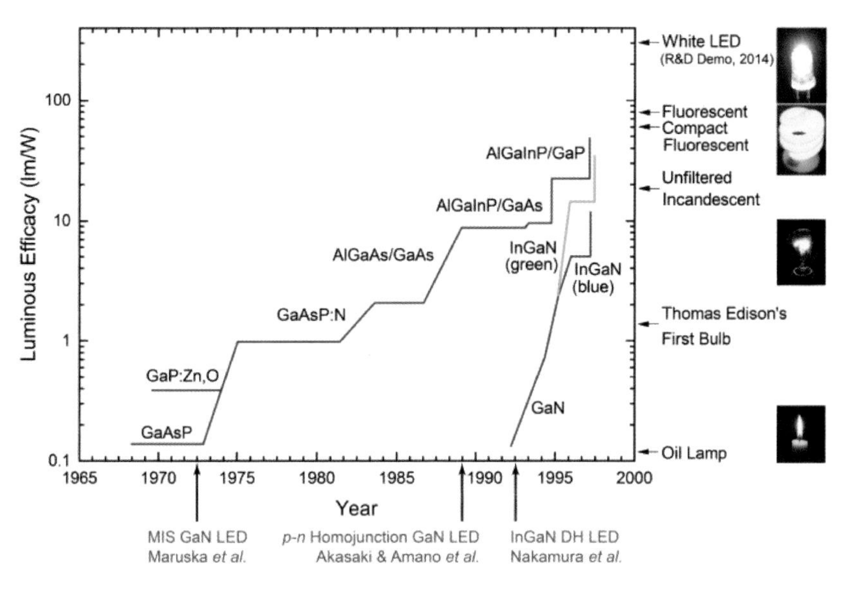

FIGURE 15. Evolution of luminous efficacy of red, green, and blue LEDs (After [34]).

blue and green LEDs in the GaN material system. In 1996, Nichia Chemical Corporation developed the first white LED using the efficient blue InGaN QW DH LED and YAG phosphors [27]. Rapid progress after the first commercialization of the white LED by Nichia Chemical Corporation has led to current state-of-the-art white LED, with a stunning 303 lm/W peak efficacy at RT as announced by Cree Inc. in March 2014 [40].

5.2 Significant Milestones on the Journey to the White LED

Table 1 provides an overview of significant developments that ultimately led to the first high efficiency white LED developed by Nichia Chemical Corporation in 1996 [27]. The first demonstrated growth of GaN was performed by Maruska *et al.* in 1969 using hydride vapor phase epitaxy (HVPE) [41]. A great thanks goes out to Dr. Maruska for opening the door to GaN research and crystal growth. Not only did Maruska *et al.* perform the first GaN growth, he also demonstrated the first violet GaN-based LED using a MIS structure (due to the lack of *p*-type GaN) by doping the GaN with Mg, which acted as a color center, in 1973 [42].

The next significant development was the demonstration by Yoshida *et al.* of growing GaN films using an AlN buffer on a sapphire substrate by reactive molecular beam epitaxy (MBE) in 1983, providing a pathway to improved quality

GaN material [43]. Amano *et al.* applied the AlN buffer layer idea to MOCVD growth of GaN by growing it at low-temperatures in 1986 [6]. The mirror-like surface morphology and reduced residual carrier concentration, on the order of 10^{17} cm^{-3} for unintentionally doped GaN, was an important milestone.

Subsequently, Amano *et al.* achieved the first *p*-type Mg-doped GaN using a post-growth LEEBI treatment in 1989, though they could not explain the origin of the mechanism by which it was achieved [10]. The hole concentration and mobility was $2x10^{16}$ cm^{-3} and 8 $cm^2/V{\cdot}s$, respectively, and output power of the *p-n* homojunction GaN LED was not reported.

The next major milestone in the development of the blue LED occurred when I was working for Nichia Chemical Corporation and developed a novel MOCVD, which I called two-flow MOCVD, in 1991 [7]. Introducing a new subflow to the system provided a significant improvement in reproducibility and uniform growth over large area substrates.

Shortly thereafter, in 1991, the GaN buffer layer was developed for GaN growth on sapphire using MBE [44] and MOCVD [9]. Lei *et al.* developed it for the MBE method, though the achieved crystal quality was poor with a rough surface [44]. I developed it for MOCVD growth by employing a low-temperature growth step, yielding high quality GaN films on sapphire [9]. Improvements to the growth of the low-temperature GaN buffer layer resulted in electron mobilities of 900 $cm^2/V{\cdot}s$ at RT in 1992 [45].

With the availability of high crystal quality material, the next major step, was my demonstration of *p*-type GaN films using post thermal annealing under a NH_3 free ambient gas, as shown in Figure 6, in 1992 [13]. The hole concentration and mobility was $3x10^{17}$ cm^{-3} and 10 $cm^2/V{\cdot}s$, respectively. Following this act, I was able to clarify the observed hole compensation in GaN:Mg being due to hydrogen passivation [11], a mechanism which had been a mystery since the early 1970s. A hydrogenation model, wherein acceptor-hydrogen (Mg-H) complexes were formed in *p*-type GaN, was proposed and theoretically confirmed by Neugenbauer and Van de Walle in 1995 [12].

Having overcome the major hurdles for GaN, the next major advance occurred for InGaN materials, arguably the most important layer in the efficient blue LED. Its first growth using electron beam plasma occurred in 1972, with Osamura *et al.* revealing some of its properties [17]. The first growth by MOCVD occurred in 1989, with Nagatomo *et al.* [19], opening the door for development using the same growth technique with which high quality GaN could be grown. It was not until 1992, when Mukai and myself were able to demonstrate the first InGaN layer which exhibited strong band-to-band emission at RT [21], that the option of forming a high efficiency blue LED became

TABLE 1. Significant developments ultimately leading to the white LED for GaN and InGaN on sapphire.

Material	Year	Achievement	Reference
GaN	1969	GaN epitaxial layer by HVPE	[41]
	1973	1st blue Mg-doped GaN MIS LED	[42]
	1983	High quality GaN using AlN buffer by MBE	[43]
	1985	**High quality GaN using AlN buffer by MOCVD**	[6]
	1989	***p*-type GaN using LEEBI** (low hole concentration)	[10]
		1st *p-n* homojunction GaN LED	
	1991	**Invention of Two-Flow MOCVD**	[7]
		GaN growth using GaN buffer by MBE	[44]
		High quality GaN using GaN buffer by MOCVD	[9, 45]
	1992	***p*-type GaN using thermal annealing** (high hole concentration)	[13]
		Discovery of hydrogen passivation	[11]
InGaN	1972	InGaN growth using electron beam plasma	[17,18]
	1989	InGaN growth by MOCVD	[19]
	1992	**InGaN layers with RT band-to-band emission**	[21]
	1994	**Efficient Blue InGaN DH LED (1 Candela)**	[22, 23, 24]
	1995	Efficient yellow, green and blue InGaN DH QW LEDs	[25, 26]
	1996	1st Pulsed Violet InGaN DH MQW LDs	[28]
		1st CW Violet InGaN DH MQW LDs	[29]
		Commercialization ***White LED using InGaN DH blue LED***	[27]

realistic. Incorporating the material in a DH LED structure, rapidly lead to slew of developments, most notably, the first highly efficient blue InGaN DH LED in 1994 [22, 23], the first high efficiency green InGaN DH QW LED in 1995 [26], and the first blue/green/yellow InGaN DH QW LEDs in 1995 [25].

While our group at Nichia Chemical Corporation worked hard on developing the first white LED, I was able to further pursue improving GaN based optoelectrontic devices, and eventually demonstrated the first pulsed [28] and CW operation [29] violet InGaN DH QW laser diode in 1996. Shortly thereafter, Nichia Chemical Corporation released the first white LED, changing the world forever [27].

A more detailed account of the history of the GaN based LED can be found in reference [46].

5.2 Contributions to the Efficient Blue LED

Of all the contributions that eventually led to the efficient blue and white LED, only a select few were highlighted in the Nobel Prize announcement. Figure 16 provides a visual summary of the key inventions and contributions to the efficient blue LED. While the demonstration and explanation of *p*-type GaN and the inclusion of buffer layers were important, the development and incorporation of high quality InGaN material was just as, if not even more, vital to achieving the efficient blue LED. A pure GaN *p-n* homojunction LED structure is too inefficient and would never have led to the development of the high efficiency LED, especially when also considering the high dislocation densities that are present due to the use of a sapphire substrate (see also Figure 13).

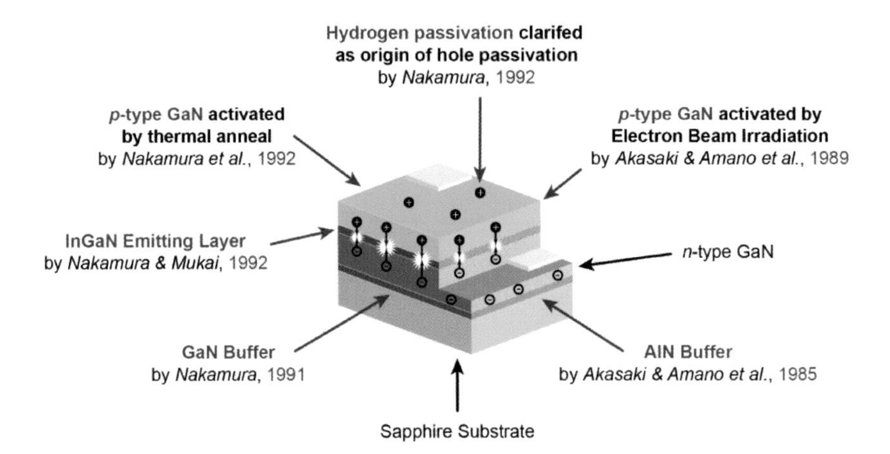

FIGURE 16. Summary of key contributions to the efficient blue LED.

6. FORWARD LOOKING

While I have been talking primary about LEDs, there is an intrinsic problem that cannot be easily overcome. Given the highly efficient nature of LEDs, it is of interest to maximize the light output per device. This is most easily achieved by increasing the current density, which runs through the device. Higher efficiencies permit higher current densities, as the device does not heat up as much. With increasing current density, though, the carrier densities within the QW increases. We can now push the current densities so high that we are seeing a reduction in efficiency with increasing current density (see Figure 17). This phenomena, referred to as efficiency droop, forces LED manufactures to operate LEDs at lower current densities (and hence reduced light output) than would be possible, in order to prevent excess heating of the device. To maintain high light output, manufactures may use multiple LEDs in parallel, effectively increasing the overall active area and hence reducing current density. The primary origin of efficiency droop, which was not discovered until recently, is the Auger recombination process [44, 48]. If you recall the contributions to the IQE of an LED (see Equation 1), the cubic term in carrier concentration ($C \cdot n^3$) was the Auger recombination process. Given the exceedingly high densities of carriers, this term starts to dominant thereby reducing the value of the IQE. Since the Auger recombination process is due to intrinsic properties of GaN, it is exceptionally challenging to overcome.

An alternative method to produce white light is by using a blue laser, as opposed to an LED, in combination with a phosphor. Above the lasing threshold, the carrier density is clamped at threshold, fixing its density. Increases in carrier density beyond the threshold density immediately contribute to stimulated emission, or lasing. Thus, the carrier density is maintained at the lower, threshold density, prohibiting it from reaching densities where the Auger recombination process becomes the dominant recombination process. Auger recombination, with the resulting efficiency droop, does not appreciably occur in blue laser diodes [49].

Current commercial blue lasers have already demonstrated comparable external quantum efficiencies to those of blue LEDs at significantly higher current densities, and hence light output (see Figure 17). It is therefore of great interest to further pursue lasers as they have the potential of operating at high current densities, resulting in white light sources with staggering light output.

While laser based lighting has the potential of being more efficient with smaller chip sizes with a very high current density region, it also offers intrinsic directionality of the light output—a feature that car manufactures have already

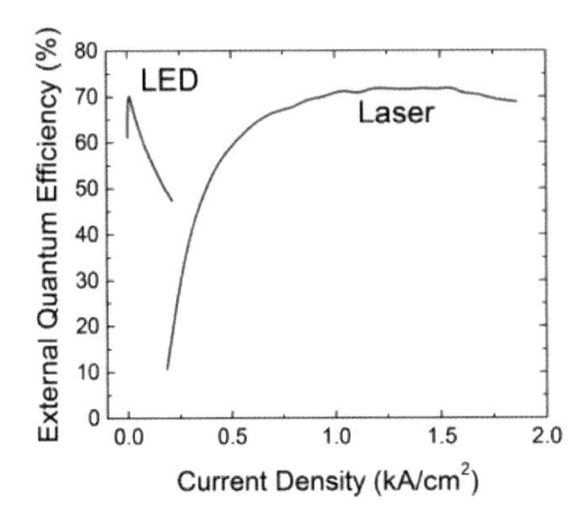

FIGURE 17. Comparison of external quantum efficiency (EQE) of a commercial LED and laser with increasing current density [49].

leveraged in their high-end vehicles. Currently, high-end vehicles already use laser based white light sources for their headlamps, allowing them see further ahead on the road without blinding oncoming traffic. Future modifications to the laser based lighting technology may well enable the next generation of white lighting with higher efficiencies at lower cost.

7. ACKNOWLEDGEMENTS

While it has been a wonderful, though sometimes rocky, journey and experience for me to develop and ultimately achieve a blue LED, I would like to acknowledge the many individuals that helped and supported me along the way. Most notably, I would like to thank the founder of Nichia Chemical Corporation, Nobuo Ogawa, who trusted me and invested a lot of money in my research despite the initial lack of results. I would also like to acknowledge the current president, Eiji Ogawa, my colleagues of the R&D department during 1989–1999 and all current employees of Nichia for their help as I would not have been able to make the high efficient blue LED without them.

I would also like to stress my gratitude towards Chancellor Henry Yang of the University of California, Santa Barbara (UCSB), for his personal support and encouragement to come to UCSB and further pursue my research there. During my years at UCSB, I have been able to experience and appreciate the tremendous support provided to me by Dean Rod Alferness and former Dean Matthew Tirrell. Last, but certainly not least, I would like to send my sincere gratitude and warmest thank to my colleagues and friends at UCSB, most notably Professors Steve DenBaars, Jim Speck, and Umesh Mishra, who make it a great pleasure and experience to work and excel at UCSB. I would also like to thank Dr. Siddha Pimputkar for his substantial help in writing this article. Thank you very much.

REFERENCES

1. S. Nakamura, T. Mukai, M. Senoh, "Candela-class high-brightness InGaN/AlGaN double-heterostructure blue-light-emitting diodes," *Appl. Phys. Lett.* 1994, **64**, 1687–1689.
2. S. Nakamura, S. Pearton, G. Fasol, *The Blue Laser Diode*, 2nd Edition, Springer, 2000, p. 230–235.
3. S. Pimputkar, J. S. Speck, S. P. DenBaars, and S. Nakamura, "Prospects for LED lighting," *Nature Photonics* 2009, **3**, 180–182.
4. *Energy Savings Forecast of Solid-State Lighting in General Illumination Applications*, DOE Report, August 2014, http://apps1.eere.energy.gov/buildings/publications/pdfs/ssl/energysavingsforecast14.pdf
5. K. Ohkawa, "Blue LEDs: Helping to Save Earth," *AAPPS Bulletin*, 2014, **24**, 11–13.
6. H. Amano, N. Sawaki, I. Akasaki, Y. Toyoda, "Metalorganic vapor phase epitaxial growth of a high quality GaN film using an AlN buffer layer," *Appl. Phys. Lett.* 1986, **48**, 353–355.
7. S. Nakamura, Y. Harada, M. Seno, "Novel metalorganic chemical vapor deposition system for GaN growth," *Appl. Phys. Lett.* 1991, **58**, 2021–2023.
8. H. Amano, I. Akasaki, K. Hiramatsu, N. Koide, N. Sawaki, "Effects of the buffer layer in metalorganic vapour phase epitaxy of GaN on sapphire substrate," *Thin Solid Films* 1988, **163**, 415–420.
9. S. Nakamura, "GaN Growth Using GaN Buffer Layer," *Jpn. J. Appl. Phys.* 1991, **30**, L1705–L1707.
10. H. Amano, M. Kito, K. Hiramatsu, I. Akasaki, "p-Type Conduction in Mg-Doped GaN Treated with Low-Energy Electron Beam Irradiation (LEEBI)," *Jpn. J. Appl. Phys.* 1989, **28**, L2112–L2114.
11. S. Nakamura, N. Iwasa, M. Senoh, T. Mukai, "Hole Compensation Mechanism of P-Type GaN Films," *Jpn. J. Appl. Phys.* 1992, **31**, 1258–1266.
12. J. Neugebauer, C. G. Van de Walle, "Hydrogen in GaN: Novel Aspects of a Common Impurity," *Phys. Rev. Lett.* 1995, **75**, 4452–4455.
13. S. Nakamura, T. Mukai, M. Senoh, N. Iwasa, "Thermal Annealing Effects on P-Type Mg-Doped GaN Films," *Jpn. J. Appl. Phys.* 1992, **31**, L139–L142.
14. S. Nakamura, T. Mukai, and M. Senoh, "High-power GaN p-n junction blue-light-emitting diodes," *Jpn. J. Appl. Phys.* 1991, **30**, L1998–L2001.
15. "MIS type Blue LEDs with a brightness of 200 mcd were developed by Toyoda Gosei," October 20, 1993, Nikkan Kogyo Shinbun (Japanese newspaper press release).
16. K. Kumakura, T. Makimoto, N. Kobayashi, T. Hashizume, T. Fukui, H. Hasegawa, "Minority carrier diffusion length in GaN: Dislocation density and doping concentration dependence," *Appl. Phys. Lett.* 2005, **86**, 052105.
17. K. Osamura, K. Nakajima, Y. Murakami, P. H. Shingu, A. Ohtsuji, "Fundamental absorption edge in GaN, InN and their alloys," *Solid State Communications* 1972, **11**, 617–621.

18. K. Osamura, S. Naka, and Y. Murakami, "Preparation and optical properties of $Ga_{1-x}In_xN$ thin films," *J. Appl. Phys.* 1975, **46**, 3432–3437.
19. T. Nagatomo, T. Kuboyama, H. Minamino, O. Omoto, "Properties of $Ga_{1-x}In_xN$ Films Prepared by MOVPE," *Jpn. J. Appl. Phys.* 1989, **28**, L1334–L1336.
20. N. Yoshimoto, T. Matsuoka, T. Sasaki, A. Katsui, "Photoluminescence of InGaN films grown at high temperature by metalorganic vapor phase epitaxy," *Appl. Phys. Lett.* 1991, **59**, 2251.
21. S. Nakamura, T. Mukai, "High-Quality InGaN Films Grown on GaN Films," *Jpn. J. Appl. Phys.* 1992, **31**, L1457–L1459.
22. S. Nakamura, M. Senoh, T. Mukai, "p-GaN/n-InGaN/n-GaN Double-Heterostructure Blue-Light-Emitting Diodes," *Jpn. J. Appl. Phys.* 1993, **32**, L8–L11.
23. S. Nakamura, T. Mukai, M. Senoh, "Candela-class high-brightness InGaN/AlGaN double-heterostructure blue-light-emitting diodes," *Appl. Phys. Lett.* 1994, **64**, 1687–1689.
24. "p-n junction DH Blue LEDs with a brightness of more than 1000mcd were developed by Nichia Chemical Industries Ltd.," November 30th, 1993, Nikkei Sangyo Shinbun (Japanese newspaper press release).
25. S. Nakamura, M. Senoh, N. Iwasa, S.-I. Nagahama, "High-Brightness InGaN Blue, Green and Yellow Light-Emitting Diodes with Quantum Well Structures," *Jpn. J. Appl. Phys.* 1995, **34**, L797–L799.
26. S. Nakamura, M. Senoh, N. Iwasa, S.-I. Nagahama, T. Yamada, T. Mukai, "Superbright Green InGaN Single-Quantum-Well-Structure Light-Emitting Diodes," *Jpn. J. Appl. Phys.* 1995, **34**, L1332–L1335.
27. K. Bando, Y. Noguchi, K. Sakamoto and Y. Shimizu, "Development and application of high-brightness white LEDs," *Tech. Digest. Phosphor Res. Soc.*, 264th Meeting, November 29, 1996 (in Japanese).
28. S. Nakamura, M. Senoh, S.-I. Nagahama, N. Iwasa, T. Yamada, T. Matsushita, H. Kiyoku, Y. Sugimoto, "InGaN-Based Multi-Quantum-Well-Structure Laser Diodes," *Jpn. J. Appl. Phys.* 1996, **35**, L74–L76.
29. S. Nakamura, M. Senoh, S. Nagahama, N. Iwasa, T. Yamada, T. Matsushita, Y. Sugimoto, H. Kiyoku, "Room-temperature continuous-wave operation of InGaN multi-quantum-well structure laser diodes," *Appl. Phys. Lett.* 1996, **69**, 4056–4058.
30. S. D. Lester, F. A. Ponce, M. G. Craford, D. A. Steigerwald, "High dislocation densities in high efficiency GaN-based light-emitting diodes," *Appl. Phys. Lett.* 1995, **66**, 1249–1251.
31. S. Chichibu, T. Azuhata, T. Sota, S. Nakamura, "Spontaneous emission of localized excitons in InGaN single and multiquantum well structures," *Appl. Phys. Lett.* 1996, **69**, 4188–4190.
32. S. Nakamura, "The Roles of Structural Imperfections in InGaN-Based Blue Light-Emitting Diodes and Laser Diodes," *Science* 1998, **281**, 956–961.
33. D. A. Browne, B. Mazumder, Y.-R. Wu, J. S. Speck, "Electron Transport in Unipolar InGaN/GaN Multiple Quantum Well Structures Grown by NH3 Molecular Beam Epitaxy," *J. Appl. Phys.* 2015, **117**, 185703.
34. G. B. Stringfellow, M. G. Craford, "High Brightness Light-Emitting Diodes," *Semiconductors and Semimetals* 1997, **48**, 47–49.

35. G. Wolff, R. Hebert, J. Broder, "Electroluminescence of GaP," *Phys. Rev.* 1955, **100**, 1144–1145.
36. D. A. Holt, G. F. Alfrey, C. S. Wiggins, "Grain Boundaries and Electroluminescence in Gallium Phosphide," *Nature* 1958, **181**, 109–109.
37. H. Grimmeiss, H. Koelmans, "Analysis of p-n Luminescence in Zn-Doped GaP," *Phys. Rev.* 1961, **123**, 1939–1947.
38. J. Starkiewicz, J. W. Allen, "Injection electroluminescence at p-n junctions in zinc-doped gallium phosphide," *J Phys. Chem. Solids* 1962, **23**, 881–884.
39. M. Gershenzon, R. M. Mikulyak, "Electroluminescence at *p-n* Junctions in Gallium Phosphide," *J.Appl. Phys.* 1961, **32**, 1338–1348.
40. "Cree First to Break 300 Lumens-Per-Watt Barrier," March 26, 2014, Cree Inc. (Press Release), http://www.cree.com/News-and-Events/Cree-News/Press-Releases/2014/March/300LPW-LED-barrier
41. H. P. Maruska, J. J. Tieljen, "The Preparation and Properties of Vapor-Deposited Single-Crystal-Line GaN," *Appl. Phys. Lett.* 1969, **15**, 327–329.
42. H. P. Maruska, D. A. Stevenson, J. I. Pankove, "Violet luminescence of Mg-doped GaN," *Appl. Phys. Lett.* 1973, **22**, 303–305.
43. S. Yoshida, S. Misawa, S. Gonda, "Improvements on the electrical and luminescent properties of reactive molecular beam epitaxially grown GaN films by using AlN-coated sapphire substrates," *Appl. Phys. Lett.* 1983, **42**, 427–429.
44. T. Lei, M. Fanciulli, R. J. Molnar, T. D. Moustakas, R. J. Graham, J. Scanlon, "Epitaxial growth of zinc blende and wurtzitic gallium nitride thin films on (001) silicon," *Appl. Phys. Lett.* 1991, **59**, 944–946.
45. S. Nakamura, T. Mukai, M. Senoh, "In situ monitoring and Hall measurements of GaN grown with GaN buffer layers," *J. Appl. Phys.* 1992, **71**, 5543–5549.
46. S. Nakamura, M. R. Krames, "History of Gallium-Nitride-Based Light-Emitting Diodes for Illumination," *Proc. IEEE* 2013, **101**, 2211–2220.
47. Y. C. Shen, G. O. Mueller, S. Watanabe, N. F. Gardner, A. Munkholm, M. R. Krames, "Auger recombination in InGaN measured by photoluminescence," *Appl. Phys. Lett.* 2007, **91**, 141101.
48. J. Iveland, L. Martinelli, J. Peretti, J. S. Speck, C. Weisbuch, "Direct Measurement of Auger Electrons Emitted from a Semiconductor Light-Emitting Diode under Electrical Injection: Identification of the Dominant Mechanism for Efficiency Droop," *Phys. Rev. Lett.* 2013, **110**, 177406.
49. A. Pourhashemi, R. M. Farrell, M. T. Hardy, P. S. Hsu, K. M. Kelchner, J. S. Speck, S. P. DenBaars, S. Nakamura, "Pulsed high-power AlGaN-cladding-free blue laser diodes on semipolar (20-2-1) GaN substrates," *Appl. Phys. Lett.* 2013, **103**, 151112.

Physics 2015

Takaaki Kajita and Arthur B. McDonald

"for the discovery of neutrino oscillations, which shows that neutrinos have mass"

The Nobel Prize in Physics

Speech by Professor Olga Botner of the Royal Swedish Academy of Sciences.

Translation of the Swedish text.

Your Majesties, Your Royal Highnesses, Esteemed Nobel Laureates, Ladies and Gentlemen,

This year's Nobel Prize in Physics is all about neutrinos, perhaps the most puzzling particles in the universe. Like ghosts, they pass through the thickest walls—indeed, they penetrate straight through the entire globe. They have no electric charge, race at almost the speed of light and weigh almost nothing at all—for a long time, we actually thought they were massless! Studying neutrinos is a real challenge!

This year's Laureates, Takaaki Kajita and Arthur B. McDonald, led two large research groups which discovered that these remarkable particles are even more puzzling than we imagined. They behave like chameleons: while travelling through space, they are continuously switching between different identities.

Neutrinos are everywhere. They are constantly created—inside the Sun, when the "cosmic rain" from space collides with the atmosphere, through the decay of atomic nuclei in the Earth's crust—and even in our muscles.

Every second, billions of neutrinos pass through our bodies, unseen and unnoticed. These elusive particles rarely interact and are almost impossible to capture. Yet their ghostlike nature seems very useful: while the light generated in the centre of the Sun never reaches the Earth, neutrino radiation from the fusion processes in the solar core passes straight out. Measuring the number of neutrinos from the Sun therefore provides a means of measuring the temperature in its core. The first experiment aimed at recording solar neutrino radiation was initiated fifty years ago and was followed by many more. It soon turned out—to everyone's surprise—that up to two thirds of the expected number of neutrinos was missing. The neutrinos seemed to be disappearing along the way!

One theory, among many attempts to explain this remarkable phenomenon, was that during its journey a neutrino could change its identity and be transformed into another type of neutrino that was not visible in the detector. This is because there are three types of neutrinos, of which the Sun only produces one—and the experiments were designed to detect only this one type.

But can a particle actually change identity? This mystery was solved with the help of two enormous underground detectors.

The Super-Kamiokande detector was located in a zinc mine in Japan, 1,000 metres below the Earth's surface, and contained 50,000 tonnes of water—more than enough to fill all the bathtubs in Stockholm! This detector recorded the neutrinos created when cosmic radiation collides with the Earth's atmosphere. To their great surprise, scientists discovered that many fewer neutrinos reached the detector from below—that is, from the other side of the globe—than from the atmosphere above. The neutrino types that are created in the atmosphere also seemed to be disappearing along the way!

Neutrinos pass unimpeded through the Earth. So Kajita-san and his team realised that the great distance between the production point on the other side of the Earth and the detector gave the neutrinos extra time to change identity. This finding was presented around the turn of the millennium.

Around the same time, on the other side of the Earth, a huge detector was built for observation of the neutrinos that are produced in the solar core. Unlike earlier detectors, the Sudbury Neutrino Observatory in Canada—two kilometres below the Earth's surface—could effectively record all three types of neutrinos. McDonald and his research team were then able to show that the total number of neutrinos from the Sun reaching the Earth corresponded closely with expectations, but that the type of neutrinos produced in the solar core was too few. They did not seem to have disappeared, though; instead, they must have been transformed into another type of neutrino and switched identity.

The results of both these experiments have one explanation in common: quantum theory describes particles travelling through space as waves. If the three types of neutrinos have different masses, they are described by waves with different frequencies. The interaction between these waves while they travel through space underlies the neutrinos' metamorphosis. The phenomenon is called neutrino oscillations and can only occur if neutrinos have mass.

The number of neutrinos in the universe is enormous. The discovery that they are not massless, as we believed for a long time, is therefore of crucial importance to our understanding of the structure of the universe—and also has far-reaching consequences for cosmology.

Professor McDonald, Professor Kajita,

You have been awarded the Nobel Prize in Physics for the discovery of neutrino oscillations, which shows that neutrinos have mass. On behalf of the Royal Swedish Academy of Sciences it is my honour and great pleasure to convey to you the warmest congratulations. I now ask you to step forward to receive your Nobel Prizes from the hands of His Majesty the King.

Takaaki Kajita. © Nobel Media AB. Photo: A. Mahmoud

Takaaki Kajita

I was born on March 9, 1959, in Higashi-Matsuyama, a small city located about an hour's train ride north of Tokyo. My house was located in the countryside, surrounded by rice fields on the north, east, and south. I grew up in such a peaceful environment.

I went to Kawagoe High School, a rather typical small-town school. This school had a tradition of allowing students to do whatever they liked rather freely. Therefore, I spent a lot of time practicing Kyudo (Japanese archery). I was not particularly good at Kyudo, but I liked it. During one's time as a high school student, you have to decide what you intend to study as an undergraduate once you are admitted to a university. Since I was interested in physics as a high school student, my choice was rather clear: I decided to learn physics in the undergraduate course at Saitama University, a local university near Tokyo.

During my four years of undergraduate education at Saitama University, I continued to play Kyudo, even more seriously than during the high school. I regret that I should have learned more physics during my undergraduate studies, because these studies in undergraduate courses form the basis of everyday research. In any case, I found that physics was indeed interesting. So I decided to continue to studying physics at the graduate level.

I was particularly interested in experimental particle physics. Very fortunately, Professor Masatoshi Koshiba accepted me as a graduate course student in his group at the University of Tokyo. My life as a graduate course student began in April of 1981. Katsushi Arisaka was also a student in Prof. Koshiba's group. He had just finished his Master's thesis based on a Monte Carlo study of a nucleon decay experiment. This was the Kamioka Nucleon Delay Experiment (Kamiokande). He was the only student working on Kamiokande in early 1981. Just when I started my studies, production of newly developed photomultiplier tubes (PMTs) with a diameter of 50 centimeters began. Katsushi Arisaka convinced me that Kamiokande would be a very interesting experiment and asked me to work on it, which I started to do.

At that time the other main members of Kamiokande were Teruhiro Suda from the Institute for Cosmic Ray Research (ICRR) of the University of Tokyo, and Atsuto Suzuki and Kasuke Takahashi from the High Energy Accelerator Research Center (KEK). Soon after I joined the Kamiokande experiment, Yoji Totsuka returned from Deutsches Electron Synchroton (DESY) and started to help us. Soon he too joined Kamiokande. Kazumasa Miyano from Niigata University and Tadashi Kifune from ICRR joined during the preparation stage of the experiment. Also, Masayuki Nakahata, who was an undergraduate course student, worked with us.

I enjoyed the preparation work for Kamiokande. In early spring of 1983, we started the construction work on the Kamiokande proton decay detector in Kamioka. It took almost four months to finish building the detector. I liked the construction work, watching the detector being assembled slowly but steadily. After it was filled with water, data taking with the Kamiokande experiment began in early July of 1983.

Soon after the beginning of the data gathering, we came back to Tokyo to search for proton decays in the Kamiokande data. I enjoyed the data analysis as well. I decided to be a professional physicist as a result of experiencing the complete process of preparation, construction, and data analysis of an experiment through my work with Kamiokande.

I received my Ph.D. in March of 1986; my thesis was titled "Search for nucleon decays into anti-neutrino plus mesons." No evidence for proton decay was observed. In the Japanese system at that time, one way to get a postdoctoral position was to be selected as a postdoctoral researcher by the Japan Society for the Promotion of Science (JSPS). I was not selected. Fortunately, Professor Koshiba offered me a position as a research associate at the International Center for Elementary Particle Physics (ICEPP) of the University of Tokyo for a fixed term of one year. However, I was unable to find a new position in a year, and so I stayed in this position for two years with the very kind understanding of the ICEPP people. I really appreciate their kind decision, because this two year period was the moment that I found the deficit of atmospheric ν_μ and carried out the initial studies.

In April of 1988, I moved to ICRR as a research associate working on the successor experiment to Kamiokande, the much larger Super-Kamiokande. From that point on I have been a member of ICRR. I was able to work on the atmospheric neutrinos and the preparation of Super-Kamiokande without worrying about finding a job.

The construction of Super-Kamiokande was approved by the Japanese government in 1991. People from the USA, most of them from the IMB experiment, joined Super-Kamiokande in 1992. (IMB was a large water Cherenkov detector operated in the USA in the 1980s and in the early 1990s.) Since then, Super-Kamiokande has been an international collaboration. The onsite construction of the Super-Kamiokande detector started in April of 1995. I moved to Kamioka in March of 1995 for the construction of the detector and worked underground for a year. The construction was completed at the end of March of 1996.

After years of planning and construction, the Super-Kamiokande experiment started taking data precisely on schedule, at the stroke of midnight that began April 1, 1996. Since then, I have worked as a convener of the atmospheric neutrino analysis. In the initial stage of the Super-Kamiokande data analysis, analyses were carried out by two independent groups. After confirming that the analyses gave quite similar results, it was decided to merge the two groups into one. Since then, Ed Kearns from Boston University and I have led the atmospheric neutrino analysis. I stepped down from this role when I was appointed the director of ICRR in April 2008.

When I joined Kamiokande, underground experiments were just a very small sub-field of particle physics experiments. At present, after more than 30 years, these underground experiments have become some of the most promising, powerful, versatile, and efficient ways to explore both particle physics and the Universe itself. This research underground continues to stimulate my interest. I look forward to what new discoveries the future will hold.

Discovery of Atmospheric Neutrino Oscillations

Nobel Lecture, December 8, 2015

by Takaaki Kajita
Institute for Cosmic Ray Research, The University of Tokyo, Japan.

1. INTRODUCTION

Neutrinos are a type of fundamental particles, like electrons and quarks. They have no electric charge and they come in three types (flavors), namely the electron neutrino (ν_e), the muon neutrino (ν_μ) and the tau neutrino (ν_τ). They are produced in various places such as in the Earth's atmosphere and in the center of the Sun. The interactions of neutrinos with matter are so weak that they easily penetrate through the Earth and even the Sun. They do interact with matter, however, but only very rarely. Because a charged current (CC) ν_e (ν_μ, ν_τ) interaction produces an electron (muon, tau), physicists can determine a neutrino's flavor by observing the lepton produced when it interacts.

The Standard Model of elementary particle interactions was established in the early 1970s and it describes the strong, weak and electromagnetic interactions very well. However, in the Standard Model these interactions are not unified and are treated as independent entities. Neutrinos are assumed to be massless within the Standard Model. However, physicists have wondered whether neutrinos really have no mass. If, on the other hand, neutrinos do have mass, they can change their type from one flavor to the other(s). This phenomenon is called "neutrino oscillations," and was theoretically predicted by Maki, Nakagawa and Sakata [1] and also by Pontecorvo [2]. For example, imagine that neutrino oscillations occur between ν_μ and ν_τ. Assuming that only these two neutrinos oscillate, so-called two flavor oscillations, the oscillation probability can be written as,

$$P(\nu_\mu \to \nu_\mu) = 1 - \sin^2 2\theta \cdot \sin^2\left(\frac{1.27\Delta m^2 (eV^2) L(km)}{E_\nu(GeV)}\right) \quad (1)$$

where $P(\nu_\mu \to \nu_\mu)$ is the probability that a ν_μ remains a ν_μ after travelling a distance, L, with energy, E. Here θ is the neutrino mixing angle and Δm^2 is the difference in the square of the neutrino masses, $|m_3^2 - m_2^2|$. A ν_τ is generated when the ν_μ disappears so that the probability that the neutrino flavor being either ν_μ or ν_τ is one. As can be seen from Equation 1, the neutrino oscillation length is longer for smaller neutrino masses.

In the late 1970s new theories that unify the strong, weak and electromagnetic forces were proposed. These theories predicted that protons and neutrons, collectively known as nucleons, should decay with lifetimes between 10^{28} and 10^{32} years. To test these predictions, several proton decay experiments began in the early 1980s. These experiments had fiducial masses ranging from about 100 tons to several thousand tons in order to detect proton decays in the predicted lifetime range. One of these was the Kamioka Nucleon Decay Experiment, Kamiokande.

Kamiokande was a 3,000 ton water Cherenkov experiment with a fiducial mass of about 1,000 tons. It was located at a depth of 1,000 m underground in the Mozumi mine in Kamioka, Japan. When charged particles propagate through the detector's water at relativistic speeds, they emit Cherenkov photons. These photons are then detected by photomultiplier tubes installed on the inner surfaces of the Kamiokande water tank. Since the photons are emitted in a forward cone along the particle direction, they form a ring pattern (Cherenkov ring) on the detector walls.

2. ATMOSPHERIC NEUTRINO ANOMALY

The Kamiokande experiment began in July 1983. As a graduate student I was involved in the experiment from its early stages. In March 1986, I received a Ph.D. based on a search for proton decays into an anti-neutrino and meson final states using its data. No evidence for proton decay was observed.

At that time I felt that the analysis software was not good enough to select the proton decay signal most efficiently from its dominant background, atmospheric neutrino interactions. Therefore, as soon as I submitted my thesis I began to work on improving the software. One of the pieces of software was to identify the type of particle seen in events with multiple Cherenkov rings. For each ring in these events, we wanted to estimate whether it was produced by an electron or by a muon. Since the rings are often overlapping it was a non-trivial task to identify the particle types that produced them. Typically one has to verify the performance of the software step-by-step, starting from the simplest cases to the most difficult ones.

Indeed, the software worked very well for the simplest case, namely for simulated atmospheric neutrino events with only a single Cherenkov ring. Muons (electrons) produced by simulated atmospheric ν_μ (ν_e) interactions were correctly identified approximately 98% of the time. With this knowledge, the flavor of atmospheric neutrino interactions observed in Kamiokande was checked. The result was strange. The number of ν_μ events was far fewer than predicted by the simulation. At the same time, no such discrepancy was seen in the number of ν_e events. At first I thought that I had made some serious mistake. In order to find where I had made the mistake, I decided to check the events in the data by eye. Immediately I realized that the analysis software was correctly identifying the particle types. Unfortunately, I thought that this meant that the problem must not be simple. It seemed very likely that there were mistakes somewhere deep in the simulation, data selection, or the event reconstruction software. Together Masato Takita and I embarked on various studies to try and find such mistakes late in 1986.

After studies for a year, we did not find any serious mistake and concluded that the ν_μ deficit could not be due to a major problem with the data analysis or the simulation. It should be mentioned that the ratio of the atmospheric ν_μ and ν_e fluxes can be accurately predicted independently of the absolute flux values due to the neutrino production mechanism. Atmospheric ν_μ and ν_e are generated in the decay chain of pions produced by cosmic ray interactions in the atmosphere. A pion decays into a muon and a ν_μ and the muon subsequently decays into an electron in conjunction with another ν_μ and a ν_e. These neutrinos have almost the same energy and therefore the ratio of the atmospheric ν_μ to ν_e fluxes is expected to be about two, regardless of the details of the cosmic ray flux. In fact, detailed calculations predicted that this flux ratio is approximately two for neutrino energies around 1 GeV. In 1988 we estimated the uncertainty of this ratio to be about 5%. However, the absolute flux was only predicted to a precision between 20% and 30%. These predictions indicated that the data from Kamiokande were very difficult to explain by uncertainties in the flux calculation.

A paper was written and published in 1988 [3]. The observed numbers of μ-like events (mostly due to ν_μ interactions), and e-like events (mostly due to ν_e interactions), were compared with the corresponding numbers from a simulation.

TABLE 1. Comparison of the number of observed events in Kamiokande with the expectation from simulations. The detector exposure was 2.87 kiloton years.

	Data	Prediction
e-like (mostly CC ν_e interactions)	93	88.5
μ-like (mostly CC ν_μ interactions)	85	144.0

The results of the comparison are summarized in Table 1. Note that though the simulation used the calculated atmospheric neutrino flux and assumed standard neutrino properties, it did not include the effect of neutrino oscillations.

From these considerations Kamiokande concluded in [3] that "We are unable to explain the data as the result of systematic detector effects or uncertainties in the atmospheric neutrino fluxes. Some as-yet-unaccounted-for physics such as neutrino oscillations might explain the data." In fact, I was most excited by the possibility of neutrino oscillations with a large mixing angle. Namely, ν_μ seemed to be almost completely oscillating into some other neutrino type. At that time a large mixing angle was not expected. So this gave me strong motivation to continue the study of atmospheric neutrinos.

Subsequently, another large water Cherenkov experiment, IMB, published an observed deficit of ν_μ events [4], [5]. Kamiokande then published a second paper on the atmospheric neutrino deficit including a detailed evaluation of systematic errors in the ratio of μ-like to e-like events for both data and simulation as well as allowed regions for oscillation parameters assuming neutrinos oscillate [6]. The results suggested that the atmospheric neutrino data might indicate neutrino oscillations. However, neutrino oscillation was still only one of the possible explanations for the data. This was due partly to the fact that the observed effect was only a small ratio of μ-like to e-like events. Indeed, the deficit of μ-like events seen in Kamiokande and IMB at energies below about 1 GeV (sub-GeV) did not show a strong dependence on the angle of the event direction relative to the vertical axis of the detector (zenith angle) nor on the event momentum.

The atmospheric neutrino flux is predicted to be up-down symmetric due to the isotropic nature of the cosmic ray flux. Note that while this is not exactly true for sub-GeV atmospheric neutrinos due to the effects of the geomagnetic field on the cosmic ray flux, the flux of the multi-GeV neutrinos is nonetheless very nearly up-down symmetric. If neutrinos have very small masses, then their oscillation length could be 100 kilometers or even longer. If this is the case, vertically downward-going neutrinos, which typically travel tens of kilometers before reaching an underground detector, may not oscillate before they interact within it. On the other hand, vertically upward-going neutrinos, which can travel up to about 12,800 kilometers, may have enough time to oscillate into another flavor. Observing a disappearance effect in the upward-going events but not in the downward-going events would therefore be a very strong indication of neutrino oscillations.

Soon after submitting the first paper on the atmospheric ν_μ deficit in 1988, we started to select and study atmospheric ν_μ events with energies larger than 1 GeV (multi-GeV) in the data. A multi-GeV ν_μ interaction typically produces a multi-GeV muon. However, multi-GeV muons produced in the detector will often penetrate through it and exit into the surrounding rock, called partially contained (PC) events. They were selected for analysis. The angular correlation between the muon and its parent neutrino gets substantially better with increasing neutrino energy and therefore the zenith angle distribution of multi-GeV muons should represent the neutrino zenith angle distribution fairly well. Consequently, we studied the zenith angle distribution for the multi-GeV events.

Since the flux of the atmospheric neutrinos decreases rapidly with increasing energy, the rate of multi-GeV ν_μ events was only about 20 events per year in Kamiokande and it therefore took several years to collect a statistically meaningful number of them. Finally, Kamiokande published a study of the multi-GeV atmospheric neutrino data in 1994 [7]. The μ-like data showed a deficit of events in the upward-going direction, while the downward-going events did not show any such deficit. On the other hand, the corresponding distribution from the e-like data did not show any evidence for a deficit of upward-going events (Fig. 1). The ratios of upward- to downward-going events (the up/down ratio) for the multi-GeV μ-like and e-like data were $0.58^{+0.13}_{-0.11}$ and $1.38^{+0.39}_{-0.30}$, respectively. The

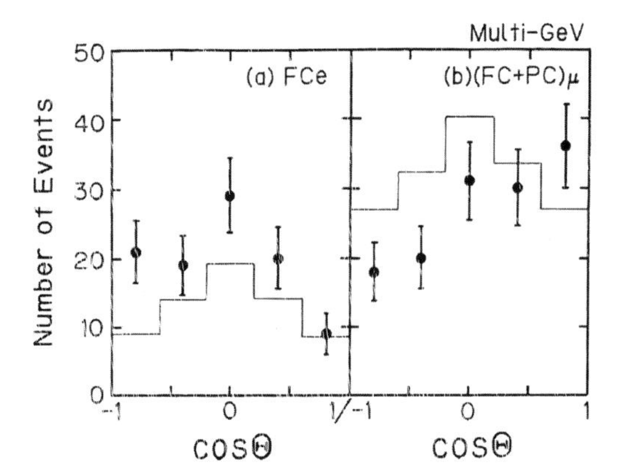

FIGURE 1. Zenith angle distributions for multi-GeV (a) e-like and (b) μ-like events observed in Kamiokande [7]. Solid histogram shows the predicted distributions without oscillations. Absolute normalization had an uncertainty of 20 to 30%.

statistical significance of the observed up-down asymmetry in the μ-like data was equivalent to 2.8 standard deviations. In other words, the probability that the observed result could be due to a statistical fluctuation was less than 1%. It was an interesting observation which showed for the first time that the ν_μ deficit depended on the neutrino flight length as predicted by neutrino oscillations. However, the statistical significance of the observation was not strong enough to be conclusive and this prompted the need for an even larger data set, namely an even larger detector.

3. DISCOVERY OF NEUTRINO OSCILLATIONS

The Super-Kamiokande detector is a large, cylindrical water Cherenkov detector; 41.4 meters high, 39.3 meters in diameter, and it has a total mass of 50,000 tons. Super-Kamiokande is divided into two parts, an inner detector that studies the details of neutrino interactions and an outer detector that identifies incoming and exiting charged particles. The fiducial mass of the detector is 22,500 tons, which is about 20 times larger than that of Kamiokande. Figure 2 shows a schematic of the Super-Kamiokande detector.

Super-Kamiokande is an international collaboration. A collaborative agreement between research groups from the USA and Japan was signed in October 1992. Many members of the Kamiokande and IMB collaborations joined in the experiment. The Super-Kamiokande detector was designed based on the

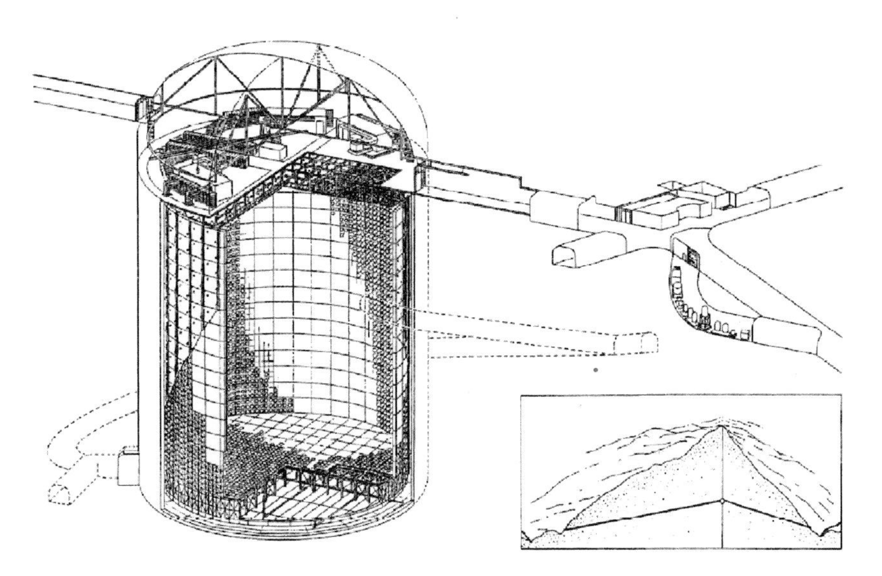

FIGURE 2. Schematic view of the Super-Kamiokande detector.

experience of these experiments and included various technological improvements. As of 2015, about 120 people from seven countries are members of the collaboration.

The Super-Kamiokande experiment started in the spring of 1996 after a five year period of detector construction. Due its larger fiducial mass, Super-Kamiokande accumulates neutrino events approximately 20 times faster than Kamiokande. Furthermore, Cherenkov rings are observed by 11,200 photomultiplier tubes making it possible to study the detailed properties of neutrino events. Methods for analyzing atmospheric neutrino interactions had been well established by studies performed in previous experiments. Therefore, from the very beginning of the experiment, Super-Kamiokande analyzed various types of atmospheric neutrino events, including fully contained events, which have no charged particle exiting the inner detector and partially contained events, which has at least one charged particle exiting the inner detector [8], [9]. In addition, upward-going muon events induced by neutrino interactions in the rock below the detector and that traverse it entirely [10] as well as those that stop within it [11] were analyzed. The topologies and features of these events types differ significantly from one another. Therefore the collaborative work of many researchers, especially young scientists, was essential for the analysis of the data. Super-Kamiokande developed simulations and analysis programs based on those from Kamiokande and IMB. For this reason Super-Kamiokande was able to produce reliable results relatively quickly after the start of the experiment.

By the spring of 1998 Super-Kamiokande had analyzed 535 days of data, which is equivalent to a 33 kiloton · year exposure of the detector. In total there were 5,400 atmospheric neutrino events, which was already several times larger than the data sets of previous experiments. At the 18th International Conference on Neutrino Physics and Astrophysics (Neutrino '98), Super-Kamiokande announced evidence for atmospheric neutrino oscillations [12], [13]. The zenith angle distributions shown at Neutrino '98 have been reproduced in Fig. 3. The top and bottom panels of the figure show the zenith angle distributions of multi-GeV e-like and multi-GeV μ-like (fully contained and partially contained events have been combined) data, respectively. While the e-like data did not show any statistically significant up-down asymmetry, a clear deficit of upward-going μ-like events was observed. The statistical significance of the effect was more than six standard deviations, implying that the deficit was not due to a statistical fluctuation. Figure 4 shows the summary of the oscillation analyses from Super-Kamiokande and Kamiokande that were presented at the Neutrino '98 conference. The allowed regions for the neutrino oscillation parameters obtained from the two experiments overlapped, indicating that the data could be consistently

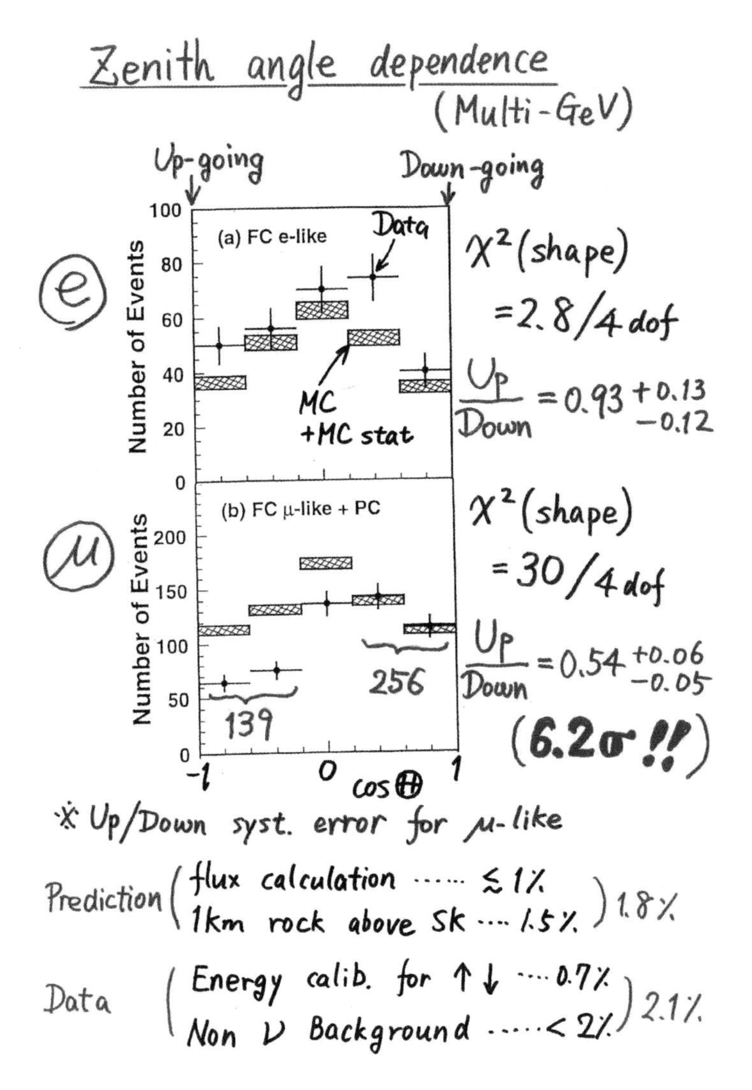

FIGURE 3. Zenith angle distributions for multi-GeV atmospheric neutrino events presented at the 18th International Conference on Neutrino Physics and Astrophysics (Neutrino '98) by the Super-Kamiokande collaboration [12].

explained by neutrino oscillations. Super-Kamiokande concluded from the analysis of these data that muon neutrinos oscillate into other types of neutrinos, most likely into tau neutrinos.

There were two other experiments, Soudan-2 and MACRO, which were observing atmospheric neutrinos at that time. Soudan-2 was a 1 kiloton iron tracking calorimeter detector which had been gathering data since 1989. This

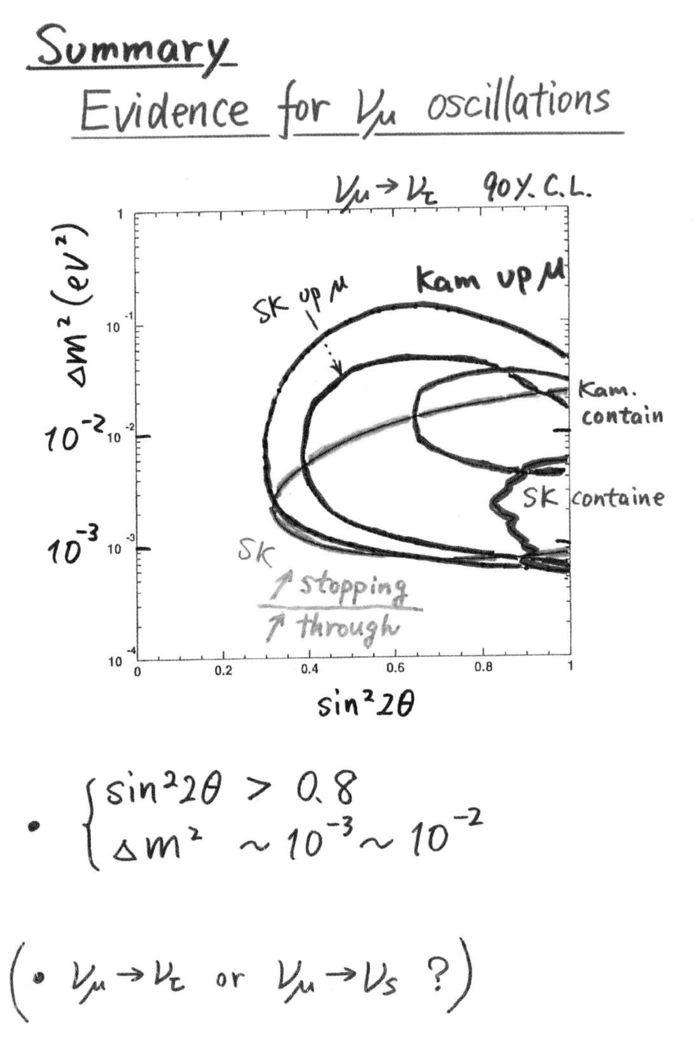

FIGURE 4. The final slide (summary slide) of the presentation by the Super-Kamiokande collaboration at Neutrino '98 [12].

experiment confirmed the zenith angle dependent ν_μ deficit [14]. Similarly, MACRO was a large underground detector which was able to measure upward-going muons as well as partially contained neutrino events. This experiment also observed a zenith angle dependent deficit of both upward-going muons [15] and partially-contained ν_μ events [16]. The results from these experiments were completely consistent with those from Super-Kamiokande and consequently, neutrino oscillations were quickly accepted by the neutrino physics community.

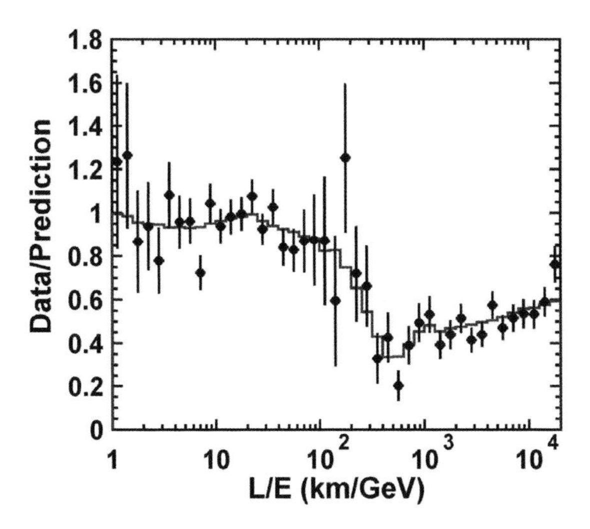

FIGURE 5. Data/Prediction as a function of L/E_ν from Super-Kamiokande [17].

4. RECENT RESULTS AND THE FUTURE

Super-Kamiokande's data showed that approximately 50% of muon-neutrinos disappear after traveling long distances, an effect which was commonly interpreted as neutrino oscillations. However, there were still several unanswered questions, such as "what are the values of the neutrino mass squared difference (Δm^2) and the neutrino mixing angle (θ)?", "does the ν_μ disappearance probability really oscillate as predicted by the theory of neutrino oscillation?", and "is it possible to confirm $\nu_\mu \rightarrow \nu_\tau$ oscillations by detecting ν_τ interactions?" These questions have been answered experimentally.

4.1. Observing "oscillation"

According to the neutrino oscillation formula shown in Equation 1, the neutrino survival probability should be sinusoidal. Specifically, at a given energy the probability should be smallest at a certain value of L/E_ν then come back to unity if twice the distance is traversed, and then continue oscillating back and forth in this way over longer distances. In Fig. 3 atmospheric neutrino events with a variety of L/E_ν values were included in each zenith angle bin so only an averaged survival probability could be observed.

Super-Kamiokande carried out a dedicated analysis that used only events whose L/E_ν value could be determined with good precision. In short, in this analysis Super-Kamiokande did not use neutrino events whose direction was near the horizon, since the estimated neutrino flight length changes significantly for even small changes in the estimated arrival direction in this regime. Similarly, the analysis did not use low energy neutrino events because the scattering angle at these energies is large, and consequently, the uncertainty in the estimated neutrino flight length becomes large. Using only the high L/E_ν resolution events, Super-Kamiokande showed that the measured ν_μ survival probability has a dip corresponding to the first minimum of the theoretical survival probability near L/E_ν= 500 km/GeV [17] as shown in Fig. 5. This was the first evidence that the ν_μ survival probability obeys the sinusoidal function predicted by neutrino oscillations.

4.2. Detecting tau neutrinos

If the oscillations of atmospheric neutrino are indeed between ν_μ and ν_τ, it should be possible to observe the charged current interactions of ν_τ generated by these oscillations. A charged current ν_τ interaction typically produces a tau lepton accompanied by several hadrons, most of which are pions. Due to the heavy tau mass (1.78 GeV/c^2), the threshold for this interaction is about 3.5 GeV. Since this threshold is rather high and the atmospheric neutrino flux at these energies is rather low, the expected event rate is only about one per kiloton per year. The rate of the charged current ν_τ interactions is therefore only about 0.5% of the total atmospheric neutrino interaction rate. It should be noted that the lifetime of the tau lepton is only 2.9×10^{-13} sec, hence, any tau lepton produced in an atmospheric neutrino interaction decays almost immediately into several hadrons and a neutrino. Therefore, a typical ν_τ interaction has many hadrons in the final state. However, high energy neutral current interactions also produce many hadrons. Searching for ν_τ events in a water Cherenkov detector is therefore complicated due to these backgrounds.

Nonetheless Super-Kamiokande searched for charged current ν_τ interactions in the detector. The search was carried out using various kinematic variables and advanced statistical methods [18], including an artificial neural network. Figure 6 shows the zenith angle distribution for candidate ν_τ events [19]. Even with these advanced methods many background events remain in the final sample. However, there is an excess of upward-going events that cannot be explained with the background events alone. The significance of the excess after taking various systematic uncertainties into account is 3.8 standard deviations [19]. These data are indeed consistent with the appearance of tau neutrinos due to atmospheric $\nu_\mu \rightarrow \nu_\tau$ oscillations.

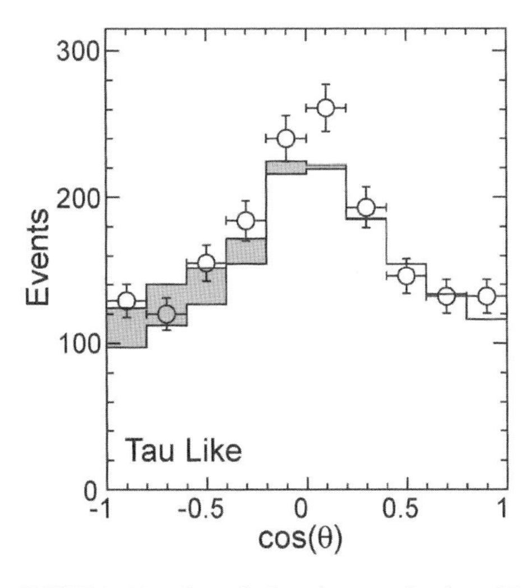

FIGURE 6. Zenith angle distributions for the τ-like events selected from the data observed in Super-Kamiokande [19]. Circles with error bars show the data. Solid histograms show the Monte Carlo prediction with $\nu_\mu \rightarrow \nu_\tau$ oscillations but without the charged current ν_τ interactions. The gray histograms show the fit result including the ν_τ interactions.

4.3. Data updates, neutrino masses and mixing angles

As of 2015 Super-Kamiokande has been taking data for about 5,000 days, which implies that approximately a factor of 10 larger data set than that in 1998 has already been obtained. Figure 7 shows the zenith angle distribution of these events. It is clear that the statistical error on the data sample has improved significantly relative to the 1998 data set (shown in Fig. 3). Neutrino oscillation parameters have been measured using these events and the data indicate that the

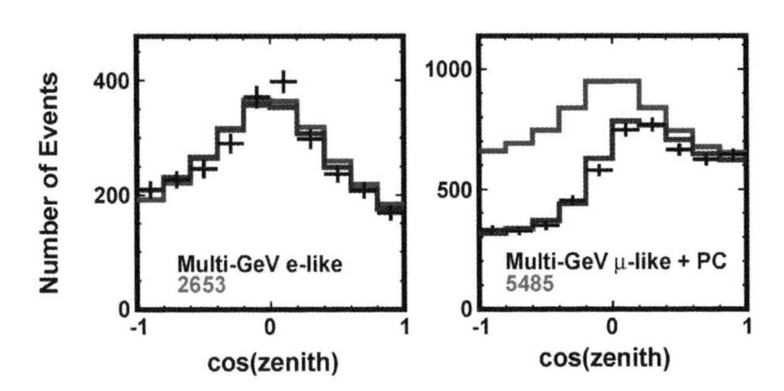

FIGURE 7. Zenith angle distribution for multi-GeV e-like (left) and μ-like (right) atmospheric neutrino events observed in Super-Kamiokande (2015).

neutrino mass squared difference (Δm^2) is approximately 0.0024 eV^2. Assuming that the neutrino masses are not degenerate, the heaviest neutrino mass is approximately 0.05 eV, which is 10 million times smaller than the electron mass (or more than one trillion times smaller than the top quark mass), suggesting that neutrino masses are extremely small compared to the masses of other elementary particles. These extremely small neutrino masses are naturally explained by the seesaw mechanism [20], [21], [22], implying that the small neutrino masses are related to the physics at an extremely high energy scale.

The measured mixing angle is consistent with maximal mixing, $\sin^2 2\theta = 1.0$. Compared to the measurements made in 1998, these parameters have been determined much more precisely. Note that the neutrino mixing angles are very different from the quark mixing angles. For instance, $\sin^2 2\theta \sim 1$ corresponds to $\theta \sim 45$ degrees, while the analogous quark mixing angle is only about 2.4 degrees. Before the discovery of neutrino oscillations this difference was not expected. Indeed, the difference in these sets of mixing angles may give us a hint to understand the profound relationship between quarks and leptons.

4.4. Neutrino oscillation experiments: past, present and future

As discussed above, the deficit of ν_μ events in early data from atmospheric neutrino experiments circa 1990 were confirmed to be the effect of neutrino oscillations in 1998 by the subsequent generation of experiments. Note that the atmospheric neutrino flux has a wide energy spectrum and also provides neutrinos with a wide range of path lengths. These features made it possible to probe neutrino oscillations over an equally large range of Δm^2 values and in fact led to the discovery of neutrino oscillations.

Early atmospheric neutrino data and the discovery of neutrino oscillations motivated accelerator based long-baseline neutrino oscillation experiments. In a long-baseline neutrino oscillation experiment the neutrino flight length is fixed to a single value since the neutrino beam is produced by an accelerator and observed in a detector located a fixed distance away. It should also be noted that the beam in such experiments has a high ν_μ (or anti-ν_μ) purity, while the atmospheric neutrino flux is a mixture of ν_μ, anti-ν_e, ν_μ and anti-ν_μ. For these reasons long-baseline experiments are well suited to carry out precision measurements.

The first generation long-baseline experiments were carried out in the 2000s. K2K and MINOS are the experiments of this generation and confirmed the neutrino oscillation phenomenon, independently measuring the neutrino oscillation parameters [23], [24]. OPERA was also a long-baseline experiment of this generation and observed tau leptons produced by the interactions ν_τ's generated by neutrino oscillations [25].

Oscillations between ν_μ and ν_τ have been well studied by both long-baseline and atmospheric neutrino experiments. Therefore, the next stage of oscillation studies is focused on three flavor oscillation effects. The first step towards realizing this goal was the establishment of the mixing angle, θ_{13}. Several reactor neutrino experiments (Daya Bay, RENO, and Double-Chooz) and long-baseline experiments (T2K and NOνA) have been carried out to search for evidence of this parameter. These experiments discovered and measured θ_{13} [26], [27], [28], [29], [30].

Including measurements by solar neutrino experiments [31] and a long-baseline reactor experiment (KamLAND) [32], all mixing angles (θ_{12}, θ_{23}, and θ_{13}) and the absolute values of the neutrino mass squared differences (Δm^2_{12} and $\Delta m^2_{23(13)}$) within the three flavor neutrino oscillation framework have been measured. It is clear that our understanding of neutrino oscillations has improved tremendously since 1998. However, there are still unmeasured, but important parameters to be probed by future neutrino oscillation experiments. Notably the measurement of the neutrino mass ordering and the potential establishment of CP violation in neutrino oscillations are measurements being targeted by upcoming experiments. The order of the neutrino masses is usually assumed to be $m_{\nu 1} < m_{\nu 2} < m_{\nu 3}$. In fact we know through measurements of solar neutrinos that $m_{\nu 1} < m_{\nu 2}$. However, we do not yet know if ν_3 is the heaviest neutrino mass state. This must be measured. If CP is violated in the neutrino sector, the probabilities of oscillating from ν_μ to ν_e and from anti-ν_μ to anti-ν_e will not be identical. Discovering CP violation in the neutrino sector could have a profound impact on our understanding of the baryon asymmetry of the Universe [33]. Consequently, several long-baseline [34], [35], atmospheric [36], [37], [38] and reactor [39], [40] neutrino experiments are currently planned or are under construction in order to measure these properties. I expect that neutrino oscillation experiments will continue producing results of fundamental importance to our deeper understanding of the elementary particles and the Universe itself.

SUMMARY

An unexpected muon neutrino deficit was observed in the atmospheric neutrino flux by Kamiokande in 1988. At that time neutrino oscillation was considered as a possible explanation for the data. Subsequently, in 1998, through the studies of atmospheric neutrinos, Super-Kamiokande discovered neutrino oscillations, establishing that neutrinos have mass. I feel that I have been extremely lucky, because I have been involved in the excitement of this discovery from its very beginning.

The discovery of non-zero neutrino masses has opened a window to study physics beyond the Standard Model of elementary particle physics, notably physics at a very high energy scale such as the Grand Unification of elementary particle interactions. At the same time, there are still many things to be observed in neutrinos themselves. Further studies of neutrinos might give us information of fundamental importance for our understanding of nature, such as the origin of the matter in the Universe.

ACKNOWLEDGEMENTS

I would like to thank the collaborators of the Kamiokande and Super-Kamiokande experiments. In particular, I would like to thank Masatoshi Koshiba and Yoji Totsuka for their continued support and encouragement of my research throughout my career. Additionally, I would like to thank the following people for their contributions. Ed Kearns worked with me on the analysis of atmospheric neutrinos in Super-Kamiokande for many years. Masato Takita and Kenji Kaneyuki worked with me during the Kamiokande analysis. Yoji Totsuka, Yoichiro Suzuki and Masayuki Nakahata have been leading the Super-Kamiokande experiment. Hank Sobel and Jim Stone have been leading the US effort on Super-Kamiokande. Kenzo Nakamura and Atsuto Suzuki played very important roles in the early stages of Super-Kamiokande. Hard work by many young Super-Kamiokande collaborators was essential for the discovery of neutrino oscillations. Also, I would like to thank Morihiro Honda for his neutrino flux calculation.

Finally, Super-Kamiokande acknowledges the Kamioka Mining and Smelting Company. Super-Kamiokande has been built and operated from funds provided by the Japanese Ministry of Education, Culture, Sports, Science and Technology, the U.S. Department of Energy, and the U.S. National Science Foundation. These efforts were partially supported by various funding agencies in Korea, China, the European Union, Japan, and Canada.

REFERENCES

1. Z. Maki, M. Nakagawa, and S. Sakata, *Prog. Theor. Phys.* **28** (1962) 870–880.
2. B. Pontecorvo, Zh. *Eksp. Teor. Fiz.* **53** (1967) 1717–1725 [*Sov. Phys. JETP* **26** (1968) 984–988].

3. K. Hirata *et al*, *Phys. Lett.* B **205** (1988) 416–420.
4. D. Casper, *et al.*, *Phys. Rev. Lett.* **66** (1991) 2561–2564.
5. R. Becker-Szendy, *et al.*, *Phys. Rev.* D **46** (1992) 3720–3724.
6. K. S. Hirata, *et al.*, *Phys. Lett.* B **280** (1992) 146–152.
7. Y. Fukuda, *et al.*, *Phys. Lett.* B **335** (1994) 237–245.
8. Y. Fukuda *et al.* (Super-Kamiokande Collaboration), *Phys. Lett.* B **433** (1998) 9–18.
9. Y. Fukuda *et al.* (Super-Kamiokande Collaboration), *Phys. Lett.* B **436** (1998) 33–41.
10. Y. Fukuda *et al.* (Super-Kamiokande Collaboration), *Phys. Rev. Lett.* **82** (1998) 2644–2648.
11. Y. Fukuda et al. (Super-Kamiokande Collaboration), *Phys. Lett.* B **467** (1999) 185–193.
12. Takaaki Kajita, for the Kamiokande and Super-Kamiokande collaborations, talk presented at the 18th International Conference in Neutrino Physics and Astrophysics (Neutrino '98), Takayama, Japan, June 1998: Takaaki Kajita (for the Kamiokande and Super-Kamiokande collaborations), *Nucl. Phys. Proc. Suppl.* **77** (1999) 123–132.
13. Y. Fukuda, *et al.* (Super-Kamiokande collaboration), *Phys. Rev. Lett.* **81** (1998) 1562–1567.
14. W. W. M. Allison, *et al.*, (Soudan-2 collaboration), *Phys. Lett.* B **449** (1999) 137–144.
15. M. Ambrosio, *et al.* (MACRO collaboration), *Phys. Lett.* B **434** (1998) 451–457.
16. M. Ambrosio, *et al.* (MACRO collaboration), *Phys. Lett.* B **478** (2000) 5–13.
17. Y. Ashie, *et al.* (Super-Kamiokande collaboration), *Phys. Rev. Lett.* **93** (2004) 101801.
18. K. Abe, *et al.* (Super-Kamiokande collaboration), *Phys. Rev. Lett.* **97** (2006) 171801.
19. K. Abe, *et al.* (Super-Kamiokande collaboration), *Phys. Rev. Lett.* **110** (2013) 181802.
20. P. Minkowski, *Phys. Lett.* B 67, (1977) 421–428.
21. T. Yanagida, in *Proceedings of the Workshop on the Unified Theory and Baryon Number in the Universe*, edited by O.Sawada and A.Sugamoto (KEK Report No. 79–18) (1979) p. 95–98.
22. M. Gell-mann and P. Ramond, and R. Slansky, in *Supergravity*, edited by P. van Nieuwenhuizen and D. Z. Freedman (North-Holland, Amsterdam) (1979) p. 315–321.
23. E. Aliu, *et al.* (K2K collaboration), *Phys. Rev. Lett.* **94**, (2005) 081802.
24. P. Adamson *et al.* (MINOS Collaboration), *Phys. Rev. Lett.* **101** (2008) 131802.
25. N. Agafonova, *et al.* (OPERA collaboration), *Phys. Rev.* D **89** (2014) 051102.
26. F. P. An, *et al.* (Daya Bay collaboration), *Phys. Rev. Lett.* **108** (2012) 171803.
27. J. K. Ahn, *et al.* (RENO collaboration), *Phys. Rev. Lett.* **108** (2012) 191802.
28. Y. Abe, *et al.* (Double Chooz experiment), *Phys. Rev.* D **86** (2012) 052008.
29. K. Abe, *et al.* (T2K collaboration), *Phys. Rev. Lett.* **107** (2011) 041801.
30. P. Adamson, *et al.* (NOvA collaboration), *arXiv:1601.05022.*
31. A. McDonald, Nobel Lecuture (2015).
32. K. Eguchi, *et al.* (KamLAND collaboration), *Phys. Rev. Lett.* **90** (2003) 021802.
33. M. Fukugita and T. Yanagida, *Phys. Lett.* B 174, (1986) 45–47.
34. DUNE collaboration, Long-Baseline Neutrino Facility (LBNF) and Deep Underground Neutrino Experiment (DUNE) Conceptual Design Report, Volume 1 to 4, *arXiv:1512.06148, arXiv:1601.02984, arXiv:1601.05471, arXiv:1601.05823.*
35. K. Abe, *et al.* (Hyper-Kamiokande working group), *arXiv:1109.3262.*

36. M. S. Athar, *et al.* (INO Collaboration), "India-based Neutrino Observatory: Project Report," INO-2006-01.
37. M. G. Artsen, *et al.* (IceCube PINGU collaboration), *arXiv:1401.2046*.
38. V. Van Elewyck (for the KM3NeT collaboration), *J. Phys. Conf.* Ser. **598** (2015) 1, 012033.
39. Z. Djurcic *et al.* (JUNO collaboration), *arXiv:1508.07166*.
40. S. B. Kim, *Nucl. Part. Phys. Proc.* **265–266** (2015) 93–98.

Arthur B. McDonald. © Nobel Media AB. Photo: A. Mahmoud

Arthur B. McDonald

I was born in 1943 in Sydney, Nova Scotia, Canada, a city of about 30,000 people on Cape Breton Island. My mother's and father's families were Scottish and French settlers who had come to Atlantic Canada in the 1700s and early 1800s. My father was a Lieutenant in the Canadian Army and left for Europe when I was about a year old to participate in the battles related to the liberation of Holland. He received the Military Cross, one of the highest decorations for bravery and was wounded, returning to Canada in 1946. He and my mother were hard-working people who appreciated the value of a good education and encouraged me throughout my school days. I also had lots of support from my extended family and enjoyed my childhood, with a balanced mix of studies, fun, sports, family activities and work (I had a 104 house paper route that I remember as being almost all uphill, particularly in winter). My father was very active in the local community, serving as a City Councillor and together with my mother in many service and charitable organizations. I have one sister, ten years younger than I am, but we have a wonderful relationship in spite of the fact that I left home for university when she was just 7 years old.

FIGURE 1. With my father, Bruce and my mother Valerie McDonald.

Sydney was a great community in which to grow up, safe, social and supportive. I remember a positive educational environment in the schools,

FIGURE 2. With my parents and my sister Faith.

with many helpful teachers. I and a number of my classmates in high school were particularly influenced by Mr. Bob Chafe, our math teacher, who went out of his way to engage us in the subject, including extra classes going beyond the normal curriculum. A number of my classmates went on to careers in academia, including several mathematics professors.

There was also a very healthy social environment in the community. I belonged to a service club for teenagers called HI-Y, associated with the local YMCA. This was the late 50s, Rock and Roll was King, and we had a dance for the boys and girls clubs on Friday nights after our meetings. Our clubs ran a community dance on Saturday night at the YMCA as a fund raiser for our service work. That was where I met my future wife, Janet, and I am very pleased to note that we will celebrate our 50th wedding anniversary in 2016. We have 4 children and 8 grandchildren who are a great joy to us. We have also loved to dance throughout our lives, undoubtedly because of our positive high school experience on the dance floor.

At the age of 17, I went to Dalhousie University in Halifax, 400 km away, to study Science, but with very little feeling for which area of Science I would like. I was strongly influenced again by my teachers, particularly Professor Ernie Guptill, my first year Physics teacher, who showed me how you could use mathematics to figure out how the world works in great detail. I also found that I was very good at solving physics problems and enjoyed doing so. These days when I am asked by young people how to choose what to follow as a career, I suggest that they try a number of areas to see which ones interest them and also to see which ones they are good at. The combination of those two features will enable them to have a successful career in an area where they are happy to go to work day after day. I feel that physics has been that way for me. I had my first experience of "engineering physics" by working in the summer for Prof. Ewart Blanchard, carefully measuring gravity on the roads of Nova Scotia, where we found an anomaly that was later developed into a profitable gypsum mine. I stayed at Dalhousie for my Master's degree, working with Prof. Innes MacKenzie, studying the lifetimes of positrons in metals. A paper that arose from that work indicating the relationship of positron lifetimes to defects in materials is still one of my most highly cited papers. I spent one year obtaining a M.Sc. degree and then knew that I really wanted to be an experimental physicist.

As we were considering where to go for graduate school, my long-time friend, Peter Nicholson and I developed an idea that we put to the Department Chair at Dalhousie. We said that if he would sponser a trip to potential graduate schools on the east coast of the US, we would take careful notes and provide a lecture and guide for other undergraduates considering where to apply. He agreed and we had a great time, visiting several of the Ivy League Schools among others. We then proceeded to apply to several of these schools for graduate study and were accepted but we were also accepted at CalTech and Stanford (Operations Research for Peter) and California just seemed like a great place to experience, so we accepted those offers.

FIGURE 3. With Janet prior to a High School dance and prior to the Nobel ceremonies and Ball.

CalTech was a marvellous experience. There was a Van de Graaff accelerator in the basement of the Kellogg Laboratory and we had as much beam time on that as we could want. My thesis supervisor, Prof. Charlie Barnes was an excellent mentor and very encouraging of the measurements that we wanted to pursue, using the nucleus to study fundamental symmetries and processes that could enable us to learn more about the basic laws of physics. I was very privileged to work with two close colleagues, Eric Adelberger and Hay Boon Mak. Eric went on to a very productive career in nuclear physics and in the study of the force of gravity at short distances with extremely beautiful and sophisticated experiments. Hay Boon became a Professor at Queen's University, Kingston, Canada and made major contributions to parity violation measurements and the Sudbury Neutrino Observatory (SNO) project. At CalTech, we studied the properties of nuclear energy levels related to the symmetries of the Coulomb interaction, known as studies of Isospin Symmetry in nuclei.

Another interesting aspect of my time at CalTech (1965–69) was that I was working in the Kellogg laboratory headed by Prof. Willy Fowler, who later received the Nobel Prize in Physics for the understanding of how the elements are produced in the sun and other stars. It was an exciting place to be a graduate student as the latest developments in physics were the subject of discussion every day. Willy had a very sunny personality and there was a very collegial atmosphere in the laboratory. The seminars were at 7:30 pm on a Friday night, often followed by a party at one of the Professor's houses. An Indication of Willy's personality is the subtitle of his Nobel Lecture: "Ad astra per aspera et per ludum" that he translated as "To the stars through hard work and fun." This is a spirit that I have always respected for research work, education and collegiality.

John Bahcall was a junior faculty member at CalTech when I was there and Ray Davis visited for several periods. They were working on the theory and experimental design for the measurement of neutrinos from the sun with Davis' proposed large tank of cleaning fluid. Some of my fellow graduate students were measuring nuclear reactions related to neutrino detection in chlorine and this experiment was a big topic of conversation in the lab. The discrepancy observed between Davis' experiment and Bahcall's theory became known as The Solar Neutrino Problem and ultimately was the impetus for the establishment of the Sudbury Neutrino Observatory almost 20 years later. Ray received the Nobel Prize in 2002 for his pioneering work in solar neutrino detection.

Following CalTech, I accepted a postdoctoral position at Atomic Energy of Canada (AECL) Chalk River Nuclear Laboratories, working on basic research at the accelerator facility there in Doug Milton's Nuclear Physics Branch. This was another very productive period with an in-house Tandem Van de Graaff accelerator, lots of beam time and very skilled colleagues. A year after I arrived, I obtained a permanent position there because a vacancy opened up as Dr. George Ewan moved to Queen's University. After completing a series of measurements providing detailed information on isospin-forbidden decays in light nuclei at Chalk River, Princeton, Washington and Michigan State, Eric Adelberger and I published a paper concluding that the data was consistent with the effects expected from the basic Coulomb interaction with no evidence for any unusual distortions of this symmetry.

I then turned my interest to studies of parity violation in nuclei. One part of the Standard Electroweak Model associated with the exchange of neutral Z Bosons between quarks could only be probed effectively for up and down quarks, as the process was supressed for the other quarks. Therefore this process must be studied in weak interaction processes between quarks in nuclei where it is in competition with the strong interaction (about a million times stronger). The way to study the weak interaction in this case was to measure processes that violated parity symmetry. Basically we looked for the difference between a nuclear reaction and its mirror image. In most cases this would be expected to be a difference of only a part in a million, but in some nuclear processes it could be increased by factors of up to 1000 by the particular properties of nuclear levels.

During the 1970s I worked on a number of experiments of this nature with a number of collaborators, including a major experiment at Queen's University involving many of the scientists who would eventually become colleagues on SNO. At Chalk River I concentrated finally on a measurement of parity violation in the disintegration of deuterium by circularly polarized gamma rays produced by a high intensity polarized electron beam. This experiment was carried out in collaboration with Dr. Davis Earle and formed the basis for our later long-time collaboration on SNO. The continuous-beam polarized electron source that we developed for this experiment was eventually transferred to the electron accelerator at MIT and used for further experiments by others there.

In 1982 I moved to Princeton University as a Full Professor and began work on polarized targets that were developed in collaboration with Prof. Will Happer, Prof. Frank Calaprice and Prof. Tim Chupp. The most interesting of these polarized targets with which I was associated was ^{3}He polarized by spin transfer from optically polarized rubidium vapor. Through the development of samples at atmospheric pressure or greater, we extended the number of polarized nuclei by several orders of magnitude over previous techniques. This enabled a variety of experiments, including the use of polarized ^{3}He to produce polarized epithermal neutrons for extensive measurements of parity violation in heavy nuclei at the Los Alamos spallation neutron facility. Following these initial developments, others used ^{3}He polarized by this technique for a variety of nuclear and particle physics measurements and eventually for medical imaging.

In the summer of 1984, while at Chalk River finishing the analysis of our parity violation experiment, I became involved in the development of the SNO Collaboration, led by Herb Chen and George Ewan. Herb had an excellent idea for the resolution of the Solar Neutrino Problem if it was possible to borrow over 1000 tonnes of heavy water from Canada's reserves. George had been seeking the best location for an underground laboratory as would be required for this measurement. Under their leadership, the original group of 16 collaboration members began work to develop a detector based on heavy water, to be sited underground and built with ultra-low radioactivity content.

It was a highly motivated group of scientists because we knew that the properties of deuterium in the heavy water could enable the simultaneous measurement of the electron flavor neutrinos produced in the sun and also the sum of all neutrino flavors. The comparison of these two measurements could provide a clear indication of whether any electron neutrinos had changed to another flavor before reaching our detector. However, the challenges associated with making these measurements were major, particularly with respect to restricting the radioactive gamma ray background that could break apart the deuterium leaving a free neutron, mimicking the second reaction that could be caused by any flavor of neutrino. The design had many challenges and required detailed simulation of the detector properties in order to be certain that it would be possible to observe both reactions.

I began work at Princeton on the measurement of radon gas emanated from materials and extracted from water. The tragedy of Herb's death from leukemia in 1987 shocked and saddened us all, but we carried on, with George Ewan as Canadian spokesman and David Sinclair as UK Spokesman. I became the US Spokesman and was joined later in 1987 by Prof. Gene Beier of the University of Pennsylvania, an experienced neutrino physicist who had worked on the conversion of the Kamiokande detector to detect solar neutrinos. In 1988–89 I took a sabbatical year from Princeton at Queen's University and worked with the international team on the development of a final design and detailed costing of the experiment for submission to the funding agencies in the three countries.

I was offered a faculty position at Queen's University which I took up in the summer of 1989. Ironically, I was once again following in the footsteps of my long-time colleague and mentor, George Ewan, who was scheduled to retire several years later. I have often said that if you want a successful career in science, follow George Ewan's lead. In December 1989 we received funding for the project from agencies in the three countries.

I became Director of the SNO Institute formed to take international responsibility for the project and also became Director of the international SNO Scientific

Collaboration. By 1989, the collaboration had expanded to 14 institutions with a large number of experienced scientists and technical people with broad capabilities. The responsibility for various parts of the project had been accepted by groups within the collaboration, a number of our collaborators accepted responsibilities as Group Leaders and they carried out those responsibilities in a very dedicated way through the design, construction and operation of the project.

During the 1990s we followed our plans, overcame many obstacles by collaborative efforts and began data acquisition in 1999. Our first scientific results were published in 2001 and 2002. In those publications we demonstrated that we had in fact built the detector to meet the specifications that had been put forward in the 1980s. We had restricted the interfering radioactivity so that we could make accurate measurements of both of the reactions on deuterium that measured separately the number of electron neutrinos and the total number of all neutrino flavors reaching the detector from ^{8}B decay in the sun. Our results showed that the calculations by Bahcall and others were very accurate for the initial rate of ^{8}B electron neutrinos but that about two thirds of those neutrinos had changed into other active neutrino flavors (muon or tau) before reaching the detector. That conclusion was in agreement with the initial, less accurate results that we obtained by combining our results for electron neutrinos with the results of SuperKamiokande for solar neutrinos, where there is a small sensitivity for all neutrino flavors. We are honored to share the Nobel Prize with Prof. Kajita and the SuperKamiokande collaboration for their detailed measurements of atmospheric neutrinos and their observed decrease of the numbers of muon neutrinos while traversing the earth, explainable through the change of flavor of muon neutrinos.

These flavor changes for solar and atmospheric neutrinos could not occur unless the neutrinos have a non-zero mass. Those neutrino properties are outside the predictions of the Standard Model of Elementary particles and require extensions to that model. Finding those extensions that match the observed properties of neutrinos from these measurements and from the ever increasing number of new neutrino measurements will enable a fuller understanding of the laws of physics at a very fundamental level and an understanding of the many ways in which neutrino properties influence the evolution of our universe.

I was very fortunate to have such a dedicated and skilled group of colleagues, technical people and construction workers who worked wonders to build and operate a unique detector. When we obtained the data for neutrino interactions in our detector we were very satisfied to observe that the simulations that had been made back in 1987 were very accurate for the case where electron neutrinos were changing their flavor before reaching the earth. That meant that we had been able to accomplish the very stringent requirements that we had set for the project, including many aspects that had never been done before. It was a true team effort and I am very grateful to all members of the team and to their spouses and families that supported them throughout.

We were also very fortunate to receive strong international support from funding agencies, educational and research institutions, federal, provincial and local governments and the people of the Sudbury region who made us all feel at home there. We are also grateful to the management of AECL who arranged the loan of the heavy water and INCO/Vale who provided the underground location over many years, in what continues to be one of their most productive mines.

I truly enjoyed working with the SNO team and consider the very positive team effort on SNO to be a highlight of my many years of research in physics. Our results are significant for the basic understanding of neutrinos and that is what we set out to do, for some of us almost twenty years earlier. I am very pleased with the very large number of young people who had the opportunity to have a "Eureka" moment with us and who have gone on to productive careers beyond SNO. This was a very significant scientific result and a very substantial educational experience for all of us, with which I am very satisfied.

CURRICULUM VITAE

Academic Experience

Position	Institution	Year
Professor Emeritus	Queen's University	2013–Present
Director	Sudbury Neutrino Observatory Collaboration	1989–Present
Gordon and Patricia Gray Chair in Particle Astrophysics	Queen's University	2006–2013
University Research Chair	Queen's University	2002–2006
Director	SNO Institute	1991–2003, 2006–2009
Associate Director	SNOLAB Institute	2009–2013
Professor	Queen's University	1989–2013
Professor	Princeton	1982–1989
Sr. Research Officer	Atomic Energy of Canada (Chalk River, Ontario)	1980–1982
Assoc. Research Officer	Chalk River	1975–1980
Assist. Research Officer	Chalk River	1970–1975
Postdoctoral Fellow	Chalk River	1969–1970

Education

Dalhousie University, Halifax, Nova Scotia—B.Sc. Physics (1964)

Dalhousie University, Halifax, Nova Scotia—M.Sc. Physics (1965)

California Institute of Technology, Pasadena, CA, USA—Ph.D. Physics (1969)

Awards

Governor General's Gold Medal, Dalhousie, 1964

Rutherford Memorial Fellowship, (1969–1970)

Fellow of the American Physical Society, 1983

LL.D., *honoris causa*, Dalhousie, 1997

Fellow of Royal Society of Canada, 1997

Honorary Life Membership at Science North, Sudbury, Ontario, 1997

Killam Research Fellowship, 1998

LL.D., *honoris causa*, University College of Cape Breton, 1999

D. Sc., *honoris causa*, Royal Military College, 2001

T.W. Bonner Prize in Nuclear Physics from the American Physical Society, 2003

Canadian Association of Physicists Medal for Lifetime Achievement in Physics, 2003

Natural Sciences and Engineering Research Council of Canada Award of Excellence, 2003

Gerhard Herzberg Canada Gold Medal for Science and Engineering, 2003

Sigma Xi Fund of Canada Award for Scientific Achievement, 2004

Bruno Pontecorvo Prize in Particle Physics, JINR, Dubna, 2005

D. Sc., *honoris causa*, University of Chicago, 2006

NSERC John C. Polanyi Award to the SNO team, 2006

Officer of the Order of Canada, 2006

Co-recipient, Benjamin Franklin Medal in Physics, 2007

LL.D., *honoris causa*, Saint Francis Xavier University, 2009

Fellow of the Royal Society of the UK and the Commonwealth, 2009

Member of Canadian Science and Engineering Hall of Fame, 2009

Killam Prize in the Natural Sciences, 2010

Member of the Nova Scotia Discovery Centre Hall of Fame, 2010

D. Sc., *honoris causa*, University of Alberta, 2011

Henry Marshall Tory Medal, Royal Society of Canada, 2011

D. Sc., *honoris causa*, University of Waterloo, 2012

Member of the Order of Ontario, 2012

Co-recipient, European Physics Society HEP Division Giuseppe and Vanna Cocconi Prize, 2013

Co-recipient, Nobel Prize in Physics, 2015

Co-recipient, Breakthrough Prize in Fundamental Physics (with the SNO Collaboration), 2015

Companion of the Order of Canada, 2015

Present Memberships

1964–Present: Member of Canadian Association of Physicists

1969–Present: Member of American Physical Society (Fellow 1983–present)

1998–Present: Fellow of the Royal Society of Canada

2004–Present: Member of the Professional Engineers of Ontario

2000–Present: CIAR Cosmology and Gravity Program Associate (Chair of the Advisory Board, 2000–05)

2009–Present Fellow of the Royal Society of the UK and the Commonwealth

Visiting Positions

CERN, Geneva (2004), University of Hawaii (2004, 2009), Affiliate Graduate Faculty, University of Hawaii (2010–), Oxford University (2003, 2009), University of Washington, Seattle (1978), Los Alamos National Laboratory (1981), Queen's University (1988)

Personal

Citizenship: Canadian

Married (four children, eight grandchildren)

The Sudbury Neutrino Observatory: Observation of Flavor Change for Solar Neutrinos

Nobel Lecture, December 8, 2015

by Arthur B. McDonald
Queen's University, Kingston, Canada.

1. SOLAR NEUTRINOS

The nuclear fusion processes that power the sun take place at such high temperatures that the nuclei of atoms are able to fuse together, a process that results in the creation of very large numbers of fundamental particles called neutrinos. As you heard from my friend and scientific colleague Professor Kajita, neutrinos only interact through the weak interaction and gravity and therefore can penetrate out from the core of the sun and through the earth with little or no interaction.

It is these neutrinos from the sun that are the subject of our measurements with the Sudbury Neutrino Observatory (SNO), 2 km underground in a mine near Sudbury, Canada. With the use of heavy water as a central element in the design of SNO it was possible to determine clearly that electron neutrinos change to one of the other active flavors before reaching our detector, a property that requires that they have a mass greater than zero. Both of these fundamental neutrino properties are beyond the predictions of the Standard Model for elementary particles. Extensions of the Standard Model to include these neutrino properties can give us a more complete understanding of our Universe at a very basic level.

The study of the sun and the processes that power it has been the subject of strong interest for many years and it is clear that in our work we "see further because we stand on the shoulders of giants" as was said by Isaac Newton. Nobel

Laureates Hans Bethe (1967) and Willy Fowler (1983) were pioneers in the study of the physics of nuclear reactions in the sun: Bethe for his work on energy production in stars via nuclear reactions and Fowler for working out the details of the proton-proton (pp) reactions and others that are responsible for the creation of the majority of the elements in stars and supernovae. The general conclusion of their work was that the "pp cycle" shown in Figure 1 was the principal source of energy generation in the sun.

I was fortunate to be a graduate student in Fowler's Kellogg Laboratory at Caltech in the 1960s, an intellectual center for the understanding of solar and stellar physics. During my time there, I met two other pioneers in the study of the sun and neutrinos, Ray Davis (Nobel Laureate, 2002) who made the first measurements of neutrinos from the sun and John Bahcall, who pioneered complete and accurate calculations of the numbers of neutrinos emitted by the sun. Their pioneering work established what came to be known as "The Solar Neutrino Problem" in that the measurements of Davis showed that the number of electron neutrinos reaching the earth were about three times smaller than the calculations of Bahcall. [1]

Possible reasons for the discrepancy could have been that the experiment or the theory was incorrect. Another possibility was put forward at the same time that Davis began his experiment in 1968 by other pioneers in the field of neutrino physics. Gribov and Pontecorvo [2] proposed that perhaps the electron neutrinos from the sun were oscillating into other flavors (muon neutrinos were known at that time) and escaping detection by Davis' experiment that was only sensitive to electron flavor neutrinos. This was a variation on a theoretical prediction that Pontecorvo had made back in 1958 [3] that electron neutrinos might oscillate into electron anti-neutrinos.

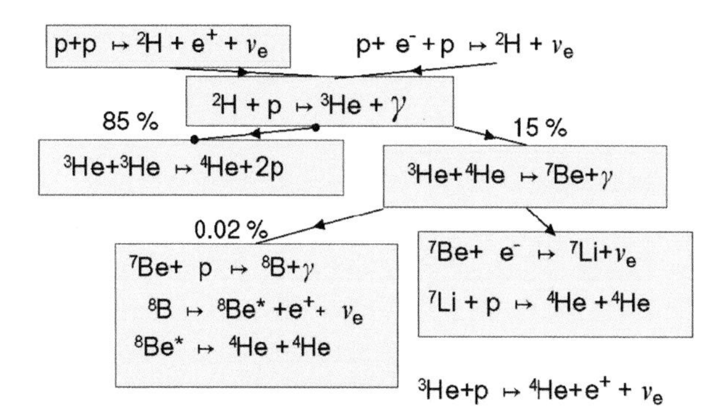

FIGURE 1. The "pp cycle" of nuclear reactions that dominate the energy production in the sun and produce neutrinos with varying numbers and energies as shown in Figure 11.

2. THE SUDBURY NEUTRINO OBSERVATORY ORIGINS

The Sudbury Neutrino Observatory Scientific Collaboration was established in 1984 with Professor Herb Chen of the University of California, Irvine, USA and Professor George Ewan of Queen's University, Kingston, Ontario, Canada as Co-Spokesmen. In 1985 the UK joined the collaboration with Dr. David Sinclair of Oxford University as UK Spokesman. The motivation for the experiment was described in a paper by Chen [4] wherein the deuterium nuclei in heavy water would enable two separate reactions to be observed, one sensitive only to the electron flavor neutrinos created in the sun and the other sensitive to all three active neutrino flavors. A comparison of the rates for these detection reactions for neutrinos from ^{8}B decay in the sun would enable a clear determination of whether electron neutrinos were changing into other flavors independent of any calculations of initial solar neutrino fluxes. In addition, the second detection reaction could be used to obtain an accurate measure of the flux of electron neutrinos from ^{8}B decay, independent of whether neutrino oscillation was occurring.

Reference 5 shows the list of the original 16 collaboration members and Figure 2 shows members at a collaboration meeting at Chalk River Nuclear Laboratories in Canada in 1986. By that time, tentative approval for the loan of 1000 tonnes of heavy water from Canada's reserves had been obtained from Atomic Energy of Canada Limited (AECL) and tentative approval had been obtained from INCO Limited, the owner of the Creighton mine near Sudbury, Ontario, a location that had been previously identified by Professor Ewan as an ideal location for a deep, low radioactivity laboratory. Tragically, Professor Chen passed away from leukemia in 1987, about six months after the picture in Figure 2 was taken.

This was a great loss for the collaboration, but they respected Herb's memory and scientific objectives and carried on with their work on experimental design and requests for funding. At the time I was at Princeton University and took over as US Co-Spokesman, joined shortly thereafter by Professor Eugene Beier of the University of Pennsylvania, Philadelphia. The collaboration grew significantly over the next few years and by 1989 comprised 14 institutions in Canada, the US and the UK. Considerable work was carried out to complete the design for the experiment and the underground location and in late 1989, funding was obtained from government agencies in the three countries. I had moved that year from Princeton to Queen's University and became Director of the SNO Institute with responsibility for the international project, and also Director of the SNO Scientific Collaboration.

The original collaboration included Atomic Energy of Canada Ltd. (Chalk River Laboratories), Carleton University, Laurentian University, the National

FIGURE 2. The picture shows some of the collaboration members at a Collaboration meeting in Chalk River, 1986. From the left: Davis Earle, Mort Bercovitch, David Sinclair, John Simpson, Doug Hallman, Hay Boon Mak, Peter Doe, Henry Lee, Cliff Hargrove, Hugh Evans, Peter Skensved, Herb Chen, Dan Kessler, George Ewan, Richard Allan, Art McDonald. Original collaboration members missing from this picture include Walter Davidson, Barry Robertson, Robert Storey.

Research Council of Canada, Oxford University, Princeton University, Queen's University, the University of California at Irvine, and the University of Guelph. By 1989, those institutions had been joined by University of Pennsylvania, Los Alamos National Laboratory, Lawrence Berkeley National Laboratory, the University of British Columbia, and Brookhaven National Laboratory and the collaboration had grown to about 70 scientists. The eventual number of authors on SNO scientific papers numbered 274. Since the majority of the additional authors were graduate students and postdoctoral fellows, this shows clearly the major educational aspect of this scientific work.

3. NEUTRINO DETECTION IN SNO

With deuterium contained in the heavy water molecules (>99.92 % D_2O) in the SNO detector, it was possible to observe three separate interactions of neutrinos in the detector, the first two mentioned before and the third, elastic scattering from electrons that takes place in any medium:

$\nu_e + \mathrm{d} \rightarrow \mathrm{e}^- + \mathrm{p} + \mathrm{p} - 1.44$ MeV (Charged Current (CC) Reaction)

$\nu_x + \mathrm{d} \rightarrow \nu_x + \mathrm{n} + \mathrm{p} - 2.2$ MeV (Neutral Current (NC) reaction)

$\nu_x + e^- \rightarrow \nu_x + e^-$ (Elastic Scattering (ES) reaction),

where x = electron, mu or tau

The first (CC) reaction is sensitive only to electron flavor neutrinos and produces an energetic electron that creates a cone of light in the detector via the Cerenkov process, observable with an array of photo-sensors. The second (NC) reaction is equally sensitive to all flavors of neutrino and produces a free neutron that was observed in different ways in the three phases of the SNO detector operation. By comparing appropriately calibrated rates for these two reactions it was possible to determine whether solar electron neutrinos had changed into other flavors before reaching the detector.

The third reaction (ES) is much weaker than the other two and is mostly sensitive to electron neutrinos (six times more sensitivity than to the other two flavors for solar neutrino energies). It produces an energetic electron that is strongly peaked in the forward direction relative to the incident neutrino and therefore can be distinguished from the other two reactions by reference to the direction from the sun.

The SNO experiment was carried out in three distinct phases. In Phase 1, pure heavy water was used and the free neutron from the NC reaction was observed as it was captured by a deuterium nucleus, producing a 6.25 MeV gamma ray that in turn generated Compton-scattered electrons producing Cerenkov light. In Phase 2, about 2 metric tons of ultra-pure NaCl salt was added to the heavy water, so that the free neutron would predominantly capture in the Cl, producing a cascade of gamma rays with energies summing to about 8.6 MeV. This increased the neutron capture efficiency from 14% to 40% and provided such an isotropic distribution of light that the events from the NC reaction could be separated statistically from the cone-shaped light emission events from the CC reaction. In Phase 3, an independent array of ^{3}He-filled neutron detectors [6] was inserted into the heavy water, providing a clear measurement of free neutrons from the NC reaction.

4. THE SNO DETECTOR

Figure 3 shows an artist's conception of the SNO detector [7] situated in a barrel-shaped cavity 34 m high by 22 m diameter, 2 km underground in INCO/Vale's Creighton mine near Sudbury, Ontario, Canada.

1000 tonnes of heavy water enriched to 99.92% deuterium (worth $300 million Canadian) are held in a transparent acrylic vessel 12 meters in diameter and 5.6 cm thick, viewed by 9438 light sensors (photomultipliers, or PMTs) mounted

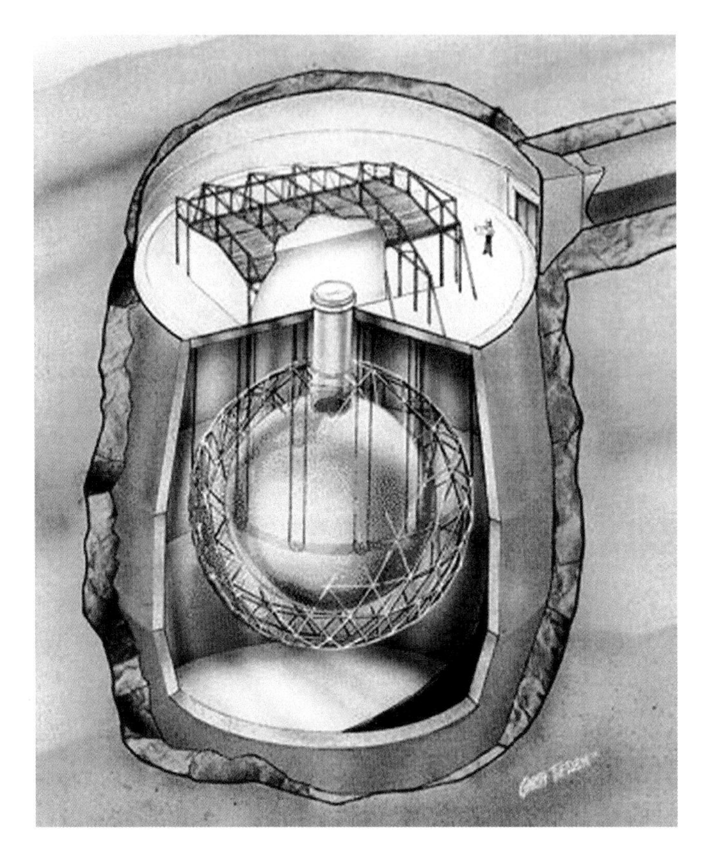

FIGURE 3. Schematic view of the Sudbury Neutrino Observatory.

on a geodesic frame made from stainless steel. The cavity is lined with water- and radon-impermeable Urylon plastic. The entire cavity outside the acrylic vessel is filled with ordinary water purified to be more than a billion times purer than tap water for the content of uranium and thorium decay chain elements.

The design and construction of this massive, complicated detector 2 km underground in ultra-clean conditions was a major engineering accomplishment. We were very fortunate to have a skilled team of engineers, technicians and construction workers to carry out this one-of-a-kind project. Scientific teams distributed across the collaboration accepted responsibilities for major parts of the experiment, led by Group Leaders for elements such as the Acrylic Vessel, PMT's, PMT Support Structure, Water Systems, Electronics, Data Acquisition, Calibration, Simulation and Analysis. These groups took responsibility for their detector elements from design through construction and operation to meet the scientific and engineering requirements, with coordination among groups through regular meetings and discussions.

All parts of the detector were carefully chosen and sampled to be as low as possible in radioactivity and the whole laboratory area was maintained at better than Class 2000 air quality (fewer than 2000 dust particles of diameter greater than 0.5 micron per cubic foot of air). It was determined that less than one gram of mine dust was present on the entire detector after construction. It was extremely important to maintain ultra-low radioactivity levels in the detector and in the heavy water because any gamma ray with energy greater than 2.2 MeV (such as from the uranium or thorium decay chains) could possibly cause the disintegration of the deuterium nucleus, producing a free neutron and mimicking the NC reaction from neutrinos. By keeping the radioactivity very low (less than 3×10^{-15} grams of Th per gram of heavy water) and measuring accurately the content it was possible to keep the numbers of neutrons from this gamma disintegration process on deuterium well below those from neutrino reactions. This background contribution was measured accurately by sampling the water for Th decay chain content and by analyzing the low energy data that was dominated by radioactivity. Similar care was taken to restrict U and measure contamination in the heavy water and the ordinary water.

Figure 4 shows the detector during construction, after the acrylic sphere had been bonded together in place from 122 pieces small enough to fit within the underground hoist used for access.

FIGURE 4. SNO detector during construction.

Figure 5 shows the completed detector prior to water fill and Figure 6 shows a wide angle camera shot looking up from the bottom of the sphere of photomultipliers (PMT's).

Figure 7 shows part of the water purification systems, including systems for the measurements of ultra-low levels of radioactivity by recirculation of the water through filters specially designed to measure the important daughter products of U, Th decay. Radon gas was another important radioactive component that had to be restricted strongly and measured carefully. The restriction was accomplished by covering the heavy water with nitrogen gas that was obtained from the boil-off of large liquid nitrogen dewars. This gas was very low in radon content and the system was designed to avoid contamination by mine air by requiring positive flow of pure nitrogen over the heavy water even during mine pressure excursions. The ordinary water was de-gassed and then re-gassed with similar "boil-off" nitrogen with ultra-low radon content. All of these precautions resulted in less than one radioactive decay from the Th or U chains per day per

FIGURE 5.

FIGURE 6.

FIGURE 7. Part of the water purification systems.

metric ton of heavy water, as required to avoid interference with the neutrino event rates.

5. SNO EXPERIMENTAL MEASUREMENTS

Photons of light generated by neutrino interactions in the heavy water were converted to electronic pulses by the photomultipliers, shaped via specially designed circuits and collected on computer systems, along with the pulses generated by radioactive background. These data were carefully analyzed to extract the pulses from a neutrino interaction, using information on the magnitude of pulses and time of arrival at all of the photomultipliers triggered simultaneously by the neutrino interaction. This set of information was collectively referred to as an event. This information was carefully analyzed to differentiate neutrino events from those generated by radioactivity or other instrumental artifacts.

Figure 8 shows a collection of events obtained from Phase 1 of the experiment with pure heavy water in the detector. The data shown here comes from the innermost 11 meter diameter of the heavy water during 306 days of data accumulation. The expected shapes of event data from the CC, NC and ES reactions and from the radioactive background pulses extrapolated from lower energies are also shown on the figure. These shapes were calculated from a very detailed Monte Carlo simulation of the expected pulses from the detector, calibrated

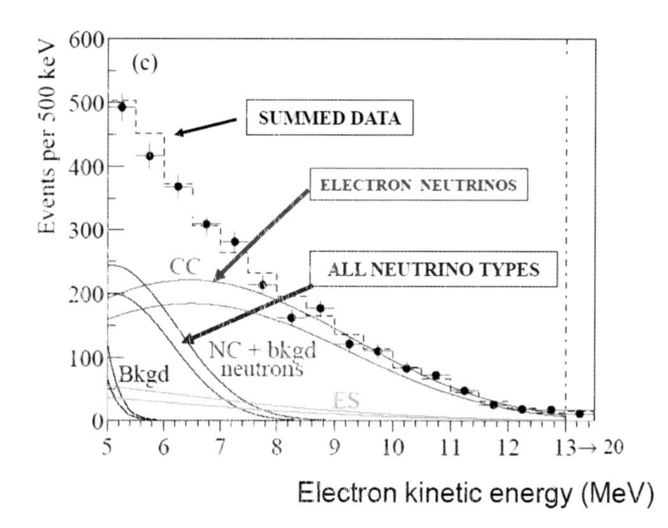

FIGURE 8. Data from Phase 1 of the SNO experiment showing the number of events versus the effective kinetic energy for an electron.

accurately by a series of subsidiary measurements using calibration sources emitting known energies and quantities of gamma rays, neutrons and electrons. These sources could be moved throughout over 70% of the volume of the detector in two perpendicular planes with a calibration source manipulator, providing a detailed mapping of the detector sensitivity.

The shape of events from the NC reaction as a function of energy was very well defined by the emission of 6.25 MeV gamma rays from the capture of free neutrons in deuterium and could be calibrated accurately by the use of a source of 6.13 MeV gamma rays from the decay of ^{16}N. The ^{16}N was generated in a shielded nearby location in the underground laboratory and transported via capillary tubes to the heavy water volume [8]. The contributions of background events from radioactivity producing Cerenkov light in the detector, as shown by the black lines in the figure were calculated from measurements made with encapsulated U and Th sources moved within the detector with the calibration source manipulator. The number of free neutrons created by gamma rays from U and Th breaking apart deuterium nuclei was calculated from measurements of U, Th decay chain elements in the heavy water and ordinary water volumes of the detector as well as smaller contributions from the known radioactivity of other detector materials. The amount of U and Th decay chain elements in the water volumes was determined by direct sampling of the recirculated water with manganese-dioxide or hydrous-titanium-oxide loaded filters and also by analysis of the U and Th events observed in lower energy regions, including differentiation between these elements using the isotropy of the light patterns on the detector. The contributions for Cerenkov light events and neutrons from gammas breaking up deuterium were respectively 8% and 12% of the total number of neutrinos observed by the NC reaction and the total combined systematic uncertainty in that quantity was only 4%.

A hypothesis test was made for the assumption that electron neutrinos were not changing their flavor before reaching the SNO detector. The test used the data shown in Figure 8, together with the additional information obtainable for each neutrino interaction, particularly the inferred direction and the location within the heavy water volume. The hypothesis of no neutrino flavor change was ruled out with a combined statistical and systematic accuracy of 5.3 standard deviations, corresponding to less than one in 10 million chance that there is no flavor change. The best fit fluxes of electron neutrinos and the combined muon and tau neutrino flavors inferred from the data, assuming no distortion of the energy distribution of ^{8}B electron neutrinos, were (in units of 10^6 neutrinos per square cm per second):

$$\phi_e = 1.76^{+0.05}_{-0.05}(\text{stat.})^{+0.09}_{-0.09}(\text{syst.})$$
$$\phi_{\mu\tau} = 3.41^{+0.45}_{-0.45}(\text{stat.})^{+0.48}_{-0.45}(\text{syst.})$$

This "best fit" implied that about two thirds of the electron neutrinos produced in the core of the sun had changed into other active flavors before reaching the SNO detector.

More detailed scientific discussion of these results is contained in the two SNO papers published in 2002 [9]. These results were consistent with a previous SNO paper in 2001 [10] where data for ^8B neutrinos detected via the CC reaction in SNO was compared with results from SuperKamiokande for ^8B solar neutrinos detected by the ES reaction in light water. This comparison provided evidence for violation of the no flavor change hypothesis with 3.3 standard deviation significance.

The results for Phase 2 of the experiment, with 2 metric tons of NaCl dissolved in the heavy water are shown in Figure 9. Comparison with Figure 8 shows that the contribution of events from the NC reaction has increased due to the increase in detection efficiency for free neutrons from 14% to 40%. In addition, the difference in the isotopy of the light emission between NC and CC events was used as an additional means to separate the events from the two reactions, providing additional improvement in accuracy.

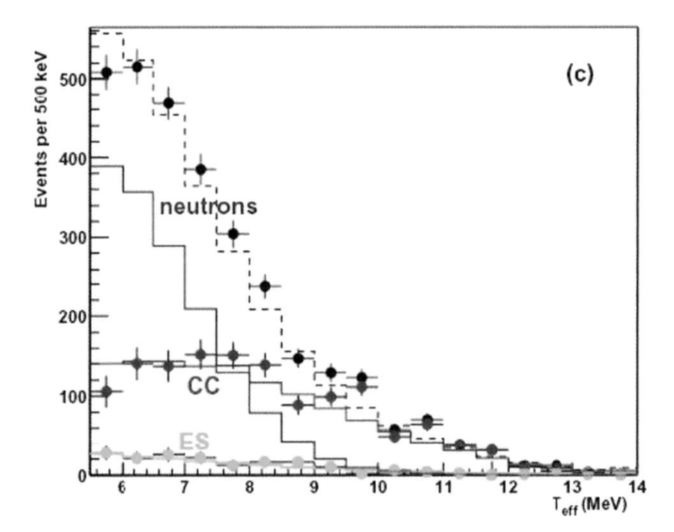

FIGURE 9. Data from Phase 2 showing the number of events versus the effective kinetic energy for an electron.

The results for Phase 2 [11] are consistent with and more accurate than those from Phase one, with a significant improvement in the accuracy of the NC measurement of the total flux of ^{8}B neutrinos. In addition, the separation of NC and CC events on the basis of the isotropy of light emission allowed a separate extraction of the shape of the spectrum of electron neutrinos via the CC detection reaction. The observed shape was consistent with the expected shape for neutrinos from ^{8}B decay in the sun, consistent with the assumption made to extract the "best fit" values for neutrino fluxes in the Phase 1 analysis. The hypothesis of no neutrino oscillation was now excluded with greater than 7 standard deviations or one in 400 billion. The total flux of all neutrino flavors was defined with substantially improved accuracy and found to be in excellent agreement with the solar model calculations of Bahcall and others.

In Phase 3 of the experiment, an array of 400 meters of ultra-low background ^{3}He-filled neutron detectors were installed in the heavy water volume using a remotely controlled submarine [6]. Figure 10 shows the submarine being used to install one of the final neutron detectors. You may notice that the submarine is a dull green color, not the first choice for anyone having a bit of whimsy in their soul, as most scientists do. Of course, our first choice was a Yellow Submarine and that was the original color. However, as luck would have it, the yellow paint

FIGURE 10. A remotely controlled submarine installing one member of the array of ^{3}He-filled neutron detectors for Phase 3 of the SNO experiment.

was much too radioactive for us to consider using it and so we had to strip it off and be satisfied with the prosaic green color shown. This is just one example of the detailed process that we went through for many of the materials used in the detector to achieve our stringent radioactivity requirements and also of the surprises that you sometimes have to deal with in a complex project like this.

The neutron detectors were used to make further measurements of the CC and NC reactions with completely different systematic uncertainties from the first two phases. The results [12] of Phase 3 were also in agreement within errors with the previous results from Phase 1 and 2, increasing the accuracy of the overall result and providing added confidence in the clear separation of CC and NC events. The analyses of SNO data through the three phases of the experiment were carried out with several approaches applied to "blind" the final result for those performing the analysis until all of the parameters to be used had been fully defined.

A final combined analysis of the three phases was obtained with the following result for the ratio of the fluxes of electron neutrinos to all neutrino flavors: 0.317 ± 0.016 (statistical) $\pm$ 0.009(systematic). This final result [13] shows clearly that over two thirds of the electron neutrinos have changed into other flavors before reaching the SNO detector. The total observed flux of neutrinos from ^8B decay in the sun was determined to be: 5.25 ± 0.16 (statistical) $+0.11$ -0.13 (systematic) million neutrinos per square cm per second. This is in agreement with and more accurate than calculations of the ^8B electron neutrino flux produced in the sun. [14] The accuracy of this measurement is being used to refine models of the sun [15], in combination with many other observations, including helioseismology.

6. COMPARING SNO RESULTS WITH OTHER SOLAR NEUTRINO MEASUREMENTS

Following Davis' pioneering measurements of solar neutrinos using the interaction of electron neutrinos with chlorine [1, 16], and prior to the SNO results, several other experiments observed solar neutrinos with exclusive sensitivity to electron neutrinos (gallium-based radiochemical detectors) [17] or predominant sensitivity to electron neutrinos through the use of the ES reaction on electrons in light water (Kamiokande and SuperKamiokande detectors) [18]. The thresholds for neutrino detection are indicated in Figure 11, which shows the calculated fluxes from various pp cycle reactions in the sun. The results from these other measurements of solar neutrino flux are shown graphically in Figure 12, where it is apparent that the measurements are factors of two or three lower than the expectations from the solar model calculations.

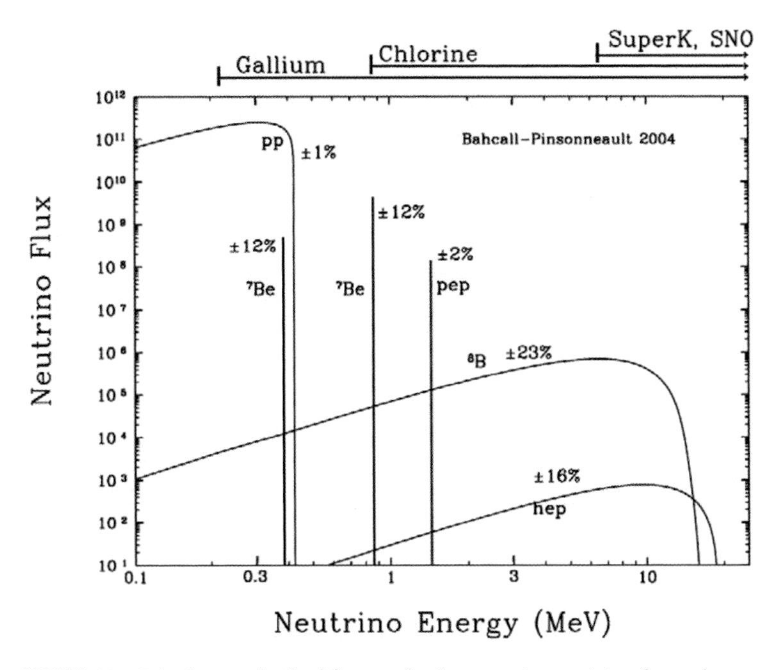

FIGURE 11. Calculations [19] of fluxes of solar neutrinos arising from the set of reactions in the pp cycle shown in Figure 1. Also shown at the top are thresholds for electron neutrino detection in experiments using chlorine, gallium and light and heavy water as discussed in the text.

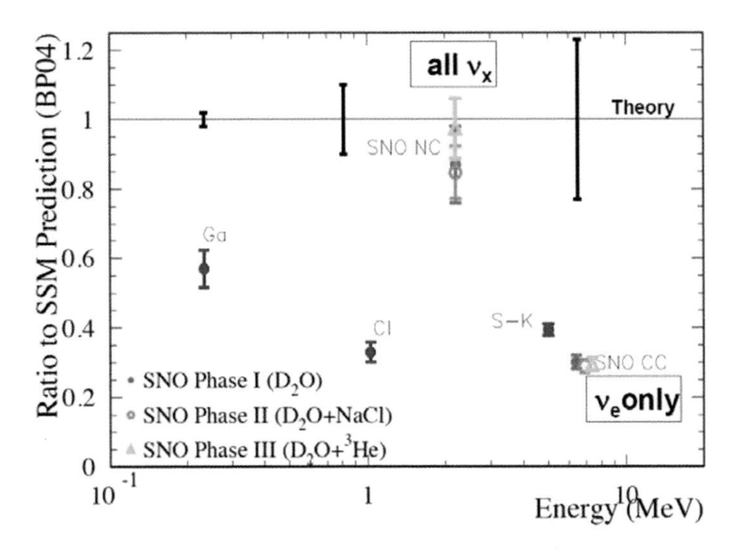

FIGURE 12. Ratios of measurements of solar neutrino fluxes to calculations of electron neutrino fluxes in the core of the sun using the Standard Solar Model (SSM) [19].

Also shown in Figure 12 are the results from the SNO measurements for the three phases, indicating the agreement between the three sets of SNO measurements. The total neutrino flux is seen to be in good agreement with solar models and the observed electron neutrino flux is about a factor of three lower than solar model predictions due to neutrino flavor change. Note also that the differences between the ratio plotted for the SuperKamiokande (SK) experiment and the SNO CC measurements is primarily due to the fact that the ES reaction measurement in SK has some sensitivity to all neutrino flavors as used in ref [10] to extract the total neutrino flux with more limited sensitivity.

7. NEUTRINO OSCILLATIONS AND FLAVOR CHANGE

The exhibition of flavor change and oscillation by neutrinos implies that they have a non-zero mass. If they had zero mass as predicted by the Standard Model of elementary particles then they would be traveling at the speed of light and would not have a measure of time in their frame of reference by which to define the process of oscillation. The presently accepted model for neutrino oscillation is based on work by Maki, Nakagawa and Sakata [20] and the work of Pontecorvo [3, 2] wherein the quantum mechanical states of flavor generated when a neutrino is created can be expressed as superpositions of mass states as shown schematically in Figure 13. As the originally pure flavor states travel through space there are changes in the descriptions of the flavor states in terms of the mass components. Then when a measurement is made, the neutrino state will look partly like the mass composition of an electron neutrino, partly like a muon neutrino and partly like a tau neutrino. This determines the fraction of events that are observed in measurements that are specific to these flavors.

For solar neutrinos, there is another effect adding to this process that was identified by Mikeyev and Smirnov, extending ideas of Wolfenstein [21] (referred to as the MSW effect). As the electron neutrinos pass through dense parts of the sun containing large numbers of electrons, the MSW effect can change the oscillation process. For ^{8}B electron neutrinos originating in the solar core, the MSW effect changes them to a pure mass 2 state and they then stay in that state until detected on earth. This provides the observed fractions of flavors in the SNO results. The determination that this effect is occurring actually arises from detailed calculations of neutrino oscillation and MSW effect in the sun, using predicted fluxes from solar models and the results from the chlorine, gallium, ordinary water and heavy water measurements, including the SNO results and results from the oscillation of reactor anti-neutrinos on Earth [22]. For more

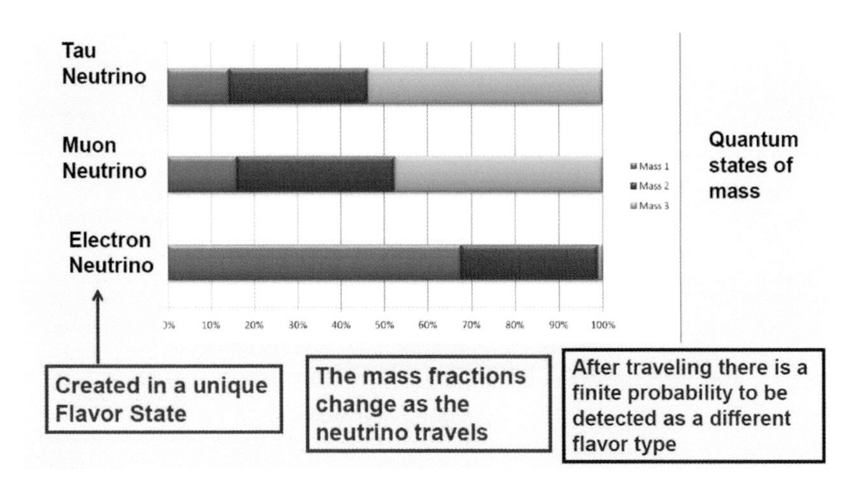

FIGURE 13. Schematic description of neutrino oscillations.

detail see, for example, reference 13. The MSW effect in the sun also determines that mass 2 is larger than mass 1.

8. FUTURE MEASUREMENTS

Since the initial results from the SuperKamiokande and SNO experiments, there have been many further neutrino detection experiments that have helped to determine the properties of neutrinos and of neutrino oscillations. There are a number of questions left to be answered about neutrino properties and experiments are underway or being planned to answer these questions.

The SNO detector itself is undergoing a conversion to a new experiment known as SNO+ which will seek to observe a rare radioactive process known as neutrino-less double beta decay for Tellurium nuclei dissolved in a liquid scintillator that will replace the heavy water that has been removed from the detector. If this rare radioactivity is observed it will confirm that neutrinos are their own anti-particles (known as Majorana particles) and the absolute mass of neutrino flavors can be observed. At present, the oscillation measurements have only determined the differences in mass of the three active mass flavors. The SNO+ detector is also planning to make measurements of lower energy solar neutrinos as well as neutrinos from the earth and nuclear reactors and will search for neutrinos from supernovae in our galaxy.

In addition to this future experiment the laboratory was enlarged under the Directorship of Professor David Sinclair to create SNOLAB and house a number of other experiments that presently include direct searches for Dark Matter particles and neutrinos from supernovae. The total excavated volume of the new laboratory is about three times the volume of the original SNO research area and cavity and the whole laboratory is maintained at Class 2000 air quality or better to control local radioactivity. For more details see www.snolab.ca.

ACKNOWLEDGEMENTS

In conclusion I would like to express my sincere thanks to the large number of collaborators who contributed to the success of the SNO project. The full set of authors on the SNO scientific papers are as follows:

263 SNO Physics Paper Authors: Adam Cox, Aksel L. Hallin, Alain Bellerive, Alan Smith, Alan Poon, Alexander Wright, Allan Myers, Alysia Marino, André Krüger, André Roberge, Andre Krumins, Andrew Ferraris, Andrew Hime, Anett Schülke, Anthony Noble, Araz Hamian, Arthur McDonald, Aubra Anthony, Azriel Goldschmidt, Barry Robertson, Bassam Aharmim, Bei Cai, Benjamin Monreal, Bernard Nickel, Berta Beltran, Bhaskar Sur, Blair Jamieson, Brandon Wall, Brent VanDevender, Brian Morissette, Bruce Cleveland, Bryan Fulsom, Bryce Moffat, Carsten Krauss, Catherine Mifflin, Charles Currat, Charles Duba, Charlotte Sims, Christian Nally, Christian Ouellet, Christine Kraus, Christopher Kyba, Christopher Howard, Christopher Jillings, Christopher Tunnell, Christopher Waltham, Clarence Virtue, Colin Okada, Darren Grant, David Anglin, David Sinclair, David Waller, David Wark, Davis Earle, Diane Reitzner, Dimpal Chauhan, Doug Hallman, Douglas Cowen, Douglas McDonald, Duncan Hepburn, Ed Frank, Edward Clifford, Michael Dragowsky, Emmanuel Bonvin, Eric Norman, Erik Saettler, Etienne Rollin, Eugene Guillian, Eugene Beier, Fabrice Fleurot, Feng Zhang, Ferenc Dalnoki-Veress, Fraser Duncan, Gabriel D. Orebi Gann, Geoffrey Miller, George Doucas, George Ewan, Gerhard Bühler, Gersende Prior, Gordana Tešić, Gordon McGregor, Gregory Harper, Guy Jonkmans, Gwen Milton, Hadi Fergani, Hamish Robertson, Hans Bichsel, Hans Mes, Hardy Seifert, Hay Boon Mak, Heidi Munn, Helen M. O'Keeffe, Hendrick Labranche, Henry Lee, Hok Seum Wan Chan Tseung, Huaizhang Deng, Hugh Evans, Hui-Siong Ng, Ian Lawson, Ilan Levine, Ira Blevis, Jacques Farine, James Cameron, James Hall, James Loach, James Leslie, Jaret Heise, Jason Detwiler,

Jason Hewett, Jason Pun, Jason Goon, Jeanne Wilson, Jeffrey Secrest, Jeremy Lyon, Jerry Wilhelmy, Jessica Dunmore, Jian-Xiong Wang, Jimmy Law, Jocelyn Monroe, John Amsbaugh, John Boger, John Orrell, John Simpson, John Wilkerson, Jon Hykawy, Jose Maneira, Joseph Formaggio, Joseph Banar, Joseph Germani, Joshua Klein, Juergen Wendland, Kai Zuber, Kara Keeter, Kareem Kazkaz, Karsten Heeger, Katherine Frame, Kathryn Schaffer, Keith Rielage, Kennneth McFarlane, Kevin Graham, Kevin Lesko, Kevin McBryde, Khalil Boudjemline, Klaus Kirch, Laura Kormos, Laura Stonehill, Laurel Sinclair, Louise Heelan, Malcolm Fowler, Manuel Anaya, Marc Bergevin, Marcus Thomson, Maria Isaac, Marie DiMarco, Mark Boulay, Mark Chen, Mark Howe, Mark Kos, Mark Neubauer, Martin Moorhead, Masa Omori, Melin Huang, Melissa Jerkins, Michael Bowler, Michael Browne, Michael Lay, Michael Lowry, Michael Miller, Michael Thorman, Michal Shatkay, Mike Schwendener, Miles Smith, Minfang Yeh, Miriam Diamond, Mitchell Newcomer, Monica Dunford, Morley O'Neill, Mort Bercovitch, Myung Chol Chon, Naeem Ahmed, Nathaniel Tagg, Neil McCauley, Nicholas Jelley, Nicholas West, Nikolai Starinsky, Nikolai Tolich, Noah Oblath, Noel Gagnon, Nuno Barros, Olivier Simard, Patrick Tsang, Paul Keener, Peter Wittich, Peter Doe, Peter Watson, Peter Skensved, Peter Thornewell, Philip Harvey, Pierre Luc Drouin, Pillalamarr Jagam, Ranpal Dosanjh, Reda Tafirout, Reena Meijer Drees, Reyco Henning, Richard Allen, Richard Ford, Richard Helmer, Richard Hemingway, Richard Kouzes, Richard Hahn, Richard Lange, Richard Ott, Richard Taplin, Richard Van Berg, Richard Van de Water, Rizwan Haq, Robert Black, Robert Boardman, Robert Stokstad, Robert Heaton, Robert Komar, Robin Ollerhead, Rushdy Ahmad, Ryan MacLellan, Ryan Martin, Ryuta Hazama, Salvador Gil, Sarah Rosendahl, Scott Oser, Sean McGee, Shahnoor Habib, Sherry Majerus, Simon Peeters, Stanley Seibert, Steffon Luoma, Steven Elliott, Steven Biller, Steven Brice, Teresa Spreitzer, Thomas Andersen, Thomas J. Radcliffe, Thomas J. Bowles, Thomas Kutter, Thomas Sonley, Thomas Steiger, Timothy Van Wechel, Tom Burritt, Tudor Costin, Tyron Tsui, Vadim Rusu, Vladimir Novikov, Walter Davidson, William Frati, William Handler, William Heintzelman, William Locke, William McLatchie, Xin Chen, Xin Dai, Yaroslav Tserkovnyak, Yasuo Takeuchi, Yekaterina Opachich, Yuen-Dat Chan. **And 11 who have passed away:** Herbert Chen, John C. Barton, John Cowan, Andre Hamer, Clifford Hargrove,

Barry C. Knox, Jan Wouters, Peter Trent, Robert Storey, Keith Rowley and Neil Tanner.

I was truly privileged to be the Director of a project with such a talented, dedicated, cooperative and hard-working group of scientists. I also want to acknowledge the very talented and hard-working group of engineers, technical people, construction, operations and administrative personnel for SNO. Our success was truly a team effort and was strongly supported by the spouses and families of all these people. I am honored to have had the opportunity to work with these people and am very grateful for all their contributions.

I would also like to thank all of the funding agencies and institutional administrations who have supported us throughout the many years it took to complete the experiment. A special word of thanks goes to AECL for the loan of the heavy water and to INCO/Vale for their continual support throughout SNO and now continuing with SNOLAB.

REFERENCES

1. R. Davis Jr., D. S. Harmer and K. C. Hoffman, *Phys. Rev. Lett.*, **20**, 1205 (1968); J. N. Bahcall, N. A. Bahcall and G. Shaviv, *Phys. Rev. Lett.*, **20**, 1209 (1968).
2. V. Gribov and B. Pontecorvo, *Phys. Lett.*, **25B**, 493 (1969).
3. B. Pontecorvo, *Sov. Phys. JETP* **7**, 172 (1958).
4. H. H. Chen, *Phys. Rev. Lett.*, **55**, 1534 (1985).
5. D. Sinclair, A.L. Carter, D. Kessler, E.D. Earle, P. Jagam, J.J. Simpson, R.C. Allen, H.H. Chen, P.J. Doe, E.D. Hallman, W.F. Davidson, A.B. McDonald, R.S. Storey, G.T. Ewan, H. B. Mak, B.C. Robertson, *Il Nuovo Cimento* **C9**, 308 (1986).
6. J. Amsbaugh et al., *Nucl. Inst. Meth* A **A579**, 1054 (2007).
7. J. Boger et al. (SNO Collaboration), *Nucl. Instr. and Meth.* **A449**, 172 (2000).
8. M.R. Dragowsky, A. Hamer, Y.D. Chan, R. Deal, E.D. Earle, W. Frati, E. Gaudette, A. Hallin, C. Hearns, J. Hewett, G. Jonkmans, Y. Kajiyama, A.B. McDonald, B.A. Moffat, E.B. Norman, B. Sur, N. Tagg, *Nucl. Instr. and Meth.*, **A481**, 284–296 (2002).
9. Q. R. Ahmad et al. (SNO Collaboration), *Phys. Rev. Lett.*, **89** 011301 (2002); Q. R. Ahmad et al. (SNO Collaboration), *Phys. Rev. Lett.*, **89**, 011306 (2002).
10. Q. R. Ahmad et al. (SNO Collaboration), *Phys. Rev. Lett.*, **87**, 07301 (2001).
11. S. N. Ahmed et al. (SNO Collaboration), *Phys. Rev. Lett.* **92**, 181301 (2004).
12. B. Aharmim et al (SNO Collaboration) *Phys. Rev. Lett.* **101**, 111301 (2008).
13. B. Aharmim et al, (SNO Collaboration), *Phys. Rev.* C **88**, 025501 (2013).
14. A. M. Serenelli, S. Basu, J. W. Ferguson, and M. Asplund, *Astrophys. J. Lett.* **705**, L123 (2009).
15. I. Lopes and S. Turck-Chieze, *Astrophys.J.* **765**, 14 (2013).
16. B. T. Cleveland et al, *Astrophysical Jour.*, **496**, 505 (1998).

17. J. N. Abdurashitov et al. (SAGE Collaboration), *Phys.Rev.* C **80**, 015807 (2009), contains combined analysis with the following references; M. Altmann et al. (GNO Collaboration), *Physics Letters* B **616**, 174 (2005); F. Kaether, Datenanalyse der Sonnenneutrinoexperiments Gallex, Ph.D. thesis, Heidelberg (2007).
18. K. Abe et al. (Super-Kamiokande Collaboration), *Phys. Rev.* D **83**, 052010 (2011).
19. J. N. Bahcall and M. H. Pinsonneault *Phys. Rev. Lett.* **92**, (2004) 121301.
20. Z. Maki, M. Nakagawa, and S. Sakata, *Prog. Theor Phys.* **28**, 870 (1962).
21. L. Wolfenstein, *Phys. Rev.* D **17**, 2369 (1978); S. Mikheyev and A. Smirnov, *Sov. J. Nucl. Phys.* **42**, 913 (1985).
22. K. Abe et al. (Kamland Collaboration), *Phys. Rev.* C **84**, 035804 (2011).